Routledge Revivals

Heinrich Rudolf Hertz (1857-1894)

Heinrich Rudolf Hertz (1857-1894)

A Collection of Articles and Addresses

Edited by
Joseph F. Mulligan

First published in 1994 by Garland Publishing, Inc.

This edition first published in 2018 by Routledge
2 Park Square, Milton Park, Abingdon, Oxon, OX14 4RN
and by Routledge
52 Vanderbilt Avenue, New York, NY 10017, USA

Routledge is an imprint of the Taylor & Francis Group, an informa business

© 1994 by Joseph F. Mulligan

All rights reserved. No part of this book may be reprinted or reproduced or utilised in any form or by any electronic, mechanical, or other means, now known or hereafter invented, including photocopying and recording, or in any information storage or retrieval system, without permission in writing from the publishers.

Publisher's Note
The publisher has gone to great lengths to ensure the quality of this reprint but points out that some imperfections in the original copies may be apparent.

Disclaimer
The publisher has made every effort to trace copyright holders and welcomes correspondence from those they have been unable to contact.
A Library of Congress record exists under ISBN:

ISBN 13: 978-0-367-18872-6 (hbk)
ISBN 13: 978-0-367-18875-7 (pbk)
ISBN 13: 978-0-429-19896-0 (ebk)

HEINRICH RUDOLF HERTZ
(1857–1894)

GARLAND REFERENCE LIBRARY
OF THE HUMANITIES
(VOL. 1697)

Gravestone of Heinrich Hertz in the cemetery of Hamburg-Ohlsdorf. Photograph by Helmut Drubba, Hannover, Germany; courtesy of Mr. Drubba and Dr. James G. O'Hara, Leibniz-Archiv, Hannover.

HEINRICH RUDOLF HERTZ (1857–1894)

A Collection of Articles and Addresses

Edited by
Joseph F. Mulligan

GARLAND PUBLISHING, INC.
NEW YORK & LONDON / 1994

© 1994 Joseph F. Mulligan
All rights reserved

Library of Congress Cataloging-in-Publication Data

Hertz, Heinrich, 1857–1894.
    Heinrich Rudolf Hertz (1857–1894) : a collection of articles and addresses / edited by Joseph F. Mulligan.
       p. cm. — (Garland reference library of the humanities ; vol. 1697)
    Includes bibliographical references and index.
    ISBN 0-8153-1288-1 (alk. paper)
    1. Physics—History. 2. Electromagnetism—History. 3. Hertz, Heinrich, 1857–1894—Biography. 4. Physicists—Germany—Biography. I. Mulligan, Joseph F. (Joseph Francis), 1920– . II. Title. III. Series.
    QC7.5.H47 1993
    537—dc20   93-34637
                                                                                                                                            CIP

Printed on acid-free, 250-year-life paper
Manufactured in the United States of America

This book is dedicated
to the memory of
Heinrich Rudolf Hertz (1857–1894)
on the occasion of
the 100th anniversary
of his death

Contents

Preface   xiii
A Note about References   xvii
Chronology of Heinrich Hertz's Life   xix

Introductory Biography
Joseph F. Mulligan

I.A. Hertz's Childhood and Early Education (1857–1875)   3
I.B. Young Manhood (1875–1878)   7
II. Berlin (1878–1883)   12
III. Kiel (1883–1885)   19
IV. Karlsruhe (1885–1889)   24
V. Bonn (1889–1894)   47
VI. Heinrich Hertz's Importance in the History of Physics   78

I. Hertz's Childhood and Young Manhood (1857–1878): Hamburg, Frankfurt, and Munich

Paper No. 1.   Introduction to Heinrich Hertz's *Miscellaneous Papers* (1895) by Philipp Lenard   89

An account, by Hertz's assistant at Bonn, which contains interesting material on Hertz's early life. The quotations from Hertz's letters provide revealing insights into Hertz's character and outlook.

II. Berlin (1878–1883): Advanced Studies in Physics, Friedrich-Wilhelm University

Paper No. 2.   Hermann von Helmholtz: On 31 August 1891
Heinrich Hertz   113

This eloquent tribute to Helmholtz, prepared for the celebration of Helmholtz's seventieth birthday in 1891, reveals Hertz's lifelong gratitude and devotion to his mentor, colleague, and friend.

viii   Contents

III. Kiel University (1883–1885): Physics Instructor (*Privatdozent*)

Paper No. 3. On the Relations between Maxwell's
Fundamental Electromagnetic Equations and
the Fundamental Equations of the Opposing
Electromagnetics

                Heinrich Hertz   127

An important theoretical paper in which Hertz
derives Maxwell's equations by a novel, ingenious
procedure. This leads him to assert the superiority
of Maxwell's theory over all competing theories of
electromagnetism.

Paper No. 4. Review of Hertz's *Miscellaneous Papers* by
George Francis FitzGerald   147

FitzGerald's review provides a good summary of
Hertz's more important early papers and, despite
some probing criticism, reveals his admiration for
Hertz's research accomplishments and those of his
mentor, Helmholtz.

IV. Karlsruhe (1885–1889): Professor of Physics, *Technische Hochschule*

Paper No. 5. Introduction to *Electric Waves, Being Researches
on the Propagation of Electric Action with Finite
Velocity through Space* by Heinrich Hertz   163

An "Introductory Overview" that summarizes both
the experimental and theoretical components of
Hertz's electromagnetic research at Karlsruhe. He
concludes that his experiments have confirmed the
fundamental tenets of the Faraday-Maxwell theory.

Paper No. 6. On Very Rapid Electric Oscillations

                Heinrich Hertz   193

A description of the techniques Hertz employed to
produce and detect electromagnetic waves with
wavelengths of about six meters. Resonance effects
turned out to be of great importance and were used
in his subsequent experiments on electromagnetism.

Contents

Paper No. 7. On an Effect of Ultraviolet Light upon the Electric Discharge
Heinrich Hertz   223

The first paper ever published on the photoelectric effect. In it Hertz shows conclusively that ultraviolet light increases the electric discharge from ordinary metals. In an amazingly thorough series of experiments he then discusses the effect on the discharge of changing the metals, the light sources, and the intervening medium.

Paper No. 8. On Electromagnetic Waves in Air and Their Reflection
Heinrich Hertz   241

Hertz describes his success in producing standing electromagnetic waves and in measuring their wavelengths. He suggests that such waves may be identical with light waves and considers that his observations provide further evidence for Maxwell's theory.

Paper No. 9. On Electric Radiation
Heinrich Hertz   257

This is the most impressive of all Hertz's experimental papers on electromagnetism. He demonstrates conclusively that electromagnetic waves of 66-cm wavelength can be formed into a beam by a parabolic cylindrical reflector and can be reflected, refracted, polarized, and made to interfere. Electromagnetic waves may therefore be properly called "beams of light of very great wavelength."

Paper No. 10. On the Relations between Light and Electricity
Heinrich Hertz   273

In this semi-popular address, delivered to a conference of German scientists and physicians in Heidelberg, Hertz presents the clearest and most eloquent account of his electromagnetic experiments in Karlsruhe. It is one of the most important and frequently quoted addresses in the history of physics.

Paper No. 11. Review of Hertz's *Electric Waves* by
George Francis FitzGerald 289

This review by FitzGerald is not as comprehensive or critical as his reviews of Hertz's *Miscellaneous Papers* and *Principles of Mechanics*. He considers Hertz's experimental research the greatest scientific advance in a quarter century, but objects to Hertz's equating Maxwell's theory with Maxwell's equations.

Paper No. 12. Nomination of Heinrich Hertz as Corresponding Member of the Berlin Academy of Sciences
Hermann von Helmholtz 293

A brief address summarizing Hertz's scientific contributions at Berlin, Kiel, and Karlsruhe and pointing up the interplay between theory and experiment that characterized his work.

V. Bonn University (1889–1894): Professor of Physics

Paper No. 13. On the Passage of Cathode Rays through Thin Metallic Layers
Heinrich Hertz 301

This is the last of Hertz's experimental papers, in which he describes his unexpected discovery that cathode rays pass through thin metal films.

Paper No. 14. Preface to Hertz's *Principles of Mechanics* by
Hermann von Helmholtz 305

Most of Helmholtz's paper is devoted to a lucid discussion of Hertz's contributions to electromagnetism and to praise of Hertz as a superb scientist. The discussion of Hertz's *Mechanics* is surprisingly brief, somewhat tentative, and lacks Helmholtz's usual enthusiasm for anything Hertz did.

Paper No. 15. Heinrich Hertz's Preface to His *Principles of Mechanics* 319

Here Hertz discusses the purpose of his *Mechanics* and its dependence on the work of Helmholtz, Mach and others. He stresses that there is little original about its contents; what is new and important is "the order and arrangement of the whole."

Contents

Paper No. 16. Heinrich Hertz's Introduction to His *Principles of Mechanics* — 323

The most philosophical of all Hertz's papers, in which he compares his newly developed system of mechanics with two more traditional systems. He also develops a theory of knowledge for the physical sciences that has been highly influential in the philosophy of science to the present day.

Paper No. 17. Review of Hertz's *Principles of Mechanics* by George Francis FitzGerald — 365

FitzGerald's review, which contains an excellent summary of the complete contents of the *Mechanics*, raises important questions about the purpose of Hertz's book and how successful he was in achieving his goal.

VI. Heinrich Hertz's Importance in the History of Physics

Paper No. 18. Heinrich Rudolf Hertz: A Memorial Address by Professor Max Planck at the Meeting of the Physical Society of Berlin on 16 Feb. 1894

translated by Joseph F. Mulligan — 383

A tribute by Planck to Hertz presented a month after Hertz's death. In it Planck summarizes the events of Hertz's life, his research contributions, and his character and personality.

Bibliography

A. Books, Articles, and Addresses by Heinrich Hertz — 405

B. Books, Articles, and Addresses Relating to Hertz, His Life and His Work — 410

Index — 423

Preface

The idea for this book came to me in 1985 when I first became interested in the scientific relationship between Heinrich Hertz and Hermann von Helmholtz. Starting to work on this subject as a physicist with little scholarly acquaintance with the history-of-science literature, I was surprised to find that no full-length biography of Hertz existed in any language and that the three volumes of Hertz's *Collected Works* in English were all out of print (even in reprint editions) and were virtually unobtainable on the second-hand book market. To someone aware of the importance of Hertz in the history of physics, it seemed that something should be done about this unacceptable situation. Combining a Hertz biography of somewhat greater length than then existed with some of Hertz's more famous papers, which are scattered through the three volumes of his collected works, struck me as an attractive and realizable step in the right direction. This idea was later expanded to include papers commenting on Hertz and his work by physicists who knew him personally; in particular, Helmholtz, Max Planck, Philipp Lenard, and George Francis FitzGerald. And so the plan for the present volume came into being.

This book is being published in 1994 to commemorate the one-hundredth anniversary of Heinrich Hertz's death at the terribly young age of thirty-six. The introductory biography together with eleven papers by Hertz and seven about him are intended to highlight the importance of Hertz's contributions to physics and at the same time to serve the needs of anyone interested in doing research on this highly gifted scientist.

This book is intended for a primary audience of those professionally involved in physics, the history of science, or the philosophy of science. The secondary audience includes students in the above fields, and nonscientists interested in the history of ideas, a field to which Hertz's contributions are of considerable significance. For example, his experimental work on electromagnetic waves led directly to radio and microwave transmission and to the communications explosion of the present day; his combined theoretical and experimental work on electromagnetism established Maxwell's electromagnetic theory in the rock-solid position it still retains in physics and electrical engineering; and his *Principles of Mechanics*, especially his introduction to that work, was one of the most significant contributions to the philosophy of science made by any nineteenth-century philosopher or scientist.

Hertz's papers set a standard for inspired research and clear, vivid technical writing seldom matched in the history of physics. For this reason a person seeking to understand the nature of scientific thought and its development can find no better place to begin than the papers of Heinrich Hertz. The intermingling of these papers with those of famous physicists commenting directly on Hertz's life and work makes the present volume an ideal vehicle for conveying the methods and values of science to scientists and nonscientists alike.

Contents of the Book

The introductory biography is based in great part on Hertz's own *Memoirs • Letters •Diaries* (San Francisco: San Francisco Press, 1977), on his research papers, and on the research done in German archives by Christa Jungnickel and Russell McCormmach for their superb *Intellectual Mastery of Nature: Theoretical Physics from Ohm to Einstein* (Chicago: University of Chicago Press, 1986), the second volume of which contains a large amount of new material on Hertz. Other important sources include McCormmach's excellent article on Hertz in the *Dictionary of Scientific Biography*, Leo Koenigsberger's life of Hermann von Helmholtz, and the unpublished *Erinnerungen* of Philipp Lenard and Heinrich Kayser. Many other primary and secondary sources were also consulted. These are referred to in footnotes and are listed in the bibliography at the end of the book.

The biography is based, therefore, on the best sources available to me, many of which were primary sources or were immediately traceable to primary sources. It is more complete and, while not definitive, it is better documented than most of the brief biographies of Hertz now in existence. I hope that the present biography may serve as a transitional step to the definitive biography that still needs to be written.

This brief biography obviously cannot convey a full sense of Hertz's abilities as a scientist or qualities as a man. The papers included in this volume were chosen to fill out the picture of this

Preface

exceptional scientist by presenting his major contributions to research in his own words and by showing the effect his research had on his fellow physicists, expressed in their own words. The criteria used in selecting the papers were importance and readability. Although Hertz was equally gifted as an experimentalist and as a theorist, most of his papers included here are experimental, of which those written at Karlsruhe in the years 1886–1888 are best known. The only paper that is rather mathematical is his famous 1884 paper comparing Maxwell's theory with the opposing theories of electromagnetism. The inclusion of this paper was dictated by its importance in tracing the development of Hertz's ideas on electromagnetism from his days in Berlin as a student of Helmholtz to his later experimental work on electromagnetism in Karlsruhe. It would have been useful to include his two theoretical papers on electromagnetism written at Bonn in 1890, but they seemed too mathematically demanding and too extensive for inclusion in what was intended to be a relatively short book.

This collection is somewhat weighted toward Hertz's research in Bonn, a period that is often neglected in favor of his more successful accomplishments in Karlsruhe. His work on cathode rays, and his preface and introduction to his *Principles of Mechanics* are reprinted here, together with Helmholtz's long preface to the *Mechanics* and FitzGerald's review of the same volume. The years in Bonn reveal a great deal about Hertz the man and the scientist. The section of the introductory biography devoted to the Bonn period is therefore considerably fuller than found in previous brief biographies.

Two papers, No. 12 by Helmholtz and No. 18 by Planck, are translated here for the first time. The other papers were either originally in English (the three reviews by FitzGerald) or were available in good English translations by Daniel E. Jones (a student of Hertz in Bonn) and his collaborators in Great Britain. These translations have occasionally been modified in small ways—punctuation, paragraphing, spelling, and the rare replacement of a word by a more up-to-date equivalent—to make them more readable. In the few cases where the translation seems in error, or where sentences or phrases have been omitted, this is clearly indicated in the footnotes. I do not claim to have compared every word of the English translation with the original German, but such a comparison has been made whenever the original translation seemed awkward or confusing.

The papers are arranged for the most part in chronological order. This makes it easy to relate them to the corresponding parts of the introductory biography. Hence a reading of the pertinent papers together with the corresponding section of the biography can provide an integrated picture of each period of Hertz's life that would be unobtainable from either the papers or the biography alone.

Acknowledgments

I am most grateful to the many persons and institutions that have provided help during the course of this project. I would especially like to thank the staffs of the Eisenhower Library at the Johns Hopkins University and the Albin O. Kuhn Library at the University of Maryland, Baltimore County, where much of the research for this book was done. In addition, the staffs at the Deutsches Museum, the Badisches Generallandesarchiv in Karlsruhe, the American Institute of Physics Niels Bohr Library, the Library of Congress, the New York Public Library, the Enoch Pratt Library in Baltimore, and the Smithsonian Institution have been most helpful and generous with their advice and services.

Among my colleagues in physics and the history of physics I would like to express my sincere gratitude to James Brennan, Alfonso Campolattaro, Bruce Hunt, Russell McCormmach, James O'Hara, Kathryn Olesko, Daniel Siegel, and Alfons Weber for their help at various stages of this project. None of these scholars has seen the complete manuscript, however, and any errors that remain in it are completely my responsibility.

I am also grateful to the helpful editorial staff of Garland Publishing, Inc., and especially to Kennie Lyman, my editor, who had the courage to recommend publication of this book and who has made working on it a pleasant and rewarding experience.

My gratitude to my wife, Eleanor, for her advice and assistance is much too great to be put into words.

Joseph F. Mulligan

A Note about References

Most of the footnotes in this book refer to items in Part B of the Bibliography. In the footnotes, books and articles are cited by author (or editor) and year of publication only; the complete reference, including page numbers, will be found in the bibliography.

All citations of papers reprinted here are made by giving the number of the paper as it appears in this volume and the relevant pages. Citations of papers of Hertz not reprinted here are made by giving the name of the volume of Hertz's *Collected Works* and the pages of interest.

Throughout this book the following abbreviations always refer to the translations of Hertz's *Gesammelte Werke* indicated below:

Misc.Pprs.: Heinrich Hertz, *Miscellaneous Papers* (translated by D.E. Jones and G. A. Schott). London: Macmillan and Co., 1896.

El.Waves: Heinrich Hertz, *Electric Waves, Being Researches on the Propagation of Electric Action with Finite Velocity through Space* (translated by D.E. Jones). London: Macmillan and Co., 1893. *Reprint:* Dover Publications, New York, 1962.

Mechanics: Heinrich Hertz, *The Principles of Mechanics Presented in a New Form* (translated by D.E. Jones and J.T. Walley). London, Macmillan and Co., 1899. *Reprint*: Dover Publications, New York, 1956.

The figures from Hertz's articles included here are photocopies of those in his *Electric Waves*, and they retain the numbers assigned to them there by Hertz. This explains the gaps that occur in the numbering of the figures in the present volume, since not all articles in *Electric Waves* are included here.

Chronology of Heinrich Hertz's Life

1857	Heinrich Rudolf Hertz born in Hamburg on 22 Feb.
1863	Enrollment in private school conducted by Dr. Wichard Lange.
1872–1874	Study with private tutors.
1874	Admitted to the upper class of the humanistic *Johanneum Gymnasium* in Hamburg.
1875	Successful completion of the *Abitur*, the school-leaving examination required for admission to a German university.
1875–1876	Work as intern to a master builder in the Public Works Department in Frankfurt am Main, as preparation for a career in structural engineering.
1876	Engineering studies at the Dresden Polytechnic.
1876–1877	Year of military service with the First Railroad Guards Regiment in Berlin.
1877–1878	Studies at both the University and the *Polytechnic* in Munich; switch from engineering to physics.
1878	Beginning of doctoral studies in physics at the Friedrich-Wilhelm University in Berlin, under direction of Professor Hermann von Helmholtz.
1879	Successful completion of prize problem posed by Helmholtz, leading to Hertz's first published paper.
1880	Doctoral oral examination on 5 Feb.; reception of degree *magna cum laude*, with theoretical dissertation on "Induction in Rotating Spheres."
1880	Appointed by Helmholtz to an assistantship in Berlin Physics Institute.

1880–1883	Research leading to fourteen additional published papers on a variety of topics in physics. Among these are two papers on cathode rays.
1883	Appointment as *Privatdozent* in mathematical physics at Kiel. As *Habilitationsschrift* Hertz submits cathode-ray research done in Berlin.
1884	Important theoretical paper on the relation between Maxwell's electromagnetic theory and rival theories of electromagnetism.
1885	Appointment as ordinary Professor of Physics at the *Technische Hochschule* in Karlsruhe.
1886	Marriage to Elisabeth Doll on 31 July.
1886–1889	Hertz's major research effort on electromagnetic waves. His experiments resulted in ten important papers and the definitive confirmation of Maxwell's electromagnetic theory.
1887	Discovery of the photoelectric effect. Birth of first child, Johanna, on 2 Oct.
1888	Unambiguous proof that electromagnetic waves of 66-cm wavelength demonstrate the same properties of reflection, refraction, interference and polarization as light waves, and propagate through space at the same speed.
1889	Appointment as ordinary Professor of Physics at the Friedrich-Wilhelm University in Bonn. Address in Heidelberg on 20 Sept.: "On the Relations between Light and Electricity."
1890	Two important theoretical papers on the equations of electrodynamics for bodies at rest and bodies in motion.

	Visit to Great Britain to receive Rumford Medal from the Royal Society (28 Nov.–4 Dec.).
1891	Birth of second child, Mathilde, on 14 Jan.
1891–1893	Noticeable decline in Hertz's health; research efforts concentrated on his book, *The Principles of Mechanics*.
1892	Publication of an experimental paper on the passage of cathode rays through thin metallic foils.
1894	Heinrich Rudolf Hertz dies of blood poisoning on 1 Jan., following operation on an abscessed tooth.
1896	First successful practical applications of Hertz's electromagnetic-wave research to wireless telegraphy (radio) by A. S. Popov in Russia and Guglielmo Marconi in Italy. First wireless telegram sent by Popov reads simply "Heinrich Hertz."
1933	Official adoption by the International Electrotechnical Commission of the name hertz (abbreviated to Hz) as the unit of frequency, in honor of Heinrich Hertz.

HEINRICH RUDOLF HERTZ

Introductory Biography

Joseph F. Mulligan

> To obtain information for myself and for others directly from nature gives me so much more satisfaction than to be always learning it from others and for myself alone—so much more that I can scarcely express it.
>
> *Heinrich Hertz (1878)*

I.A. Hertz's Childhood and Early Education (1857–1875)

Heinrich Rudolf Hertz was born in Hamburg on 22 February 1857. By that time the Free and Hanseatic City of Hamburg, as it was officially known, had already become Germany's busiest port and an important industrial and commercial city. Its excellent harbor gave it a window to the world, and led to close commercial ties with Great Britain. In 1241 Hamburg had formed an alliance with Lübeck that later became the basis for the Hanseatic League. Hamburg was a city built around water, being located near the mouth of the Elbe River on the North Sea. The Alster River flowed through the city; and near the center of the city were two attractive lakes, the Binnenalster (Inner Alster) and the Aussenalster (Outer Alster). Hertz's love for vacations near large bodies of water goes back to his childhood in Hamburg, where water was seldom out of his sight.

What little we know of the Hertz family impresses us with the happiness of the life of Heinrich, his three brothers and one sister, and their parents in Hamburg.[1] Hertz's father was Gustav Ferdinand Hertz, J.D. (1827–1914), a very successful attorney. In 1877 he was appointed an Appeals Court judge, and in 1887 became the head of Hamburg's judicial system and a member of its Senate. His ancestors were Jewish, but his father Heinrich David Hertz had been "assimilated" in 1834, i.e., converted to the Evangelical Lutheran Church. Hertz's mother was

---

[1] A good picture of life in a city like Hamburg in 1857 can be obtained from Thomas Mann's novel *Buddenbrooks*, which depicts the life of the Buddenbrook family in Lübeck in the last part of the nineteenth century.

the beautiful and devoted Anna Elisabeth Pfefferkorn Hertz (1835–1910), the daughter of a Frankfurt physician and a member of the Evangelical church. Both mother and father were cultured, open-minded individuals. They encouraged their children's individual talents and interests, although they were not always capable of appreciating the technical details of their studies. Heinrich was always close to his father, even though in his younger days he only saw him at dinner every day. In later years Heinrich confided to his mother that he had come to know many outstanding men during his life, including his idol, the physicist Hermann von Helmholtz, but he had never found anyone with the sure insight and broad knowledge of his father. It was said by those who knew the family that Hertz's sense of duty and practical orientation came from his father, his amiable disposition from his mother.[2]

During the years 1857–1875, when Heinrich lived at home, his mother always called him "Heins."[3] He appears to have been a model child, who deeply loved his parents, brothers and sister, liked to study, to make things out of wood or metal, and to play. When Heins was four, his mother heard him praying at bedtime: "Dear God, make the night pass quickly, so I can get back to playing."

Throughout his life Hertz remained grateful to his parents for their devoted upbringing and care. Just three years before his death he wrote them a moving letter on 24 February 1891:

> ... I want to thank you not for mere good intentions, but also for all the good you have done me all these years as you gradually nurtured me to independence. If my external successes (for which I can really take very little credit) give you some pleasure in return, then this is the best and only reward you may expect.[4]

---

[2]More details of Hertz's early life, together with some charming photographs of the Hertz family, are to be found in Gerhard Hertz (1988). Heinrich's only sister, Melanie, was born in 1873.

[3]Hertz's early days are well described in: Heinrich Hertz: Memoirs•Letters•Diaries. Second enlarged [German-English] edition; arranged by Johanna Hertz; edited by Mathilde Hertz and Charles Susskind. (San Francisco Press, San Francisco, 1977). Johanna Hertz and Mathilde Hertz were the two daughters of Heinrich Hertz. In what follows, this volume will be referred to simply as H. Hertz, *Erinnerungen*. His mother's charming account of Hertz's childhood, which was written after Hertz's death, is on pages 1–21.

[4]H. Hertz, *Erinnerungen*, p. 313. This appreciation of his parents' loving support reminds one of Ernest Rutherford's cable to his mother after being

Introductory Biography

In 1863, when he was six, Heinrich was enrolled in a school run by Dr. Wichard Lange. There he was always near the top of his class, but somewhat embarrassed by his success. A few years later, when he was ten, he told his mother that Dr. Lange had asked his pupils who was the brightest and most ingenious boy in the class, and they had all pointed to him. Heinrich was pleased at this but reported to his mother that at first he felt like crawling under his desk. He remained in Dr. Lange's school for nine years. When he was seventeen, he set down on paper his impressions. The work was hard and the discipline strict, but still most of the students loved the school, especially "the lively spirit of competition that was kept alert in us, and the conscientiousness of the teachers who never let merit go unrewarded nor error unpunished."[5]

Many of his teachers and friends commented on Hertz's considerable artistic ability. The Hertz family doctor, Dr. Eduard Cohen, startled Mrs. Hertz one day by inquiring which of her four sons was the Michelangelo of the family. He had seen some clay models made by their eldest son and was convinced that Heinrich should be a sculptor. Heinrich had learned to draw when he was only three, and at age eight took drawing lessons under a Herr Erich. He excelled particularly in geometrical drawing—a talent evident in his line drawings for later scientific articles. He also attended classes in the industrial high school (*Gewerbeschule*), where he learned carpentry and metalworking. He was soon producing footstools, small tables and handsome cabinets for his parents and brothers and even constructed some scientific apparatus for his own use.

In later years, when Hertz was appointed Professor of Physics at Karlsruhe, the craftsman who had originally taught him carpentry exclaimed: "What a pity! That boy had the makings of a first-rate woodturner." This ability to work with his hands would become one of Hertz's great advantages as an experimental physicist: he was able to make his own equipment in much shorter time than he could purchase it commercially. Sometimes the equipment was not so

---

named a baron in 1931: "Now Lord Rutherford, more your honor than mine, Ernest."

[5]H. Hertz, *Erinnerungen*, p. 15.

elaborate or substantial as the commercial version might be, but it did the job and saved equipment money that was always in short supply.[6]

In 1872 his father withdrew Heinrich from Dr. Lange's school to allow him to study with private tutors in preparation for admission two years later to the most advanced class in the humanistic *Johanneum Gymnasium*. Both from his tutors and his *Gymnasium* instructors Hertz imbibed a healthy dose of history and philology, which were the core of the curriculum. He loved languages, literature and especially poetry. He studied Latin, Greek, Arabic, and even Sanskrit. In particular he treasured Dante, Homer and the Greek tragedians, and when he was alone in later life often recited aloud their resonant verses. For him this pleasure seems to have taken the place of music, a field in which he had no talent. He could not hold a note and was discouraged from singing in school choruses and choirs. Despite this, while a university student in Berlin he insisted on torturing his fellow students by walking up and down the corridors in the middle of the night trying to sing in the loudest possible voice.[7]

Although he loved knowledge in any form, his unique bent was towards mathematics and the sciences. His mother tells how he tried to talk to her about the advanced mathematics he was learning in school. She told him honestly that she was too untrained in mathematics to understand what he was telling her. He put his arms around her with great tenderness and said with deep emotion: "Poor Mama, that you have to miss this pleasure." Mathematics was one of the great pleasures he continued to enjoy throughout his life, especially in his last years, when he was unwell and could no longer work in the laboratory.

At the end of his *Gymnasium* studies in 1875,[8] Hertz passed the matriculation or "school-leaving" examination (*Abitur*), which qualified him to attend any German university of his choice. At that time he had to submit an autobiographical essay. Encouraged by his

---

[6]The high quality of Hertz's craftsmanship can be seen in some apparatus that remains today either in its original form or in replicas at the Deutsches Museum in Munich and the Science Museum in London. Photographs of this apparatus may be found in Appleyard (1930) or in Bryant (1988).

[7]Kayser (1936), p. 106.

[8]At this time his *Gymnasium* instructors commented on his one failing, "a monotonous enunciation which his ear is incapable of perceiving."

Introductory Biography 7

ability to design and construct useful scientific apparatus, along with his interest in mathematics and science, he wrote of his intention to become a structural engineer (*Bauingenieur*): "... only if I were to prove unsuited to this profession or if my interest in the natural sciences were to increase further, would I devote myself to pure science. May God take care that I choose whatever I am best suited for."[9]

I.B. Young Manhood (1875–1878)

Before Hertz finally gave in to his propensity for pure science, however, he was to experience some difficult years of indecision. After passing the matriculation examination, he determined to postpone his university studies and go to Frankfurt am Main to work for a year as an intern to a master builder (*Baumeister*) in the Public Works Department. Such experience was required if he wanted to obtain a license as a structural engineer.

Hertz spent the year from March 1875 to April 1876 in Frankfurt. It was his first experience living away from his family, and he was terribly homesick. He had no friends in Frankfurt (only a few older relatives of his mother), nor does he appear to have made much of an effort to make new friends. When his mother and the other Hertz children passed through Frankfurt (in October 1875) on their way back to Hamburg, he met them at the train station, walked back to his apartment afterwards and confided to his diary: "How I would have liked to have gone with them!"

The first few months in Frankfurt were especially hard. Hertz found himself bored with the make-work assigned to him by his supervisors, sketching from plaster casts of buildings, changing the scale of architectural drawings, and finally—somewhat more to his liking—helping with plans for a new stock-market building and a bridge across the Upper Main. But still he longed for something more challenging and more theoretical. His diary entries indicate the diversity of his interests during this period: he took sculpting lessons at the Staedel Institute; read Plato's *Republic* and the plays of Euripides in the original Greek; studied mathematics, political economy and

---

[9]H. Hertz, *Erinnerungen*, p. 21.

physiology; and even dissected a frog. He was losing interest in engineering, but had not discerned that this casting about was a clear sign that the field was not for him.

On 1 August 1875, Hertz wrote to his parents that he had gone with his uncle to visit a friend of his uncle's in Sachsenhausen.[10] He says that, when the son of the friend showed them his workshop and his small laboratory, he suddenly became somewhat envious and very homesick.

In late October he finally withdrew a copy of Wüllner's *Physics* from the library and dug into it with delight. It nourished his strong interest in pure science, but he still did not reverse his decision to be an engineer. He wrote to his parents on 25 October 1875: "... I cannot convince myself to give up what I have set as the most desirable goal for myself." Perhaps the root of his hesitancy to submit to his strong drive toward theoretical science was a lack of confidence in his own abilities. This lack of confidence, which seems often to have troubled him, surfaced in a letter to his parents a few years later:

> I have not forgotten what I often used to say to myself, that I would rather be a great scientific investigator than a great engineer, but I would rather be a second-rate engineer than a second-rate investigator.[11]

Hertz gradually acclimated himself to Frankfurt and returned there after Christmas better able to cope with his situation for the remaining few months. On 21 April 1876 he moved to Dresden to begin his engineering studies at the Dresden Polytechnic, a technical institute.[12]

---

[10] This Sachsenhausen was at that time a town near Frankfurt on the other side of the Main river. It is not the Sachsenhausen near Berlin made infamous by a Nazi concentration camp during World War II.

[11] Letter from Hertz in Munich to his parents, 1 November 1877. See the first page of Paper No. 1 in the present collection.

[12] In the latter half of the nineteenth century Germany had nine "technical institutes," located in Aachen, Berlin, Braunschweig, Darmstadt, Dresden, Hannover, Karlsruhe, Munich and Stuttgart. These had grown in size and status over the years. They were originally called *Polytechnics*, but by about 1880 had officially gained the status of "higher schools." A change of name from *Polytechnic* to *Technische Hochschule* accompanied this official change of status, but was implemented at different times in different places. For this reason we will refer to the Dresden and Munich technical institutes as *Polytechnics* when Hertz studied there in 1876–1877, and the one at Karlsruhe

Introductory Biography

In Dresden Hertz was lonely at first and, in an endeavor to make new friends, decided to join the Cherusker student corporation. This was a student organization something like a modern college fraternity, but famous not only for drinking but for fencing. After less than a week as a member, Hertz was upset by the heavy beer drinking and wrote to his father, requesting that he write a letter forbidding Heinrich to remain a member of the Cherusker. His father wisely replied in the way his son had requested, but wrote a second letter advising Heinrich not to make use of the first letter as an excuse for dropping out of the club. Hertz followed this advice and stayed in the club for the three months of the summer semester. He began to enjoy the fencing, which he found good for his health, and even added later in a letter "nor is the drinking quite so terrible."

Hertz attended classes in physics, descriptive geometry, drafting and comparative psychology, but found them all boring and elementary. He found the history of philosophy course, however, "incredibly interesting," as were the lectures of Professor Leo Koenigsberger[13] on the integral calculus. What he liked best about Koenigsberger as a teacher was that he did not routinely follow a textbook, and occasionally digressed to provide helpful glimpses of more advanced mathematics.

In Dresden Hertz worked hard, in particular on analytical geometry and calculus, but found time to read Kant's *Critique of Pure Reason* and Dickens' *Oliver Twist* (which he did not like as much as other books by Dickens) and gave a lecture on Darwinism at the Cheruscia. His engineering studies had not taken him too far astray from his true love—mathematics and pure science—and Koenigsberger's lectures on the calculus, supplemented by Hertz's own wide readings in the subject, prepared him well for subsequent work in physics.

---

as a *Technische Hochschule* during Hertz's stay there from 1885 to 1889. On this see Jungnickel and McCormmach (1986), vol. 2, pp. 54–56, especially footnote 83 and the references cited there. In present-day Germany the former *Technische Hochschule*n are called *Technische Universitäten*.

[13] Leo Koenigsberger (1837–1921), a well-known mathematician, had come to Dresden from Heidelberg, where he had been a colleague of Helmholtz and Kirchhoff. He and Kirchhoff had worked very closely together in a mathematical physics seminar at Heidelberg. Koenigsberger later wrote the definitive three-volume biography of Helmholtz.

But now he was forced to interrupt his studies for a while. At this period in German history, after the unification of Germany in 1871, all young men were required to contribute some time to the military service of their country. University students had their full-time military training reduced to one full year. On the successful passing of an examination at the end of this period, they were commissioned as reserve officers,[14] with the obligation of spending additional time in future summer exercises to retain their commissions. Hertz's year of duty, from 1 October 1876 to 1 October 1877, was spent with the First Railway Guards Regiment in Berlin.

Hertz adapted to the strict military regime rather well. There was little time to do anything but drill and work all day, relieved only by breaks to eat, drink and sleep. There was no leisure—to judge from his diary—even to enjoy the treasures and pleasures of a great city like Berlin. But at least these embryonic soldiers ate reasonably well, slept like logs at night, and remained in good spirits with the realization that they were all in the same boat and that their voyage was limited in duration. The exercise and strict hours improved Heinrich's stamina, and he enjoyed the camaraderie with his fellow soldiers. He made new friends whom he liked and whose opinions he respected. He also won the approval of his superior officers, who appreciated his highly motivated sense of duty.

At the end of his year of service, in September 1877, Hertz took the officers' final examination. He did moderately well: he passed the examination, but was transferred to the infantry instead of being allowed to remain with the Railway Guards. He was content with this, fairly happy with his year in Berlin, but pessimistic about the future. He wrote to his parents that he was sorry that he had accomplished so little in the twenty years of his life, and was afraid that he would not do much better in the years ahead.

He and his parents then decided that he would not return to Dresden, but would continue his studies at the *Polytechnic* in Munich. After a few weeks of needed relaxation at home, he moved to Munich at the end of October 1877.

Just when the winter semester (November through February in German universities) was about to begin, Hertz had a sudden change

---

[14]In German society at that time being a *Reserveoffizier* conferred as much prestige as would the academic title of *Privatdozent*.

of heart. On 1 November he wrote an agonizing letter to his father asking that he be allowed to switch from engineering to the natural sciences.[15] His father consented, and Hertz began the year in Munich with new joy and zest for his studies, for now he knew that he was finally where he belonged.

He liked Munich better than either Berlin or Dresden. The buildings were more impressive, the parks lovelier, and the surrounding countryside brighter than in northern Germany. He also found the art in the museums and galleries superior to that in Berlin, but this may have only been because he now had more time to enjoy such things.

Hertz took courses at both the *Polytechnic* and the University, but was unhappy that they were not more demanding. He told his parents that some of the physics courses in Munich did not differ much from those he had taken at Dr. Lange's school in Hamburg! He complained that professors often arrived one-half hour late for their classes, cancelled classes for the most flimsy reasons, and by so doing short-changed their students. Every day seemed to be a holiday in Munich, and little substantial work was done.

Hertz consulted the physics professor at the University, Philipp von Jolly, who was very friendly and advised him to take as much mathematics and mechanics as possible, and to study on his own the great works of Lagrange, Laplace and Poisson. During the summer semester he took laboratory courses at both institutions.

His letters during this year in Munich reveal a growing impatience with his lot as a student. He longed to put textbooks behind him and learn for himself in the research laboratory. He wrote to his parents on 25 November 1877: "... I am burning with impatience to reach the frontiers of what is already known and to go on exploring into unknown territory; ..." In a letter on 4 February 1878 he says he would have preferred living in earlier days when so many wonderful things were still undiscovered. He longed for the days when the

---

[15]This occurrence is well described by Philipp Lenard in the first few pages of his Introduction to Hertz's *Miscellaneous Papers* (Paper No. 1); Hertz's letter to his father can also be found there.

telescope and microscope were still new, and so many startling discoveries remained to be made.[16]

By the end of July Hertz had grown disillusioned with his studies in Munich and decided to go either to Leipzig or Berlin for his doctorate in physics, with Bonn, where Clausius[17] was the professor of physics, an outside possibility. Hertz was determined to choose the university best able to prepare him for a research career in physics. Here Berlin held the trump card, since it had not one but two distinguished physics professors, Helmholtz and Kirchhoff, and Helmholtz was widely recognized as the best physics research advisor in any German university. Hertz wisely settled on Berlin.

He spent the vacation months of August, September and October at home studying and constructing apparatus on his small lathe. He built a very sensitive tangent galvanometer, which was to be of considerable use to him in Berlin, where he was finally able to do the front-line physics research he had always craved.

II. Berlin (1878–1883)

Heinrich Hertz arrived in Berlin to begin his university studies in physics near the end of October 1878. Although he had spent his year of military service in Berlin (1876–77), this great city was still a revelation to him, for during that time he had enjoyed little leisure to appreciate its wonders and charm. Now he was seeing Berlin through the eyes of an inquisitive, sensitive student at one of the foremost universities in the world.

Berlin had emerged after the Napoleonic Wars as a center of national feeling, a rival to Vienna as the intellectual and cultural center of the German-speaking world. Its enduring symbol was the monumental Brandenburg Gate, constructed in 1788–1791 during the reign of King Friedrich Wilhelm II, ruler of Prussia from 1786 to 1797.

---

[16]H. Hertz, *Erinnerungen*, p. 81. Hertz's statement is reminiscent of the words of Victor Hugo: "Where the telescope ends, the microscope begins. Who is to say of the two, which is the grander view?"

[17]Rudolf Clausius (1822–1888), professor of physics at Bonn from 1869 to 1888, famous for his contributions to thermodynamics, and in particular for the concept of *entropy*. Hertz would succeed Clausius in the chair of physics at Bonn in 1889.

Introductory Biography

The advent of the railroads in the middle of the nineteenth century made Berlin also the industrial and commercial center of Germany. After Berlin was named capital of the new German Empire in 1871 at Versailles, its population, which was about 900,000 at that time, grew to almost three million by 1900. The influx of so many new residents fostered a simultaneous growth in its educational and cultural life.

The university in Berlin was founded in 1810 at the instigation of Friedrich von Humboldt and King Friedrich Wilhelm III, who ruled from 1797 to 1840. It was named the Friedrich-Wilhelm University. Soon the university had added professors of the caliber of Fichte, Hegel and Ranke to its faculty. After the unification of the German nation in 1871, a determined effort was made to convert it into the showpiece of the German university world. This effort attracted Hermann von Helmholtz (1821–1894) to Berlin in 1871 as professor of physics, to be followed in 1875 by his friend and colleague, Gustav Kirchhoff (1824–1887), who became "professor of mathematical physics." For ten years these outstanding physicists brought distinction to Berlin as they had to Heidelberg in the years 1858–1871. That period had been called by Emil Du Bois-Reymond (1818–1896) "an era of brilliance such as had seldom existed for any university and which will not readily be seen again." Physics at Berlin in the period beginning about 1875, however, was to rival and even outdo Heidelberg in academic stature and scientific accomplishment.

Hertz liked the university from the first day he entered its gates. He was especially impressed by the Physics Institute—by the availability of so many physics periodicals in the library, by the high quality of the equipment, by the conviction on the part of the faculty and students that physics research was important and that some of the best research in the world was being done right in their Institute. Research had always been Hertz's goal, and in Berlin he knew he would have the opportunity to test his ability as an independent investigator in the kind of intellectual environment required for top-flight research.

Hertz was able to test his research skills much sooner than he expected. On arriving at the university in October 1878, he noticed that a prize was being offered by the Philosophical Faculty (at the suggestion of Helmholtz) for the solution of a physics problem on the inertia of electric currents in conductors. Hertz felt he had the background to compete. He did so, completed the work by the end of

January 1879, and won the prize.[18] This research led to his first published paper.[19]

Helmholtz was obviously pleased by Hertz's success, since he already considered him "a student of most unusual promise." He therefore brought to Hertz's attention the following prize problem set by the Berlin Academy of Science in 1879 (at Helmholtz's instigation): "To establish experimentally any relation between electromagnetic forces and the dielectric polarization of insulators."[20] Helmholtz had suggested this problem because its solution was crucial in deciding among the competing theories of electromagnetism at that time, a subject in which Helmholtz was deeply interested.[21] Helmholtz promised Hertz all the assistance the Physical Institute could provide, were he to accept this challenge. In his usual cautious manner, Hertz considered the problem, did some detailed mathematical calculations, and concluded that "a clear-cut effect is not to be hoped for, but only one that lies at the limit of observation."[22] He informed a disappointed

---

[18]This research is described by Philipp Lenard in Paper No. 1, and is clarified by the letters of Hertz to his parents that are included in Lenard's account. The problem Hertz solved is also discussed by Helmholtz in Paper No. 14.

[19]H. Hertz, "Experiment to Determine an Upper Limit to the Kinetic Energy of an Electric Current," Wiedemann's *Annalen der Physik und Chemie* 10, 414–448 (1880). This paper is contained in Hertz's *Miscellaneous Papers*, pp. 1–34. To avoid confusion, it is worth noting here that the official name of the *Annalen* from 1824 to 1899 was *Annalen der Physik und Chemie*. It was edited from 1824 to 1877 by J.C. Poggendorff (1796–1877), and usually referred to as Poggendorff's *Annalen*. From 1877 to 1899 G.H. Wiedemann (1826–1899) was the editor, and the journal was commonly known as Wiedemann's *Annalen*. In 1900 Paul Drude (1863–1906) became the editor, and the name was officially changed to *Annalen der Physik*, with no further mention of chemistry. Henceforth in this book we shall refer to this journal simply as *Annalen*, no matter what the year of publication.

[20]Hertz's own account of this can be found in his Introduction to his *Electric Waves* (on the first page of Paper No. 5). Helmholtz had set this problem expecting that Hertz would succeed in solving it. On this see footnote 4 to Paper No. 14.

[21]In 1879, the year under discussion here and the year in which James Clerk Maxwell died, his theory of electromagnetism was still not widely accepted, even in Great Britain.

[22]J.G. O'Hara and W. Pricha have discovered a relevant unpublished manuscript in the British Science Museum in London, which was presented to the museum when Elisabeth Hertz and her two daughters emigrated to

Introductory Biography 15

Helmholtz of his negative decision and turned instead to a theoretical problem that he hoped to submit as his doctoral dissertation. (At Karlsruhe in 1887, however, he would return by a rather circuitous route to the proposed prize problem and solve it to his own and his mentor's complete satisfaction.) Hertz devoted much of his short research career (1880–1893) to theoretical and experimental work on various aspects of this 1879 prize-problem.

Hertz's doctoral dissertation concerned the currents induced when metal spheres rotate in a magnetic field. He had made some preliminary studies of the problem at home during the 1878 autumn vacation and worked at it in earnest during the winter semester of the 1879–1880 academic year in Berlin, completing it in February 1880. Anyone who doubts that Hertz was a first-rate mathematical physicist should read this dissertation.[23] It contains some powerful and (for that day) sophisticated mathematics. That he could have completed it in such a short time is remarkable.

Hertz took his doctoral oral examination on 5 February 1880. His committee consisted of Professors Helmholtz, Kirchhoff, Kummer, and Zeller.[24] In the examination Hertz was disappointed with the easy questions Zeller proposed on the philosophy of Plato and his school. Hertz had a thorough knowledge of philosophy and would have enjoyed the chance to display more of it. He found some of the physics and mathematics questions more probing, but the examiners encouraged him by sprinkling in some easy questions and by changing the subject if he ran into any difficulties.[25] Hertz was awarded his

---

Cambridge from Bonn in 1937. This represents work done by Hertz in Hamburg during the vacation period from August through October 1879, on this prize problem. The manuscript consists of a preamble, in which Hertz outlines the experiments that would be required, and appendices in which his calculations are presented. The preamble to the manuscript in both German and English has been published by O'Hara and Pricha (1987), pp. 120–128.

[23]This is in *Miscellaneous Papers*, pp. 35–126. In the rest of this book we shall refer to this volume of Hertz's collected works as *Misc.Pprs*.

[24]E. Kummer (1810–1893), professor of mathematics in Berlin, 1856 to 1883; E. Zeller (1814–1908), professor of philosophy in Berlin, 1872 to 1895. Zeller was an eminent philosopher who was persuaded to come to Berlin by Helmholtz. On the nature of the doctoral examination at German universities during this period, see Mulligan (1992).

[25]J. Kuczera quotes the actual proceedings of Hertz's examination from the University of Berlin archives. They state that Hertz answered all Helmholtz's

degree *magna cum laude*, a rare achievement in Berlin, and especially rare for students of Helmholtz and Kirchhoff, two mentors who set the highest standards for their students.

Hertz received his D. Phil. degree after only five semesters of university-level study in physics, two at Munich and three in Berlin. This was only possible because of his strong background in mathematics and physics from his earlier studies, together with all he had absorbed by reading classic treatises like those of Laplace and Lagrange on mechanics.

After his doctoral examination Hertz was tired of books and theoretical physics. He plunged himself totally into experimental research, sometimes spending the day from nine in the morning until nine at night in the laboratory. He had no lack of ideas for research; he wrote to his parents that he could think up more research projects in a day than he could complete in a year.

In August 1880 Helmholtz offered Hertz an assistantship in the Physics Institute. Hertz was delighted to accept the appointment, but it turned out to be more work than he bargained for. He was the assistant in charge of one part of the basic laboratory course and was responsible for a part of the Institute's equipment. This kept him busy from ten to three every day, stealing a huge block of time from the research he was encouraged to perform as an assistant. He retained this position until 1883, when he left Berlin for Kiel. During those three years he completed the research for fourteen additional papers on a variety of subjects.[26] He was able to accomplish so much by the wise use of weekends and vacation periods; fortunately, the academic schedule typical of German universities at that time offered March, April, August, September and October as vacation months.

But Hertz still found time for some social life, although he did not always enjoy it. His letters frequently refer to his being at the

---

questions on the wave theory of light "with great confidence and clarity," and that Zeller was very "well satisfied" (*sehr befriedigt*) with his answers on Greek philosophy. See Kuczera (1975), p. 24.

[26]These fourteen papers are all included in *Misc.Pprs.*, pp. 127–272. They are discussed briefly in Lenard's Introduction (Paper No. 1), and more critically in George Francis FitzGerald's review of *Misc.Pprs.* in *Nature* (Paper No. 4) and in Planck's tribute to Hertz after his death (Paper No. 18). Included in the published papers based on his research in Berlin were eight on various aspects of electromagnetism, which gave him an excellent experimental background for his later research on electromagnetism in Karlsruhe.

Helmholtz home for breakfast, luncheon or tea. He preferred it when only the Helmholtz family was there, or when he was accompanied by another assistant like H. Kayser or E. Hagen. He found that he was less self-conscious and more at ease in such circumstances. There is also a reference to his meeting Kirchhoff at a large party in the city, and of his gratitude to Kirchhoff for his cordiality and kindness to him on that occasion. It seems likely that he was at least occasionally a guest at the elegant dinner parties hosted by Helmholtz and his charming, socially conscious wife, to whose home professors, artists, musicians, politicians—the cream of Berlin's intellectual community—were invited.[27]

Hertz was extremely happy during his five years in Berlin. He had proved to his own satisfaction and that of renowned physicists like Helmholtz and Kirchhoff that he could do publishable research which, if not world-shaking, was solid and promised well for the future. For this reason he felt that he had gained a position at the center of his field, and, "whether it be folly or wisdom, it is a very pleasant feeling."[28]

This feeling of security in his physics career was supported by Hertz's realization that he could do either experimental or theoretical work with equal ease. His work was marked by transitions from theory to laboratory research and back again—the same pattern that had distinguished Helmholtz's career. While in Berlin, Hertz had also absorbed some of Helmholtz's feeling for the important problems of physics. Throughout his career Hertz paid very close attention to the problems that Helmholtz found important, but he approached these problems independently, bringing to them his own critical insights.[29] The problems were often suggested by Helmholtz, but it was Hertz who solved them.

There were occasional storm winds stirring up the "calm seas and prosperous voyage" of Hertz's relationship to his mentor. Although he had the greatest respect for Helmholtz, he was understandably upset when in October 1882 his monthly stipends as an assistant came to an abrupt halt because Helmholtz had failed to

---

[27]On Helmholtz's wife, Anna, see Dilthey (1961).

[28]H. Hertz, *Erinnerungen*, p. 137.

[29]For Hertz's immense respect and affection for Helmholtz, see his article on the occasion of Helmholtz's seventieth birthday (Paper No. 2). For a more detailed discussion of Helmholtz's influence on Hertz's contributions to physics, see D'Agostino (1971) and Mulligan (1987).

submit the papers required for his payment and that of another assistant (Hagen). And in 1883 Helmholtz unaccountably did something that might have lost Hertz to physics forever.[30] The administrators of the city of Berlin wanted someone to supervise the installation of electric lighting in the city. An official approached Helmholtz and asked him to suggest someone for the job. Helmholtz proceeded to recommend Hertz as his most promising young physicist with a background in electricity. It is true that there were few experts on practical electricity in 1883, and prominent physicists like Kelvin and Lodge had devoted time to practical applications of electricity without damaging their physics careers. But Hertz's situation was different: he had earlier abandoned engineering in favor of physics and his success in Berlin meant that he could now look forward to a distinguished career in the field he loved. He was reluctant to reintroduce into his life the uncertainty that had plagued his earlier studies. Out of deference for his mentor, Hertz met with the government official once and then broke off all further talks on the matter. After this incident, however, some nagging doubts may have remained in Hertz's mind about how well Helmholtz really understood his favorite student.

Despite his five wonderful years in Berlin, Hertz knew that the time had come to look elsewhere for an academic position. It would have been easy for him to qualify for the *Habilitation* (the right to teach) in Berlin, but he felt uncomfortable about this possibility. There already were too many *Privatdozenten* in physics in Helmholtz's institute, and too few regular faculty positions opening up anywhere in Germany. He might have expected that he would be favored by Helmholtz for any position that did become available, but still he was not sure. And so he decided to test the waters in a smaller and quieter sea.

---

[30]The following account is based on an article by Eugen Goldstein (1850–1930), who received his doctorate at Berlin under Helmholtz in 1881 and was a colleague of Hertz during the years 1878–1883. Goldstein is known for his research on cathode rays and first interested Hertz in work on cathode rays at Berlin. See Goldstein (1925), p. 42.

III. Kiel (1883–1885)

In February 1883, near the end of the winter semester in Berlin, Kirchhoff told Hertz that the university in Kiel was looking for a *Privatdozent* for mathematical physics, and that he and the Professor of Mathematics, Karl Weierstrass (1815–1897), would be happy to recommend Hertz for the position. Hertz at first found the idea of a change of venue appealing, but on discussing the matter with Weierstrass he was less certain. The faculty in Kiel really wanted an *Extraordinarius* (Associate Professor) in mathematical physics, but the Education Ministry could not support such a position at that time. It seemed likely, however, that the ministry would approve the appointment of a *Privatdozent*, with an unwritten agreement that the position would be upgraded to the rank of *Extraordinarius* in a few years.

Helmholtz was reluctant to express any opinion, since he felt that Hertz should decide his own future. He agreed that Hertz would be of more use in Kiel as a *Privatdozent* than he would in Berlin. He also agreed to recommend Hertz for the position, with the caveat that it would probably do little good, since Professor Gustav Karsten, the director of the Kiel Physics Institute, was not on the best of terms with the Berlin faculty. But, more important, Helmholtz pointed out that Kiel's physical facilities were very limited, and that Hertz would probably not have adequate equipment there to do any worthwhile experimental research. This was a major drawback for Hertz, since he would miss not only the laboratory space and the equipment, but also the library books and journals so abundant in Berlin. Hertz also wondered how he would react to the transition from the busy artistic and cultural life of Berlin to the small-town atmosphere of Kiel, whose entire university enrolled only 300 students.

Finally, after visiting Kiel and finding that the faculty were enthusiastic about his joining their ranks, he accepted the position. The probability of an eventual associate professorship and the advice of his parents seem to have won out over his worries that he was being appointed only as a mathematical physicist and would probably have little opportunity in Kiel to demonstrate that he was more than this. Unlike many *Privatdozenten*, he was to receive a modest yearly stipend of 500 Thaler ($375), in addition to whatever student fees he received for the courses he taught.

Hertz submitted as his *Habilitationsschrift* his research on cathode rays done at Berlin. He defended it before a faculty committee, who decided that he had indeed earned the right to teach at the university level. Hertz assumed his new position at the beginning of the 1883 summer semester, delivering an inaugural lecture on May 5 on the foundations of the mechanical theory of heat.[31]

Kiel is situated on the Baltic sea, seventy miles by train north of Hamburg, and is the chief city and port of Schleswig-Holstein, which in 1866 had become part of Prussia. Since the tenth century Kiel had been renowned for its excellent and beautiful harbor. During his two years there Hertz greatly enjoyed boat excursions around the harbor with his university colleagues and, during the brief summers, swimming at the nearby Baltic beaches.

But Hertz found Kiel a dark and lonely place. In December 1883, he wrote to his parents that at that time of the year the sun rose when he was going to lunch, and set at the time for afternoon coffee. He added: "... the mind is hard put not to succumb to the same mood." This loneliness was intensified by the death of his brother Otto in 1884. Though ten years younger than Heinrich, Otto had been his favorite sibling and the joy of the Hertz household.[32] Hertz's diary for this period is filled with indications of unhappiness: he was depressed, and unable to do any meaningful research.

In Kiel Hertz's research suffered, as he had feared, from a lack of proper experimental facilities and equipment. This forced him to devote much of his time to thinking about what experiments he would perform when he once again had the means to perform them. From this perspective the two years in Kiel may have been extremely valuable to his development as a physicist. He also found that he had much more time to devote to his own work, since he had fewer duties at the university, and there were far fewer social obligations than there had been in Berlin.

But it is unlikely that Hertz considered these circumstances advantageous at that time, which was a distressing one for him. He tried his hand at hydrodynamics, but without much success; he

---

[31]For an excellent discussion, based on archival materials, of Hertz's two years at Kiel, see Jungnickel and McCormmach (1986), vol. 2, pp. 43–48. The material presented here owes much to this account.

[32]As an indication of Hertz's closeness to his youngest sibling, see his jesting letter to Otto on 24 Feb. 1878, in *Erinnerungen*, p. 85.

Introductory Biography

attempted to find explanations for many of the meteorological phenomena he observed while sailing on the harbor, but little came of it except one minor theoretical paper on the adiabatic changes of moist air;[33] and he read literature, history and philosophy in an attempt to find something to hold his interest. His lectures, after a shaky beginning, seem to have gone well, and he was happy to report good class attendance to his parents. He also found pleasure in building some small equipment for use in demonstration lectures and simple laboratory exercises.

By the following spring Hertz had begun serious theoretical work on electrodynamics and on 19 May 1884 he confided to his diary that he had hit upon the solution to the electromagnetic problem he had been struggling with. This, his major research contribution while at Kiel, was published later that year in the *Annalen* under the title "On the Relations between Maxwell's Fundamental Electromagnetic Equations and the Fundamental Equations of the Opposing Electromagnetics."[34]

It was inevitable that, given his inability to do worthwhile experimental work at Kiel, Hertz should be attracted to Maxwell's theory (more precisely, Helmholtz's version[35] of Maxwell's theory) as a topic for theoretical research. Hertz started with the basic equations of the older action-at-a-distance theories of Weber and Neumann.[36] But

---

[33]This paper turned out to be of considerable use to meteorologists. On this see Garber (1976), especially pp. 57–60.

[34]Paper No. 3.

[35]Hertz frequently makes reference to the famous 1870 paper of Helmholtz: "On the Theory of Electrodynamics. Part l: On the Equations of Motion of Electricity in Conducting Bodies at Rest," in Helmholtz (1882–1895), vol. l, pp. 545–628.

[36]Wilhelm Weber (1804–1891) and Franz Neumann (1798–1895) were the two German physicists whose theories dominated electromagnetism in Germany in the middle of the nineteenth century. Both theories were action-at-a-distance theories, which assumed that electromagnetic forces propagated at an infinite speed through space. In other respects the two theories differed greatly. Weber assumed that electricity consisted of two fluids, one of particles with positive charges, the other of particles with negative charges. The force-law between pairs of particles had a Coulomb term, which depended only on the charges and their separation, but also contained two additional terms that depended on the relative velocities and relative accelerations of the two particles.

he introduced a new basic principle, the "unity of electric force," which asserted that the electric field generated by a changing magnetic field is in every way equivalent to an equal and equally directed electric field of electrostatic origin.[37] Using this principle, together with the known experimental facts of electricity and the principle of conservation of energy, Hertz was able to derive Maxwell's equations by an ingenious method that employed a process of successive approximations. Hertz's conclusion was that all action-at-a-distance theories were incomplete and unsatisfactory and that, in contrast, Maxwell's theory was both complete and far easier to apply. As a result, ". . . if the choice rests only between the usual system of electromagnetism and Maxwell's, the latter is certainly to be preferred. . . ."[38]

This provocative statement, especially coming from a disciple of Helmholtz, caused considerable discussion, but few German physicists were convinced by Hertz's valiant efforts.[39] His work on this paper, however, gave Hertz great insight into the beauty and power of Maxwell's ideas and suggested to him a possible line of

---

Neumann's theory, on the other hand, was a potential theory, in which electromagnetic induction was handled in terms of the vector potential. This theory was free of any assumptions about the nature of electricity, but was quite abstract and mathematical, and provided little physical insight into electromagnetic processes.

These two theories dominated electrodynamics in Germany until Helmholtz's 1870 paper (see footnote 35) appeared. For further discussion of Weber's and Neumann's contributions to electromagnetic theory, see Whittaker (1951), vol. 1, pp. 198–211; and Helmholtz in Paper No. 14, pp. 307–310.

[37]Hertz's principle for electricity was completely analogous to the "principle of the unity of magnetic force," which had been developed by Ampère and which states that the magnetic field produced by an electric current is in every way equivalent to an equal and equally-directed magnetic field produced by a magnet.

[38]See the last few pages of Paper No. 3.

[39]Both Max Planck (Paper No. 18) and Sir Edmund Whittaker (1951), vol. 1, p. 321, point out that Hertz's derivation of Maxwell's equations from his assumption of action-at-a-distance forces cannot be considered a valid proof. Buchwald (1990), p. 294, has commented that Hertz's 1884 paper "attempts to meld two theories that are incompatible at the most fundamental level," but precisely for that reason the paper turned out to be important. Even Hertz himself admitted that his deduction was "in no sense to be regarded as a rigid proof that Maxwell's system is the only possible one." See the last few pages of Paper No. 3.

Introductory Biography

experimentation to verify Maxwell's theory, if and when the necessary equipment became available.

It is remarkable that Hertz was able to complete this major paper at Kiel, since it appears that he went through a severe emotional crisis at the time. This is clear from the tone of his diary entries, and from his intense desire to get away from Kiel.

What was the source of the profound depression Hertz experienced in Kiel? We simply do not know. There have been suggestions that he had fallen in love and had been rejected, but no documentary evidence seems to exist to support this view. There are no personal letters for the period 8 December 1883 to 30 November 1884 to be found among those published, and only four from that time to the end of his stay in Kiel.[40] It seems unlikely that Hertz wrote so few letters to his parents during this period of trial, since he remained a faithful correspondent throughout his life. It appears more likely that some letters that could clarify this matter were destroyed by his parents.[41]

Hertz's daughter, Mathilde, does suggest that the fault was at least in part that of her father: "He suffered severely in his emotional life through his own errors of judgment and the mistakes of others."[42] What these errors and mistakes were, we do not know. But we do know this: there was a side to Hertz—an excessive sensitivity, a tendency to brood, an aloneness—that is often forgotten in celebrating his great triumphs.

---

[40]There are over one thousand personal letters of Hertz (mostly to his parents) in the Staatsarchiv in Hamburg. Of these, excerpts from about 15 percent are contained in the *Erinnerungen*. For the period 1883 to 1886 there are about 130 items in the Hamburg archives, of which perhaps twenty-five have been excerpted for the *Erinnerungen* [The writer is extremely grateful to Dr. James G. O'Hara for this information]. Future research in the Hamburg archives may throw more light on this question.

[41]This view is based on statements by Hertz's daughter, Mathilde, in her Preface to the second edition of Hertz's *Erinnerungen*. On page xi she writes about the period in Kiel and the first year in Karlsruhe when many letters appear to be missing: "He was very unhappy then, depressed almost to the point of mental illness...." She explains that for Hertz's parents it would have seemed natural, even a duty, to destroy letters that "disturbed the picture of an ideal relationship—expressions of differences, of morbid thoughts and human weakness."

[42]H. Hertz, *Erinnerungen*, p. xiii.

IV. Karlsruhe (1885–1889)

During the 1884 winter term in Kiel, Hertz remained unproductive and unhappy. He tried to do some work on a theoretical problem in hydrodynamics, but declared himself completely discouraged by his lack of success.

On 30 November 1884 he mentions in a letter to his parents a rumor that the *Technische Hochschule* in Karlsruhe was interested in his coming there as *Ordinarius* (Full Professor) to succeed Ferdinand Braun, who was leaving Karlsruhe for a professorship in Tübingen. There was also the likelihood that Kiel would react to such an offer by making Hertz at least an *Extraordinarius* (Associate Professor) to encourage him not to leave. Hertz was in such a bad mood that he convinced himself that neither position would eventuate. In this same letter to his parents he continues:

> ... my feelings are in part most irrational, and I could not even defend them myself, if I were not certain that what is troubling me is a longing for activity and life and change, rather than any longing for a silly title.[43]

Apparently Helmholtz had been asked by the faculty in Karlsruhe for a recommendation of Hertz.[44] Immediately rumors began to fly (as they tend to do in academic circles) that he might leave Kiel. The Kiel faculty then pressed strenuously for his promotion. A professorship was tentatively approved by the educational ministry in its 1885–1886 budget, subject to the approval of the minister of finance and the legislature. The Kiel faculty then voted to make Hertz their only candidate for this professorship.

On 20 December 1884 Hertz received a welcome—and much needed—Christmas present: he was officially called to the chair of physics at the *Technische Hochschule* in Karlsruhe. The fact that he was a student of Helmholtz and that his former mentor had recommended him, the youngest of the three candidates, as the one with the "greatest hope for the future" had much to do with Hertz's receiving the appointment. The day after Christmas he departed for Berlin to seek

---

[43]H. Hertz, *Erinnerungen*, p. 199.

[44]On Hertz's call to Karlsruhe and his research there, my account is especially indebted to the material in Jungnickel and McCormmach (1986), vol. 2, pp. 85–92.

Introductory Biography 25

the advice of Helmholtz, Kirchhoff, and Heinrich Kayser (1853–1940).[45] Kayser had been his colleague in Berlin, where both had been assistants to Helmholtz. He also met with Friedrich Althoff, the departmental director for universities in the Prussian Ministry of Culture, to discuss with him the details of the position. Althoff exerted a major influence on the implementation of university science policy in Wilhelmian Germany in the years 1882–1907, and the tremendous progress in physics at German universities during that time was significantly fostered by his enlightened decisions.[46]

On 28 December Hertz took the train to Karlsruhe to survey the situation. Karlsruhe was a city of about 60,000 people on the northern fringes of the Black Forest, near the western border of Germany. It was named for the hunting lodge ("Charles-Rest") of the former Margrave of Baden and is located five miles east of the Rhine, connected by a canal with its port, Maxau, on the Rhine. Hertz's initial impression of the city was negative; he confided to his diary that he had a "great aversion" to the city. His reasons were probably subjective, since at that time Karlsruhe, a planned city built in 1750, was considered one of the more attractive and best laid-out cities in Europe (its streets, radiating out in a semi-circle from a focal point provided by the beautiful palace of the former margrave, so impressed Thomas Jefferson that he wanted Washington, D.C., to have the same radial streets).[47]

The next day everything changed. Hertz saw the well-outfitted laboratories and up-to-date equipment available, much of which had been acquired by Ferdinand Braun, his predecessor as director of the

---

[45]On Kayser, see Mulligan (1992).

[46]Friedrich Althoff (1839–1908) received the honorary LL.D. degree from Harvard University in 1906 as the leading contributor to the development of German policy for higher education. In the degree citation President Charles Eliot referred to Althoff as "a judicious, energetic, indefatigable, perspicacious, prudent, and courageous man, the most important individual in German academic life." On Althoff see Sachse (1928) and vom Brocke (1991). Althoff was also a very controversial figure; for this aspect of his life, see McClelland (1980), pp. 288–299, McCormmach (1982), pp. 39–50, and Kayser (1936), pp. 169–175. Althoff valued Hertz's opinions very highly and, after Hertz's death, continued to ask other physicists what Hertz had thought of physicists being proposed for important university positions.

[47]Jefferson opposed Pierre L'Enfant's original plan for Washington, D.C., but in 1901 L'Enfant's drawings were resurrected and became the basic plan for present-day Washington.

Karlsruhe Physics Institute. He also chatted with Dr. August Schleiermacher, the *Privatdozent* in Physics with whom he would be working most closely. He carried away the impression that Althoff was serious about making the *Technische Hochschule*, which had been in existence since 1825, into a first-rate technical university, and would provide the means for him to do worthwhile research. Of course, the interest of the Karlsruhe faculty was more along practical engineering lines, especially in electrotechnology (electrical engineering), which was a burgeoning field at the time. But Hertz sensed an advantage in this, for the Institute already possessed much of the electrical equipment he required for his intended research, and he felt that Althoff would provide whatever else was needed.[48] That evening, when pressed for a decision, he immediately accepted the chair of physics at Karlsruhe.

But Hertz still had to return to Kiel to finish out the winter semester, which only ended at the beginning of March. As soon as he returned to his lonely life in Kiel, he was again overwhelmed by depression and uncertainty. On 7 February 1885 he wrote to his parents:

> As beautiful as I found Kiel during some fine summer months, so sad and godforsaken does it seem to me now... It is not even raining, but that is how I feel at the moment.[49]

Anyone who has ever lived through a winter in northern Germany—the constant dreary clouds and damp chill in the air—can sympathize with Hertz. But Hertz's distaste for Kiel had far deeper roots than the weather. He felt that he was terribly alone, cut off from the experimental research he loved, and therefore not living the full and interesting life he so much desired.

Finally the semester came to an end and Hertz rushed home for a vacation in Hamburg, to enjoy for a short while the warmth of life with his loved ones. On 29 March 1885 he set out for Karlsruhe to

---

[48]In April 1886, Hertz received a sum of 500 M ($125) for equipment for the electrotechnical laboratory, and a special grant of 4,100 M ($1,025) for the purchase of electrical equipment, instruments and laboratory furniture for his own research. He also had the services of a machinist, at a salary of 1,400 M ($350) a year. On this see Lehmann (1892), p. 262.

[49]H. Hertz, *Erinnerungen*, p. 203.

Introductory Biography

prepare his institute for the summer semester, which began at the end of April.

On April 20 Hertz delivered his inaugural lecture to the faculty, on "The Energy Balance of the Earth."[50] This was not a topic on which he had done much original research; rather, he appears to have presented an up-to-date summary of the state of knowledge in the field. He was not pleased with the lecture, but the audience seemed satisfied.

He found his duties as director of the physics institute rather burdensome, although there were fewer problems with space and equipment than he had anticipated. It consoled him that he no longer had to worry about his future, and hence could take time away from his research if administrative duties and committee work demanded it. He continued to worry, however, that he might become one of those who, after achieving a professorship, never contribute any more worthwhile scholarship to their fields.

Ferdinand Braun (1850–1918), another student of Helmholtz, was Hertz's predecessor at Karlsruhe. He is known as the inventor of the oscilloscope and shared the 1909 Nobel Prize in Physics with Guglielmo Marconi (1874–1937) for his contributions to radio technology. Braun had been called to Karlsruhe in 1882 to develop a teaching and research program in what we would now call electrical engineering. He introduced a lecture course on the theoretical aspects of electrotechnology and also developed a laboratory course in this field.[51] Hertz was expected to take over the same courses Braun had taught, and did so for his four years in Karlsruhe. Beginning in November 1885, he also taught, during the winter semester every year, a course on meteorology (one of his great interests) for forestry students.

---

[50]Hertz sent the fifty-page manuscript of this lecture to his father in Hamburg on 20 April 1885; it has never been published. It was purchased after World War II by Mr. David P. Wheatland of Boston and appears on a list of Hertz's papers possessed by Mr. Wheatland. [This information was obtained from Dr. Bern Dibner and from Dr. Philip Weimerskirsch, formerly of the Burndy Library in Norwalk, Conn.].

[51]On Braun at Karlsruhe, see Jungnickel and McCormmach (1986), vol. 2, pp. 85–86; and Kurylo and Susskind (1981), pp. 48–59.

Since a *Technische Hochschule* was not allowed at that time to offer the doctoral degree,[52] there were few advanced students there, either in physics or electrical engineering. As a consequence, Hertz had no students sufficiently advanced to assist him with his research. The only other instructor in physics was Schleiermacher. He was a theoretician, to whom Hertz occasionally demonstrated his new experimental results, but who on only one occasion seems to have actually helped Hertz with his experiments.[53] Thus Hertz did his research at Karlsruhe by himself, although he had a technician to help with the construction of equipment and some routine tasks.

Hertz's personal situation in Karlsruhe was initially even worse than in Kiel. He lived in an apartment and took most of his meals by himself in a restaurant. Unlike Kiel, where there had been a number of congenial *Privatdozenten* of about Hertz's age, in Karlsruhe most of the faculty were older, married and accustomed to a comfortable social life with their families and married friends. This was hard on Hertz, who on 21 April 1885 was driven to write to his parents: "... if I am not married by the end of a year, then I will be in a raging fury." His thoughts seemed to be running toward marriage at an accelerating pace.

Hertz's spirits remained at a low ebb during his first months in Karlsruhe. Even during the Christmas holidays, when he was in Hamburg with his family, he mentions his bad humor and feeling of hopelessness. Hertz's diary entry for New Year's Eve, 1885, expresses his relief that a distressing old year had passed, and his hope that the New Year would be an improvement on the old.

Fortunately for Heinrich Hertz, the New Year differed from the old as light from darkness. In the early months of 1886 he met Elisabeth Doll, whose father, Max Doll, was a lecturer in trigonometry and geodesy at the *Technische Hochschule*. Hertz seems to have fallen in love with Elisabeth almost at first sight. They became engaged in mid-April and were married on 31 July 1886, at the end of the summer semester. They spent their honeymoon in Bavaria, the Tyrol and Lake Constance, where they delighted in their love for each other and their shared love for nature. Elisabeth was seven years younger than Hein-

---

[52] The *Technische Hochschule* in Karlsruhe finally was granted the right to offer the doctoral degree in 1899, along with all similar technical institutes in Germany.

[53] H. Hertz, *Erinnerungen*, p. 269.

Introductory Biography

rich (She outlived her husband by forty-seven years, dying in Great Britain in 1941 at the age of seventy-seven). She was a lovely, graceful, devoted and intellectual woman, who especially delighted in sharing literary insights with her husband, since they were both avid readers. Hertz occasionally took his wife to the laboratory to show her some of his experiments. She, in turn, saved him research time by helping to draw technical diagrams for his scientific articles and by making copies, in a beautiful hand, of his scientific correspondence.[54]

Hertz found much happiness in his life with Elisabeth, and this seems to have resolved many of his emotional problems. Two children, both girls, resulted from the marriage. Their first daughter, Johanna, was born in October 1887 in Karlsruhe, and the second, Mathilde, in January 1891 in Bonn.[55]

Research on Electromagnetic Waves

During his first six months in Karlsruhe, Hertz had attempted experiments on a variety of topics but without much success: he tried his hand at research on gas discharges, motors and generators, fog droplets, and battery construction, but nothing seemed to be important or interesting enough to hold his attention for long. On returning from his honeymoon in early September 1886 he was still undecided about what project to pursue. Then on October 25 he writes in his diary that he has begun experiments with a spark-micrometer.[56] Ideas that he had been mulling over ever since his 1884 theoretical paper on Maxwell's theory[57] finally began to jell with the equipment he

---

[54]These copies are a godsend to scholars, since Hertz's handwriting is very difficult to decipher, especially in old German script. For an example, see H. Hertz, *Erinnerungen*, pp. 237, 242–245. Unfortunately Elisabeth did not make copies of Hertz's personal letters.

[55]Both daughters contributed significantly to the publication of Hertz's *Erinnerungen*: Johanna edited the first German edition (1927); Mathilde edited and helped translate the second combined German-English edition (1977).

[56]A spark-micrometer consists of two metal spheres (or other terminals) whose distance apart can be adjusted with a micrometer screw and read off the calibrated scale on the barrel of the micrometer. This gives an accurate measurement of the length of an electric spark between the two terminals.

[57]Paper No 3.

had available, not merely the spark-micrometer, but also a set of Riess or Knochenhauer spirals he had used for lecture demonstrations.

What happened is a perfect example of how a chance observation or a new piece of equipment can activate the well-prepared mind of a first-rate scientist. Hertz observed that sparks produced in the primary spiral could produce sparks in the secondary spiral over such considerable distances as to make direct inductive effects unlikely.[58] He virtually exploded into action, and for almost three years devoted his research time to the exploration of a previously unknown phenomenon—the propagation of electromagnetic waves through space. These years proved to be the most productive of his life. The combination of his marriage and mounting enthusiasm for his work in the laboratory made his earlier emotional problems a thing of the past.

Most discussions of Hertz's Karlsruhe research on electromagnetic waves merely paraphrase his own account in *Electric Waves*, the collection of his papers on electromagnetism.[59] This biography avoids this approach, since Hertz's papers—especially those included in the present collection—present a more lucid, carefully-reasoned, and often eloquent account of his accomplishments than could any second-hand summary of his work. For example, it would be impossible for anyone to improve on Hertz's succinct statement: "The object of these experiments was to test the fundamental hypotheses of the Faraday-Maxwell theory, and the result of the experiments is the confirmation of the fundamental hypotheses of this theory."[60]

This is not to say that Hertz's achieving the goal of his experiments was in any sense easy. There were many mistakes and miscalculations in his exploration of this new field of research. The impressive thing is that he was always the first to point out his

---

[58] An anonymous reviewer of *Electric Waves* in *The Electrician* makes the point that, although Hertz was led to begin his work on electric waves by what might be styled an accident, he was so prepared for this chance discovery that he knew exactly what to do next. The review goes on: "... in thus stumbling upon the problem, Hertz made the only stumble in connection with it. For there is not in the entire annals of scientific research a more completely logical and philosophical method recorded than that which has been rigidly adhered to by Hertz from start to finish." *The Electrician* 32 (April 13, 1894), p. 657.

[59] Some well-illustrated accounts of Hertz's experiments may be found in Bordeau (1982), Bryant (1988) and Gerhard-Multhaupt (1988).

[60] Paper No. 5, last sentence of part A.

Introductory Biography 31

mistakes and then to strive to eliminate them as he steadily, but not always directly, approached his research goal.[61]

As an example, Maxwell's theory predicted that electromagnetic waves on wires and in free space should travel at the same finite velocity. In his paper, "On the Finite Velocity of Propagation of Electromagnetic Actions,"[62] Hertz tried to establish experimentally this equality, using waves of 5.6-m wavelength. He first concluded that such waves travelled much more rapidly in air (a good approximation to free space) than on wires and, in fact, that the velocity in air might well be infinite—in complete disagreement with Maxwell's theory. Gradually he refined his experiments to a point where it was clear that the velocity was not infinite (a very important conclusion for deciding between Weber's and Maxwell's theories), but still he could not establish that the velocities in air and on wires were even approximately the same.

The experimental problems leading to these spurious results were many: the wavelength was 5.6 m, and the room used for the experiment was only 15 m long, so that the standing-wave measurements were limited to the first few wavelengths near the transmitter, where the radiated waves had not yet fully developed their final form; rows of iron pillars, and an iron stove only 1.5 m from the line of flight of the waves, undoubtedly produced reflections that caused uncertainties in the nodal positions measured; the 5.6-m wavelength was long compared to the dimensions of the apparatus used to generate and detect the waves, and therefore diffraction could be expected to be large.[63] All these problems could be simultaneously reduced by shortening the wavelength, as Hertz did in his most important Karlsruhe paper,[64] in which the wavelength used was 66 cm. He found, by setting up standing waves and measuring the wavelengths, that the speeds in air and on wires were approximately the same at this wavelength.

---

[61]Hertz wrote in his Introduction to his *Electric Waves*: "... I have carried out with the greatest possible care these experiments (by no means easy ones), although they were in opposition to my preconceived views. And yet, although I may have been lucky elsewhere, in this research I have been decidedly unlucky." (Paper No. 5, p. 171.)

[62]*El.Waves*, pp. 107–123.

[63]On this see the comments of H. Poincaré in footnote 74 below.

[64]Paper No. 9 in this book.

In the following discussion we will merely summarize some of the original ideas and experimental techniques used by Hertz to achieve his goal of confirming Maxwell's theory.[65] What factors made it possible for Hertz to succeed so spectacularly when other competent physicists either tried and failed, or were too discouraged by the difficulties of the task ever to try at all? Among the most important factors in his success were the following:

1. His thorough knowledge of electromagnetic theory that focussed and directed his experimental work.

Hertz had devoted years to the study of Maxwell's theory and those of F. Neumann, W. Weber and Helmholtz. His paper on these competing theories of electromagnetism[66] convinced him that the crucial proof needed to decide between Maxwell's theory and the opposing theories was not the relationship between electromagnetic forces and dielectric polarization, as set forth by Helmholtz in the prize problem of 1879. Rather it was to establish whether empty space behaves like all other dielectrics. To prove that it does, he set out to measure the speed of electromagnetic waves (at what we now call radio frequencies) in air—the best available approximation to free space—to determine if they propagated at the speed of light, as Maxwell's theory predicted.

2. Skill in using simple equipment to achieve important experimental results.

In addition to his ability to do excellent theoretical research, Hertz was also a gifted experimentalist. Using a pair of Riess or Knochenhauer spirals[67] for a lecture demonstration in 1886, he was

---

[65]The most important writings of Hertz describing his experiments on electromagnetic waves are the six papers (Nos. 5, 6, 7, 8, 9, 10) in the present collection. Brief summaries of what Hertz accomplished are also contained in Papers No. 11, 12, 14 and 18. In what follows, Hertz's *Electric Waves* will be abbreviated to *El.Waves*.

[66]S. D'Agostino (1975) has emphasized the importance of this 1884 theoretical paper (Paper No. 3). He rightly points out that, without this theoretical work, Hertz would have had neither the motivation nor the proper approach needed to succeed in his classic experiments.

[67]For a photograph of these spirals, see Bryant (1988), p. 14, or Appleyard (1930), p. 116. For a photograph of the spark-micrometer, see Bryant (1988), p. 20.

surprised to find that weak discharges through one spiral could be easily detected in the other, even if they were not very near each other. Hertz modified this equipment in various ways to produce what was basically a transmitting and a receiving circuit, where sparks jumping a gap in one circuit caused electric disturbances that travelled through the air and produced sparks across the gap in the other circuit. As a detector of the sparks across the gap, Hertz used his spark-micrometer. A simple horizontal telescope enabled the spark gap to be observed as the gap size was adjusted by turning the screw. The maximum gap size for which sparks were observed could then be read off the micrometer scale. This gave Hertz some rough quantitative idea of the strength of the incoming disturbance at the position of the spark-gap. By moving the receiving coil and spark-gap to different positions in what we would now call the electromagnetic field, Hertz was able to plot out the field strength at various points in space.

As Hertz pointed out in his Heidelberg address,[68] the thin thread on which the success of his Karlsruhe research hung was the fact that sparks as small as 0.01 mm in length and lasting about a microsecond can be seen by a dark-adapted eye in a perfectly dark room. He spent so many hours watching his tiny sparks merrily jumping across the gap that he was soon able to judge their strength from their appearance alone, without the need to read the micrometer. Hertz pointed out in this connection that "the success of a workman depends on whether he properly understands his tools." After three years of such experimentation, Hertz understood the behavior of his spark-gaps very well indeed; so well, in fact, that physicists who have tried to repeat his experiments with his primitive equipment have stood in awe at his great experimental technique and his unlimited patience.[69]

3. Use of resonance techniques to intensify the observed sparks.
   Hertz was familiar with the well-known Helmholtz resonators used in acoustics at that time. He decided to apply similar techniques to tune his receiving circuit to the frequency of the primary spark transmitter. By adjusting the dimensions of his receiver and thus

---

[68]Paper No. 10, p. 282.

[69]On this see J.J. Thomson (1931), pp. 43–44. Thomson's conclusion is that "Hertz's researches were one of the most marvelous triumphs of experimental skill, of ingenuity, of caution in drawing conclusions, in the whole history of physics" (p. 43).

changing the inductance of the circuit, he was able to put it into resonance with the transmitter.[70] The required tuning was not very precise: the transmitter produced a highly damped oscillatory spark that radiated a rather broad band of frequencies,[71] so that a receiver tuned to any frequency within that band would produce resonant enhancement of the signal. This resonance technique allowed him to experiment with much larger separations between source and receiver.

4. Employment of higher frequencies (shorter wavelengths).

Hertz also found that he could increase the propagation distance of the electromagnetic waves by increasing the frequency of the radiation emitted through a reduction of the inductance and capacitance of the transmitting coil. (As we know today, the energy radiated varies as the fourth power of the frequency.[72]) Using higher frequencies had the added advantage of making the wavelengths shorter, thus enabling accurate standing-wave measurements[73] to be carried out even in the limited spaces he had available. Higher frequencies also reduced significantly the disturbing effects of diffraction on his experiments.[74]

---

[70] On this see Hertz's discussion in the section "Resonance Phenomena" in Paper No. 6.

[71] Hertz did not realize this at the beginning of his research. This involved him in some discussions with E. Sarasin in Geneva and H. Poincaré in Paris about the phenomena of "Multiple Resonance," the name given to it by Sarasin and de la Rive. On this see Paper No. 5, pp. 179–180.

[72] This fourth-power dependence of radiated energy on frequency had been derived by FitzGerald in 1883, but was probably not known to Hertz until he began to communicate with the Maxwellians after 1888. For FitzGerald's derivation, see FitzGerald (1902), pp. 128–129.

[73] See Paper No. 8.

[74] Neither Hertz himself nor commentators on his research have paid much attention to the adverse effects of diffraction on his experiments. One of the few physicists to discuss this in Hertz's case was Henri Poincaré, who commented: "... we know that light does not exactly follow the laws of geometric optics, and the discrepancy due to diffraction increases proportionately to the wavelength. With the great wavelengths of the Hertz undulators, these phenomena must assume enormous importance and derange everything." Poincaré goes on to say that it was only the crudeness of Hertz's observed data that concealed these diffraction effects. (They are clearly more evident in Hertz's experiments at 9-m wavelengths than in those at 66-cm

The pre-Maxwellian theories of Weber and Neumann recognized only an induced electric field, which arose when a magnetic field changed in time. According to the Faraday induction law, this electric field was proportional to the time rate of change of the magnetic field. Maxwell's theory predicted in addition that a magnetic field would be induced by a changing electric field, and that this magnetic field would be proportional to the time rate of change of the electric field. This effect was so weak, however, that it could not be observed unless the electric field changed very rapidly in time, and this required higher-frequency oscillators and detectors capable of responding to such high-frequency signals. Hertz's genius was twofold: first, on the basis of his 1884 comparison of Maxwell's theory with earlier theories of electromagnetism, he came to see the need for higher frequencies to decide which theory was correct; secondly, his experimental skill was such that he was able to push his research to higher and higher frequencies, culminating in the 66-cm waves he used for his 1888 paper, "On Electric Radiation." These frequencies of about 500 MHz enabled him to show convincingly the overwhelming superiority of Maxwell's theory in explaining his results.

5. *Familiarity with the tools of his trade, and a unique ability to design experiments and employ apparatus to achieve his goals.*

After Hertz had convinced himself that his electromagnetic waves resembled light waves in that they propagated through space at approximately the same speed as light,[75] he saw that the convincing proof of the equivalence of his electromagnetic waves with light waves would require demonstrating that both kinds of waves were reflected, refracted, diffracted and polarized in much the same way. This would require the production of a directional beam from his transmitting spark, and the use of special reflectors, refracting prisms, and metal wire gratings to study reflection, refraction and polarization of the waves.

He first constructed a cylindrical parabolic reflector out of a zinc sheet 2 m long, 2 m wide and 0.5 mm thick, bent into the desired shape on a wooden frame of the exact parabolic curvature needed. With a

---

wavelengths, where he obtained his best results.) On this see Poincaré (1894b), p. 137, and Paper No. 9.

[75]See Paper No. 8, footnote 2 on p. 241.

linear spark-source situated at the focal line of the reflector, Hertz produced a beam of linearly polarized electromagnetic waves. He used a similar parabolic reflector to focus the incoming beam on the spark gap of the secondary circuit.

To refract the beam Hertz constructed a wooden container in the form of a prism 1.5 m high, 1.2 m on a side, and with an apex angle of about 30°. This, when filled with asphalt or hard pitch, weighed more than 1,200 pounds! To study polarization he used a rotatable octagonal frame almost 2 m high, strung with parallel conducting wires.[76]

This was large-scale and expensive equipment for a physics experiment in 1888, but Hertz had enough confidence in his own ideas to be sure the equipment would work. His gamble paid off when he was able to show conclusively that 66-cm electromagnetic waves travel in straight lines at the same speed as does light and can be reflected, refracted and polarized in the same fashion as light waves.[77] He proved that they were transverse waves (they could be polarized) and similar to light waves in almost all respects except their frequency. As Hertz stated at the conclusion of his Heidelberg lecture in 1889,

> The connection between light and electricity, of which there were hints and suspicions and even predictions in the theory, is now established.... Optics is no longer restricted to minute ether waves, a small fraction of a millimeter in length; its domain is extended to waves that are measured in decimeters, meters, and kilometers. And in spite of this extension it appears merely...as a small appendage of the great domain of electricity. We see that this latter has become a mighty kingdom. We perceive electricity in a thousand places where we had no proof of its existence before.[78]

Hertz had thus confirmed the fundamental predictions of the Faraday-Maxwell theory and proved that all action-at-a-distance theories of electromagnetism were certainly invalid. This was his most important contribution to physics. He was obviously pleased with his

---

[76]This work is all described by Hertz in Paper No. 9. Also see the photograph in this book at the beginning of the group of papers published by Hertz during his Karlsruhe period (p. 159).

[77]Hertz also tried to observe diffraction fringes of the kind observed with light waves, but did not obtain clear enough fringes to make meaningful measurements. See the section on "Rectilinear Propagation" in Paper No. 9.

[78]Paper No. 10, p. 285.

remarkable success in the laboratory. On 17 March 1888 he described to his parents how, after removing some large chandeliers from the auditorium he was using for his experiments now that classes were no longer in session, he carried out the experiments just described. He called them "brilliantly successful." He now had the "comfortable feeling" that he was ahead of the pack in the field of electromagnetic-wave research and would not suddenly read in the literature that someone else had done the same experiments. He added: "It is really at this point that the pleasure of research begins, when one is, so to speak, alone with nature and no longer bothers about the intentions, opinions, or claims of men."[79]

Research on the Photoelectric Effect

The year 1987 was the centenary of Hertz's discovery of the photoelectric effect. Hertz first noticed this phenomenon in early December 1886,[80] while engaged in his work on electromagnetic waves. Although the initial discovery was accidental, he immediately recognized its importance, since the only other effect of light on electrical phenomena known at that time was the change in the electrical resistance of selenium when exposed to light. He suspected that an interaction between light and electricity of the kind he had observed might clarify the nature of both light and electricity. He therefore interrupted his electromagnetic-wave research to explore this unexpected phenomenon. By May 1887 Hertz had achieved as much clarification of the photoelectric effect as seemed possible at that time. At the end of May he sent off a seventeen-page paper to Wiedemann, the editor of the *Annalen*, describing his observations[81] and put further work on the phenomenon aside to direct his attention once more to the main question, which in this highly productive period of his career was the nature and properties of electromagnetic waves.

---

[79]H. Hertz, *Erinnerungen*, p. 255.

[80]H. Hertz, *Erinnerungen*, p. 227. On Hertz's contributions to the understanding of the photoelectric effect, see Stuewer (1971).

[81]H. Hertz, "On an Effect of Ultraviolet Light on the Electric Discharge," Paper No. 7.

Hertz's only paper on the photoelectric effect describes how he first noticed that light falling on the pole pieces of his secondary spark gap increased the sparking across the gap. To clarify how light enhanced the sparks, Hertz first used pole pieces of copper, brass, aluminum, iron, tin, zinc and lead; next he varied the light source, using sparks, sunlight, gas flames, and electric arcs to produce the effect; lastly he interposed various gases, liquids and solids between the light source and the receiving air gap. The materials interposed included metal and glass plates of various thicknesses, paraffin, shellac, mica, agate, wood, cardboard, paper, rocksalt, and more exotic substances like topaz, amethyst, ivory, feathers and the skins of animals.

Hertz's daily involvement with these experiments gradually led him to the hypothesis that the radiation most effective in strengthening the observed spark had a wavelength close to the short-wavelength limit of the visible spectrum. He then used a quartz prism to disperse the radiation from different light sources and found that the wavelengths enhancing the sparks were indeed all in the ultraviolet.

On 7 July 1887 Hertz wrote a beautiful letter to his father in an attempt to explain, in the simplest possible terms, his work on the photoelectric effect.[82] In the course of the letter he summarizes the situation, as he saw it just after he completed his research on the subject, as follows:

> ... the effect is striking and yet totally puzzling. Naturally it would be nicer if it were less puzzling; however, there is some hope that, when this puzzle is solved, more new facts will be clarified than if it were easy to solve.

Hertz was completely right in his evaluation of the situation. The photoelectric effect was indeed puzzling and could not have been understood in 1887, when very little was known of the structure of atoms, metals, or even electricity. But later attempts to untie this knot of ignorance would have to be based on the solid foundation of Hertz's experimental research.

A year after Hertz deliberately terminated his research on the photoelectric effect, Wilhelm Hallwachs (1859–1922), a *Privatdozent* at

---

[82]H. Hertz, *Erinnerungen*, pp. 225–229.

Introductory Biography 39

Leipzig, published a paper[83] in which he extended Hertz's work by showing that when a metal is negatively charged and then exposed to ultraviolet light, the charge on the plate rapidly leaks off. This was the first convincing evidence that the effect was produced by the emission of *negative* charges from metals under irradiation with ultraviolet light. In his diary for 29 January 1888 Hertz commented:

> My consolation must be that I expected and accepted that; moreover, I have reached the goal that I wanted to reach, the goal for which I set aside those other experiments.[84]

The six months Hertz "stole" from his electromagnetic-wave research to straighten out the physics of the photoelectric effect were certainly well spent. His paper on this subject reflects his great skill in planning and carrying out a logical sequence of experiments, in recording precisely what he observed, and in reporting his results with exceptional clarity and great physical insight. Not a single flaw has been found in his results. His research, although obviously incomplete, provided a firm foundation for subsequent experiments by Hallwachs, Righi, Stoletov, Elster and Geitel, Lenard, and Millikan, whose work led to today's more complete picture of the photoelectric effect.

Other Research at Karlsruhe

While doing research on electromagnetic waves at Karlsruhe, Hertz made a number of related discoveries that were to be important for the development of both physics and electrical engineering. For instance, he appears to have been the first scientist to produce resonance curves for tuning what we would now call a primitive radio receiver to a transmitter.[85] In his graph of the resonance he plotted the length of the spark he observed in the air gap of the receiving circuit as a function of

---

[83]W. Hallwachs, *Annalen* 33 (1888), pp. 301–312. Hallwachs had received his doctorate under August Kundt (1839–1894) in Strassburg in 1883, and was later Professor of Physics and Electrical Engineering at the Dresden *Polytechnic*. After Hertz yielded the field to him, he became the leading expert on photoelectric processes in Germany. For a time the photoelectric effect was known in Germany as the Hallwachs Effect.

[84]H. Hertz, *Erinnerungen*, p. 251. Hertz himself had pointed out in a number of places that the negative terminal of his spark gap was the one sensitive to the photoelectric effect. See, e.g., section 14 of Paper No. 7.

[85]For examples of such curves, see Fig. 10 in Paper No. 6.

the total length of the wire in the rectangular coil of his detector. The wire had a distributed inductance, which could be varied by changing the wire's length to produce resonance at a frequency emitted by the transmitting circuit. The resonance peak then corresponded to a maximum length of the observed sparks in the receiver's air gap.

Hertz maintained an interest in resonance phenomena in electromagnetism after he moved to Bonn in 1889 and encouraged his first research student, Dr. Vilhelm Bjerknes, to take up this research topic. Bjerknes published a series of papers on the subject that represented a considerable advance over Hertz's groundbreaking work.[86]

An 1889 paper of Hertz[87] contains a very detailed discussion of the "skin-effect," in which he demonstrated conclusively that at the frequencies used in his research (about 100 MHz) the current was confined to the exterior skin of the wires, as Oliver Heaviside and J.H. Poynting had predicted on the basis of Maxwell's equations. He extended this idea to show how electrical equipment can be shielded from radio-frequency radiation. He demonstrated that a conducting sheet only 0.05 mm thick (or even an open wire mesh) completely surrounding his spark gap was sufficient to shield the equipment from strong radiation in the 100–MHz range. At the end of his stay in Karlsruhe, Hertz was working with electromagnetic waves of 66-cm wavelength[88] and thus deserves to be called the discoverer of microwaves. He also constructed the first coaxial transmission line,[89] and designed a resonant detector to explore the electric-field configurations in the space between the inner and outer conductors. J.H. Bryant[90] has pointed out that this was the first "slotted line," although that term only came into use with the development of radar during World War II.

A few years after Hertz's death in 1894, a Russian physicist A.S. Popov (1859–1906)[91] began to use Hertz's experimental results on electromagnetic waves to explore the possibility of long-distance

---

[86]These papers were later published in Bjerknes (1923). For further information on Bjerknes see Part V of this biography, and Friedman (1989).

[87]H. Hertz, "On the Propagation of Electric Waves by Means of Wires," El.Waves, pp. 160–171.

[88]See, in this regard, Paper No. 9.

[89]H. Hertz, El.Waves, p. 169.

[90]Bryant (1988), p. 33.

[91]On Popov, see Susskind (1962).

wireless communication. He was the first to develop a practical antenna and in 1897 succeeded in sending a signal over a distance of three miles from ship to shore. In the same year Guglielmo Marconi (1874–1937), who in 1896 had obtained the first patent in the history of radio, sent a signal from land to an Italian warship anchored twelve miles off shore. Both Popov and Marconi paid tribute to Hertz, whose ideas and experimental techniques had led them to hope for great success (and perhaps enormous financial profit) in the new field of wireless telegraphy, or radio. The first wireless telegram sent by Popov in 1896 over a distance of 250 m read simply "Heinrich Hertz."[92]

It is surprising that Hertz himself had not realized the extraordinary possibilities that his electromagnetic waves presented for long-distance wireless communication.[93] He had, after all, passed such waves through a wooden door into another room.[94] The most probable reason that Hertz did not move into the development of long-distance communications was that his interest was sharply focussed on pure physics, on the understanding of nature, not on the use of that understanding for practical purposes. An additional reason may have been Hertz's happiness when he was alone with nature, especially in his own laboratory, and his reluctance to imperil that happiness by involvement with practical applications.

But this is all conjecture. The only bit of concrete evidence we have is an exchange of letters between Hertz and Heinrich Huber, an engineer from Munich, who was employed at the power station that provided electricity to The Hague. Huber wrote to Hertz on 1

---

[92]Another physicist who made major contributions to the development of wireless telegraphy was Oliver Lodge (1851–1940). There is a good discussion of Lodge's contributions in chapter 4 of Aitken (1985). Chapter 3 of the same book is devoted to Hertz, and chapter 5 to Marconi.

[93]As Arnold Sommerfeld has written, "It is amazing how much of the later development of wireless telegraphy has been anticipated in this paper." He was referring to Hertz's 1888 paper, "The Forces of Electric Oscillations, treated according to Maxwell's Theory," El.Waves, pp. 137–159, a paper that contains the first diagrams of the development of an electromagnetic wave as it leaves a dipole antenna. On this see Sommerfeld (1954), p. 6. These radiation diagrams have only recently been superceded by the development of computer graphics.

[94]See Hertz's letter of 30 Nov. 1888 to H. von Helmholtz, in Koenigsberger (1902–1903), vol. 3, pp. 7–9.

December 1889[95] asking about the possibility of using parabolic mirrors, as Hertz had, to transport magnetic forces (fields) through space. He clearly had in mind audio frequencies like those used in the telephone or in electric power lines. Hertz replied that the oscillations of a power transformer or of a telephone transmitter were much too slow. As a consequence, the parabolic mirrors, which need to be approximately the size of the wavelength, would have to be 300 km in width at frequencies of 1,000 cycles per second. This was a perfectly correct answer to the question Huber had raised.

Nowhere in either letter is there any mention of "wireless telegraphy" or even of "telegraphy." Hence Hertz's letter cannot be interpreted as denying the technological possibility of electrical communication at the higher frequencies he had been using in his experiments, although it has been misinterpreted in this way by some writers on the subject.[96]

If we assume the role of Monday-morning quarterback, it is easy to see that Hertz was right not to involve himself in the practical applications of his discoveries. As we now know, the tremendously important, but seriously flawed, process of developing a radio industry from scratch was plagued by an abundance of financial and legal strife. Hertz would have been excruciatingly unhappy in such an intellectual climate.

And yet Hertz's name still lives on in today's radio industry. The word "hertz" (abbreviated as Hz) had been used as the unit of frequency in Europe as early as the 1920s. However, it was not officially adopted by the International Electrotechnical Commission until October 1933, with Germany, France, The Netherlands, Italy, Poland, Rumania and Spain voting in favor of the proposal to so honor Heinrich Hertz.[97] The United States, Great Britain and Japan voted "nay," preferring to retain the older and more descriptive "cycles per second" as the unit of frequency. This latter group was outvoted, however, and today the "hertz" is the official frequency unit in the United

---

[95]Both Huber's and Hertz's letters, which are in the Deutsches Museum in Munich, are given in the original German and an accurate English translation in Susskind (1965). Susskind was the first to correct some of the earlier misleading interpretations of the Hertz-Huber correspondence.

[96]See, for example, Appleyard (1930), p. 140. The same author includes a photographic copy of Hertz's letter to Huber on p. 138.

[97]Boerger (1988), pp. 97–98.

Introductory Biography

States. Thus, when people worldwide tune their AM or FM radio dials to frequencies measured in kHz or MHz, they are paying tribute to Hertz, whose basic research first suggested the possibility of radio.

Helmholtz and Hertz's Research on Electromagnetism

In concluding this discussion of Hertz's remarkable success in confirming the predictions of Maxwell's theory and demonstrating that light is merely a small part of the complete electromagnetic spectrum, it is important to indicate not only how indebted Hertz was to Hermann von Helmholtz, his mentor and lifelong friend, in this regard, but also how Hertz gradually departed from the views of his mentor as his own research progressed. He probably would never have carried out his famous experiments in Karlsruhe if he had not been associated with Helmholtz in Berlin in the years 1878–1883.[98] He first learned electromagnetic theory from Helmholtz, who was at that time one of the few physicists on the continent to appreciate the importance of Maxwell's work. By 1884 Hertz had progressed far enough in his understanding of Maxwell's theory to reject all purely action-at-a-distance theories and even Helmholtz's own compromise theory. Immediately he began looking for ways to test Maxwell's theory in the laboratory.

A hint as to the way to proceed can be found in a memorandum written by Helmholtz on 16 June 1883, just a few months after Hertz left Berlin for Kiel. This memorandum outlined the research to be done at the proposed *Physikalisch-Technische Reichsanstalt (PTR)*.[99] Among the tasks he proposed for the researchers at this projected research institute was the following:

> In the theory of the magnetic action of electrical currents, a velocity that appears to be exactly equal to the velocity of light, and which W. Weber characterizes as critical, ... seems to play a fundamental role. Its identity with the velocity of light appears to me to indicate an essential and intimate relationship between optical and electrical processes. We seem to have here a clue to

---

[98]Michael Pupin (1858–1935) has expressed his doubts that anyone in Germany in the late nineteenth century could have done what Hertz did unless they had been closely associated with Helmholtz. See Pupin (1926), p. 267.

[99]On the *PTR* and on Helmholtz's position there, see Cahan (1989).

the mysterious aspects of electromagnetic phenomena, which probably may lead us to their deepest foundation.[100]

It is unknown whether Hertz ever saw this memorandum, but, if Helmholtz felt this strongly about it, it is certainly likely that he would have in some way communicated his thoughts on the subject to Hertz, his prize student. Hertz's 1884 paper made two things very clear: Hertz still retained his admiration for Helmholtz's ideas, but on the question of electromagnetic waves he much preferred Maxwell's theory;[101] his conviction was growing that electromagnetic waves travelled at the speed of light. As a consequence, he felt the need for a definitive measurement of the speed of electromagnetic waves at nonoptical frequencies, and he devoted much thought to ways of carrying out such a measurement if the necessary equipment became available (as it did in Karlsruhe in 1886).

During the course of Hertz's electromagnetic research in the years 1885–1888, letters were frequently exchanged between Hertz in Karlsruhe and Helmholtz in Berlin. Of the nine research papers published in the *Annalen* between 1885 and 1888, four were sent by Hertz to Helmholtz to be first presented by him to the Berlin Academy of Sciences for publication in the *Sitzungsberichte* of the Academy.

Hertz's letters to Helmholtz during this period are very deferential, as might be expected of a former student writing to his mentor. They frequently point out how grateful Hertz is for Helmholtz's contributions to his research success. Thus in a letter of 5 December 1886, detailing some of the results contained in his paper "On Very Rapid Electric Oscillations,"[102] Hertz writes:

> I should like to take this opportunity to let you know about some experiments that I have recently successfully completed, *since I had hoped from the beginning that they might interest you.*[103]

A very interesting letter arrived from Hertz dated 5 November 1887. It contained a request for Helmholtz to submit to the Academy the paper in which Hertz was finally able to show the effect of

---

[100]Koenigsberger (1902–1903), vol. 2, p. 349.

[101]For a fuller discussion of the gradual evolution of Hertz's ideas on electromagnetism along different lines from those of his mentor, see D'Agostino (1971) and Buchwald (1990), especially Section 6.

[102]Paper No. 6.

[103]Koenigsberger (1902–1903), vol. 2, p. 344 [italics not in original].

Introductory Biography 45

electromagnetic fields on the dielectric polarization of insulators.[104] This was the same research topic Helmholtz had suggested to him at the beginning of his university studies in Berlin. Hertz writes:

> ... it deals with a topic that you yourself once urged me to tackle some years ago, and *that I have therefore always kept in mind*, but which until now I have found no way of approaching with any clear prospect of unambiguous success.[105]

Helmholtz was always equally gracious in his replies and enthusiastic about transmitting Hertz's papers to the Academy. To his letter mailed on Friday, November 5, Hertz received a postcard reply on Tuesday, November 9, that reads: "Manuscript received. Bravo! Will hand it on to be printed on Thursday. H.v.Htz"[106] Again, in a letter of 15 December 1888, Helmholtz wrote to Hertz:

> Your latest achievements have given me much pleasure. For years I have puzzled over the possibility of getting at these things, trying to find an opening through which they could be reached, and therefore I am familiar with the entire train of thought and recognize at once its immense importance.[107]

Helmholtz officially transmitted Hertz's paper to the Academy on the same day and lectured on it to the Physical Society the next day. When Hertz sent him his paper on the production of standing electromagnetic waves in air, Helmholtz sent it on to DuBois-Reymond for the *Sitzungsberichte* of the Academy, with the comment that Hertz's research was that of a "genius."

These few excerpts make clear the mutual respect, even warm admiration, which existed between these two great physicists.

On 6 August 1889 Helmholtz's son, Robert, died, leaving his father devastated with grief. Robert had also been a physicist, and his father had expected great things of him. It must have consoled Helmholtz a little to travel to Heidelberg the following month to hear

---

[104]H. Hertz, "On Electromagnetic Effects Produced by Electrical Disturbances in Insulators," *El.Waves*, pp. 95–106.

[105]Letter of 5 Nov. 1887 from Hertz to Helmholtz; in Koenigsberger (1902–1903), vol. 2, p. 356 [italics not in original]. See also pp. 14–15 above.

[106]H. Hertz, *Erinnerungen*, p. 235. There is also a facsimile of the actual postcard on p. 236 [A present-day reader of this correspondence is in awe at the efficiency of the German postal system one hundred years ago!].

[107]Koenigsberger (1902–1903), vol. 3, p. 9.

his "second son," Hertz, deliver his famous address, "On the Relations between Light and Electricity."[108] The evening of the talk Helmholtz wrote to his wife:

> Today we had the address from Professor Hertz; it really was extraordinarily good, very polished in style, tactful and tasteful, and called out a storm of applause.[109]

This pride in his former student may have wiped away, at least for a time, the tears of Helmholtz's grief.

At the end of his tenure in Karlsruhe, Hertz had succeeded in demonstrating the unity of electricity and optics suggested in 1883 by Helmholtz as a research topic for the *PTR*. Hertz's appreciation of Helmholtz's contributions to this success is clear from his request that Helmholtz allow him to dedicate *Electric Waves* to his mentor, since "my work . . . derives rather essentially from the study of the works of Your Excellency, and since the original impetus even came from your personal suggestion."[110] For this reason A.E. Woodruff rightly concludes his article on Helmholtz's contributions to electrodynamics with the following insightful remark:

> In retrospect, the significance of the work of Helmholtz in electrodynamics is that of making Maxwell's theory intelligible to the German physicists and of inspiring the experimental work of Hertz which confirmed it.[111]

---

[108] Paper No. 10. In this address Hertz never mentioned Helmholtz, probably because he would have had to indicate that his own experiments had demonstrated the superiority of Maxwell's theory to that of Helmholtz. At a get-together of Helmholtz, Hertz, Kundt, Wiedemann and others at Koenigsberger's house afterwards, Koenigsberger praised Hertz's address and called Hertz the greatest living physicist after Helmholtz. Hertz immediately arose and in his modest fashion lauded Helmholtz to the skies as the greatest physicist of all time. He also expressed his sincere opinion that his own discoveries were only the logical outcome of Helmholtz's ideas. On this see Koenigsberger (1902–1903), vol. 3, p. 26.

[109] Koenigsberger (1902–1903), vol. 3, p. 25.

[110] Letter of 24 Feb. 1892 from Hertz to Helmholtz; in H. Hertz, *Erinnerungen*, p. 321.

[111] Woodruff (1968), p. 310.

V. Bonn (1889–1894)

On 3 October 1888 Hertz received a call to the chair of physics at Giessen, the principal university of the Grand Duchy of Hesse. He consulted Althoff and was told that it was his duty as a Prussian to decline the position and return to a university in Prussia when the opportunity arose.[112] Althoff suggested a few possible openings and implied that, if Hertz did not accept an appointment to one of these universities, he might never have the opportunity again. On the death of Kirchhoff in Berlin in 1887 and of Clausius in Bonn in 1888, two especially prestigious physics chairs in Prussian universities had become vacant. Helmholtz wrote a letter to the Bonn mathematician Rudolf Lipschitz, recommending Hertz for the chair at Bonn because he was "the most talented among the younger physicists and most replete with original ideas."[113]

On 10 December 1888 Althoff offered Hertz a choice between Bonn and Berlin.[114] Hertz chose Bonn almost immediately because the position in Berlin would be in theoretical physics,[115] and Hertz very much wanted to continue to do experimental physics. Also, he probably had grown to like the idea of being the only ordinary professor of physics at a smaller university like Bonn, instead of being surrounded by luminaries like Helmholtz, who was now the director of the *PTR*, and August Kundt (1839–1894), who had recently taken over Helmholtz's position as director of the Physics Institute in Berlin. He was happy in the knowledge that Helmholtz approved his decision. On 15 December 1888 Helmholtz sent him a letter containing the following thoughtful assessment of the situation:

> Personally I am sorry that you did not want to come to Berlin, but, as I have already told you, I do believe it is for your own interest to go to Bonn. Those who have still much scientific work in view are better away from big cities. At the end of one's life,

---

[112]On this see the very interesting letter of Hertz to his parents on 5 Oct. 1888, in *Erinnerungen*, pp. 261–263.

[113]H. Hertz, *Erinnerungen*, p. 351.

[114]For many parts of this discussion of Hertz's life and work in Bonn I am greatly indebted to the account on pages 141–144 of vol. 2 of Jungnickel and McCormmach (1986). References to primary sources on this period are provided there.

[115]The Berlin position finally went to Max Planck (1858–1947).

when it is more a question of utilizing the points of view one has arrived at for the education of the coming generation and the administration of the State, the situation is different.[116]

Three days before Christmas Hertz travelled to Bonn to meet Althoff and discuss with him the details of his appointment. He had an easy time negotiating an acceptable salary of 6,400 M ($1,600), with a housing allowance of 660 M ($165). This would be supplemented by the fees students paid to attend his lectures, and at Bonn these fees would be considerably greater than they had been in Karlsruhe, since the *Technische Hochschule* was a smaller institution. He had, as a matter of fact, been offered a considerable increase in salary to remain there, but the added fees in Bonn made it impossible for Karlsruhe to compete financially with Bonn's offer.[117] In addition, Hertz preferred to be at a full-fledged university, rather than at a technical institute.

Another agreeable outcome of this meeting was Althoff's insistence that Hertz have a light teaching load and devote as much time as possible to research, since this was the reason he had received the Bonn appointment. Althoff's understanding of the importance of Hertz's research was much appreciated, and on Christmas Day, 1888, Hertz sent his parents a long letter that reflected his happiness.[118]

Despite the advantages he saw in Bonn, Hertz found it hard to leave Karlsruhe. After all, it was there he had met Elisabeth and had experienced with her the joy of their first child. He also would leave many good friends there. Most important for his career, the *Technische Hochschule* had provided the equipment and support he had needed to do the pioneering research that had won him a prestigious place in the world of physics; for that Hertz was deeply grateful.

An indication of Hertz's growing visibility in university circles was the visit he received in early February 1889 from Professor G. Stanley Hall (1846–1924), the well-known American psychologist, who was at that time president of Clark University in Worcester, Massachusetts. Hall had been a student of Helmholtz in Berlin during the period 1879–1882, and had asked Helmholtz to recommend an

---

[116]Koenigsberger (1902–1903), vol. 3, p. 9.

[117]Hertz had also been promised a special grant of 11,000 M ($2,750) for his own research equipment, which was almost three times what he had received in Karlsruhe. On this see Konen (1933), p. 350.

[118]H. Hertz, *Erinnerungen*, p. 277.

Introductory Biography

excellent physicist who might be interested in a position in the United States. Helmholtz was reluctant to recommend Hertz, since he felt it important for physics that he remain in Germany, but he suggested that Hall meet with Hertz to see if there were some colleagues close to Hertz's age whom he could recommend. In the course of the meeting, Hall let it be known that he was able to offer a salary of $5,000 to the right man and would agree to provide $250,000 to build and equip a Physics Institute at Clark.[119]

Hertz was staggered by the size of the salary, which was three times the base salary he had accepted at Bonn. In a letter to his parents on 11 February 1889 he admitted that, if he were not married, he would certainly have accepted the position. Hall knew that Hertz was already committed to Bonn and told him "he greatly regretted that Bonn was so beautiful," making it clear that the Clark position would have gone to Hertz if he had expressed any interest in it.

This visit from Hall was only one of many during Hertz's last days in Karlsruhe. Suddenly everyone seemed to want to visit his laboratory to see his apparatus and discuss his research. They wrote him letters filled with praise or posing problems related to his now-famous experiments; they invited him to deliver important lectures; they sought him as a member of illustrious scientific societies.[120] On 10 March 1889 he gave a final public lecture to an audience of 300 at the *Technische Hochschule*. "At the end," his wife told Heinrich's parents, "there was so much applauding and cheering that I was quite embarrassed." And on March 16 he received his diploma as a corresponding member of the Berlin Academy.[121]

All this notoriety was taking its toll on Hertz's research. He found his new fame a great distraction that kept him from "the inner and outer peace needed for complete absorption in his research," which had almost come to a standstill. During February and March 1889, he also had constant trouble with his eyes (aggravated by

---

[119] H. Hertz, *Erinnerungen*, p. 281. The Clark University Archives substantiate that President Hall met with Hertz in 1889, but contain no other details of their meeting. The available salary of $5,000 or 20,000 marks may be compared with the annual income of German residents, which averaged 352 marks in 1871–1875 and 603 marks in 1896–1900.

[120] For a list of the honors Hertz received, see Helmholtz's account in Paper No. 14. A more complete account can be found in Wenig (1970), p. 117.

[121] On his nomination for this honor, see Paper No. 12.

repeated demonstrations of his experiments to well-intentioned visitors) and some painful dental problems.

For these reasons the last few months in Karlsruhe were not the satisfying, happy ones that Hertz and his wife might have desired (and certainly deserved) after his total immersion in experimental work on electromagnetic waves during the preceding three years.

On 3 April 1889 Heinrich Hertz moved to Bonn to begin what would be his last academic appointment. His wife was to join him later in the month with their one-and-a-half-year-old daughter, Johanna. The move from Karlsruhe required a trip of about 150 miles north along the majestic Rhine river to the beautiful city of Bonn.

Hertz and his wife had decided to buy the house in which his predecessor, Rudolf Clausius, had lived with his family for fifteen years, at Quantiusstrasse 13.[122] It was a spacious house with a lovely garden dominated by a beautiful old chestnut tree. Hertz had fallen in love with it, and it had the added attraction of having been the former residence of a world-famous physicist. The Hertzes paid 37,700 M ($9,425) for the house, which seemed ideal for their needs. As Hertz wrote to his parents on 6 January 1889:

> ... the entire external appearance of the house is so dignified and yet so unpretentious that it seems almost to have been built for me.[123]

Although Hertz's published memoirs seldom provide glimpses of the private life of his family, an exception is a letter to his parents on 20 January 1889, in which he describes the pleasure he and Elisabeth were having planning the arrangement of their new home.

> Our thoughts are now much occupied with the furnishing of the house, and we mentally arrange the furniture about the floors and the pictures on the walls.[124]

As was to be expected, it was easier to do this mentally than in the everyday world, and the Hertzes experienced great frustration during their first few months in Bonn trying to get workmen to meet their high

---

[122]For his last five years in Bonn, Clausius and his family had occupied a newly-renovated apartment in the Physics Institute.
[123]H. Hertz, *Erinnerungen*, p. 279.
[124]H. Hertz, *loc.cit.*

Introductory Biography 51

expectations for their new home and also complete the renovations in time for the arrival of the furniture from Karlsruhe.

Elisabeth had arrived on 26 April with daughter Johanna, and on June 7 Hertz's parents appeared, apparently to help "the children" get settled in their new home. This turmoil was too much for Hertz, a man who loved quietude and resented the constant interruptions he had to endure from workmen and visitors. Throughout June he noted that he was "nervous and tense to a high degree" and again "very depressed and overburdened." Even by early July he found "no peace either at home or at the Institute." By 31 July, the end of the summer semester, he was "quite exhausted" and discouraged by his complete lack of progress in his research since coming to Bonn.

Gradually, however, things settled down and peace returned to the Hertz household. A second daughter, Mathilde, was born on 14 January 1891. The Hertzes made many friends, for Bonn was an active social town, with many dances, dinners, luncheons and visits to the homes of friends. Although Hertz did enjoy a moderate amount of socializing and was known as a cordial host, a lively conversationalist, and a lover of good jokes, too much social life left him feeling unfulfilled and out of sorts. He had the gnawing feeling that he was living on the dividends of his work on electromagnetic waves and was anxious to demonstrate, by producing some worthwhile research results, that he remained alive as a physicist.

The Physics Institute also demanded much of his time. A large apartment, which had once been part of the medical school but more recently used as a residence,[125] was being converted into additional laboratory space. This required that he deal with university officials, building contractors and suppliers of laboratory furniture and equipment—a task Hertz found time-consuming and nettlesome. Shortly after Hertz arrived in Bonn, he had walked through the Institute and found it very spacious, but quite damp. To him it looked so deserted;

---

[125]This apartment, which had been occupied by the Clausius family for the previous five years, had been offered to Hertz as his family residence, but doctors whom he consulted told him that, since it had once been used as a clinic for contagious diseases, there might still be germs present. It had been thoroughly scrubbed and disinfected, and the Clausius family had lived there for five years with no ill effects, but still it was probably not a good place to raise a family with two young children. Hertz therefore decided to use the apartment for additional physics space, as had been advocated by some of the faculty. On this see H. Hertz, *Erinnerungen*, pp. 273–275.

there was none of the bustle of activity expected around a productive physics research laboratory like the one in Berlin. Even after a year he had only six advanced students doing research in laboratories in which fifty could have worked without crowding. Nagging doubts tormented him about his ability to attract enough good students to make his Institute a true center for physics research. Althoff had earlier warned him to put such thoughts out of his head; his responsibility was to perform and direct first-class physics research; then all the rest would fall into place. But this put all the more pressure on him to achieve some important research results—and soon.

Hertz had been invited to deliver the feature address at the Sixty-Second Conference of German Scientists and Physicians, to be held in Heidelberg at the end of September 1889. He worked hard on his lecture throughout August, at a time when he was slowly recovering from extreme exhaustion after the move and his first semester at the Bonn Institute. He found the lecture more difficult to prepare than he had expected, since it was to be in great part a popular account of his work on electromagnetic waves, intended for a mixed audience of professional physicists, other natural scientists, and medical doctors. At the end of August he was miserable about his lack of progress. By September 8 he was still upset about the talk, which was then only two weeks away. He wrote to his parents in near despair:

> ... what I am producing is, in my opinion (my *honest* opinion), incomprehensible to the layman, trivial to the expert, and nauseous to myself. Alas, there is no way out this time; I have to say something.[126]

Of course, this was Hertz the perfectionist speaking, and the lecture was in actuality very much better than he thought. He delivered it in Heidelberg on 20 September 1889; the large audience was most enthusiastic.[127] To a present-day reader this talk seems an exemplar of a popular, yet scientifically precise, lecture for a quite heterogeneous audience.

Hertz received much praise for the lecture afterwards, both from former colleagues like Helmholtz, Kundt, Koenigsberger and Wiedemann, and from younger scientists anxious to make his

---

[126] H. Hertz, *Erinnerungen*, p. 293.

[127] H. Hertz, "On the Relations between Light and Electricity," Paper No. 10.

acquaintance. In the evenings, after a day spent listening to papers and discussions, there was much conviviality and frequent beer-drinking sessions (*Kneiperei*) with old friends. By Monday Hertz had had his fill of the excitement, the talk, the constant drinking, and decided to pack up and return to Bonn before the meeting formally ended.

Research on Electromagnetic Theory

During his first half year at Bonn Hertz had very little time for research. In September 1889, he was bothered by an infected foot and unable to move around without considerable difficulty. This made laboratory work impossible and so he turned back to electromagnetic theory, just as he had done at Kiel in 1884 when he did not have the equipment needed for his experiments. This time his research resulted in two important papers on the basic equations of electrodynamics.

The research for his first paper: "On the Fundamental Equations of Electromagnetics for Bodies at Rest," was begun in late September 1889. The paper was completed by March 1890, and quickly published in the *Annalen*.[128] In his introduction to this paper Hertz indicates that his intention is "to sift Maxwell's formulae and to separate their essential significance from the particular form in which they first happened to appear."[129] For Hertz this meant removing all nonessentials like the electric displacement and the vector potential from Maxwell's formulae. He considered these part of the scaffolding Maxwell used in constructing his theory, but confusing if retained once Maxwell's theory was complete.

From February to September 1889, Hertz had carried out a lengthy correspondence with Oliver Heaviside[130] on electromagnetic waves and Maxwell's equations. In his paper he pointed out that Heaviside had been working along similar lines to his since 1885, and that the equations he had derived were, in their simplest form, the same as those derived in this paper by Hertz. In his usual scrupulously

---

[128]H. Hertz, *El.Waves*, pp. 195–240.

[129]H. Hertz, *El.Waves*, p. 196.

[130]Oliver Heaviside (1850–1925), the self-taught British physicist and electrical engineer, made many important contributions to electrical theory. On Heaviside, see Nahin (1988); on Heaviside's interactions with Hertz, see O'Hara and Pricha (1987), pp. 55–84, and Hunt (1991), pp. 180–182, 200–201.

honest fashion, Hertz then makes clear that "Mr. Heaviside has the priority."[131]

The most important result of Hertz's paper was Maxwell's equations in the symmetric form we use today, free of both scalar and vector potentials. These equations were referred to at the turn of the century as "Maxwell's equations in the Hertz-Heaviside form." Although Hertz had conceded the priority of this approach to Heaviside, the clarity of Hertz's presentation played a very important role (especially in Germany) in changing the minds of physicists about electromagnetic theory. Arnold Sommerfeld recalled that, on reading this paper of Hertz, the shades fell from his eyes, and he understood electromagnetic theory for the first time.[132] The once-neglected equations of Maxwell, stripped by Hertz of the mechanical models and auxiliary quantities on which Maxwell had based them, soon became the basis for all research on electromagnetism and optics in Germany.[133] Since Germany was the world's leader in physics at the end of the nineteenth century, this development had a major impact on physicists throughout the rest of the scientific world.

Starting in March 1890, Hertz tackled the much more difficult problem of the electromagnetics of bodies in motion.[134] In this paper he extended the approach used in the previous paper. Since Hertz had always been a firm believer in the ether, the question naturally arose: "Does the ether remain at rest, or is it dragged along with the moving body?" For purposes of simplifying the theory, Hertz assumed that the ether inside a material object is completely dragged along with the object when the object moves. This made it possible for him to proceed with his theoretical treatment of the problem, but, as Hertz admitted, "It is scarcely probable that these restrictions correspond to the actual facts of the case."[135] For this reason the second paper of

---

[131]H. Hertz, *El.Waves*, p. 197. Hertz seems to have been excessively modest in this matter, since in his 1884 paper he had derived Maxwell's equations in essentially the same form arrived at by Heaviside.

[132]Sommerfeld (1954), p. 2.

[133]On this see Planck (1906), p. 298–299.

[134]H. Hertz, "On the Fundamental Equations of Electromagnetics for Bodies in Motion," *El.Waves*, pp. 241–268.

[135]H. Hertz, *El.Waves*, p. 268.

Introductory Biography

Hertz has not played any significant role in the development of classical electromagnetic theory.[136]

By late September 1890, Hertz had mailed off the manuscript of this last theoretical paper to Wiedemann for the *Annalen* and turned back again to laboratory research on electromagnetism. This time, however, he had neither the time, the energy, nor the inspiration to repeat his great success in Karlsruhe. He did publish one small piece of research,[137] which had been planned but never completed in Karlsruhe. This involved the mechanical effects of electric waves on wires. Hertz had wanted to study such effects in "free air," but "this last hope was frustrated by the feebleness of the effects produced...."[138] For this reason he thought his results incomplete and unsatisfactory.

During late 1890 Hertz's experiments constantly encountered new difficulties; he had increasing problems with his health; and had to devote so much time to running the Institute and to social obligations that he began to grow discouraged. At this time a letter arrived that boosted his spirits. On November 8 he received word from London that the Royal Society had awarded him one of its major awards for scientific achievement, the Rumford Medal.[139] On November 28 Hertz departed for London to receive the medal. It is said that modesty prevented him from announcing this honor to his colleagues at the university; he simply slipped away from Bonn and tried to return there just as quietly afterwards. By that time, however, the news had spread around the university, and his students greeted him with applause equal to that he had received when he accepted the Rumford Medal in London.

---

[136]This was a first, admittedly tentative, attack on a difficult problem that was eventually solved only by Einstein's special theory of relativity.

[137]H. Hertz, "On the Mechanical Action of Electric Waves on Wires," *El.Waves*, pp. 186–194.

[138]H. Hertz, *El.Waves*, p. 187.

[139]Named after Benjamin Thompson, Count Rumford (1753–1814), an American who fled the United States during the American Revolution and spent most of his life in Great Britain, France and Germany. He was the founder of the Royal Institution, a scientific research and teaching institute dedicated to the application of science to the needs of society, and made important contributions to the field of thermodynamics. On Rumford, see Brown (1979).

Hertz had a wonderful time in England, was treated royally, and was happy he had agreed to receive the medal in person.[140] He had corresponded with a number of British physicists about his work on electromagnetic waves, and especially with the Maxwellians,[141] Oliver Lodge, George Francis FitzGerald and Oliver Heaviside. Now Hertz enjoyed the opportunity to have a quiet dinner with Lodge and FitzGerald, but Heaviside, who was quite a recluse, did not join them. Hertz also met Ayrton, Dewar, Crookes, Lockyer, Kelvin, Stokes, Rayleigh and other well-known British physicists.

He was shown Westminster Abbey, the Royal Institution, the National Gallery, a number of imposing London clubs, the laboratories of many of the physicists he had met, and even an underground power station that provided electricity for the lights of London. He would have preferred more time on his own to see London, but all he could squeeze in was a two-and-a-half-hour walk through the city, from London Bridge to the famous Tower. The last day of his visit he spent in Cambridge, meeting more colleagues and touring the laboratories and landmarks of the university. In all, he spent five days in England, days of pleasure and accomplishment but, he had to admit, "in many respects a great strain." He returned to Bonn on December 4, physically tired but mentally refreshed, and confident that he had made some new friends who might trade ideas with him on research projects of mutual interest.

Research on Cathode Rays

Hertz had been an associate of Eugen Goldstein (1850–1930) in the years 1878–1881 when Goldstein was doing experiments at the Berlin Physics Institute on the conduction of electricity through gases. Goldstein received his doctorate in 1881, with Helmholtz as his mentor. He first interested Hertz in "cathode rays," the name he had himself given in 1876 to the emissions from the negative electrode in evacuated Geissler tubes.

---

[140]Many details of Hertz's visit to London are contained in his long letter to his parents on 5 Dec. 1890 (*Erinnerungen*, pp. 307–311). This letter is the source of the material contained in the present section.

[141]On the Maxwellians, see Hunt (1991).

Hertz retained this interest throughout his life. Among his many papers based on work done during his years in Berlin were two relating to cathode rays.[142] Hertz had been able to deflect cathode rays in magnetic fields, but failed to do so in electric fields because of the poor vacuum his equipment could produce and the static charges built up by the ions produced from the residual gases. This and his failure to observe any magnetic effects of the rays outside the Geissler tube led most German physicists to conclude that cathode rays were not charged particles, but waves or some other kind of unusual disturbance in the ether. Hertz never changed this incorrect, though understandable, conclusion from his poor experimental data.[143]

Hertz had used this work on cathode rays as his *Habilitationschrift* when he was called to Kiel as *Privatdozent* in 1883. He then put aside all further work on the subject, since he did not have the necessary equipment at Kiel. On being appointed professor at Karlsruhe in 1885, Hertz resumed work on cathode rays. Soon, however, he turned his attention to electromagnetic waves, and his great success with this pioneering research drove all thoughts of further research on cathode rays from his mind.

On 1 April 1891 Dr. Philipp Lenard (1862–1947) arrived in Bonn to serve as assistant to Hertz at the Physics Institute, beginning in the summer semester. Lenard was an experienced experimentalist who had studied at the universities of Budapest, Vienna, Berlin and Heidelberg. His dissertation topic was suggested by Helmholtz, but the work was completed in Heidelberg under Georg Quincke (1834–1924). Before coming to Bonn Lenard had done some research on cathode rays. This common interest in cathode rays may have stimulated Hertz to return to the laboratory on 29 September 1891 to work on phenomena in rarified gases.[144] Since he was also actively engaged in

---

[142]These are "On a Phenomenon Which Accompanies the Electric Discharge," *Misc.Pprs.*, pp. 216–223; and "Experiments on the Cathode Discharge," *Misc.Pprs.*, pp. 224–254.

[143]On the cause of Hertz's mistaken views on the nature of cathode rays, see G. Thomson (1970) and Hon (1987).

[144]Bonn had a considerable tradition of cathode-ray research before Hertz went there in 1889. Heinrich Geissler (1814–1879), a skillful glassblower, opened a shop in Bonn in 1852 for the construction of scientific instruments. In 1855 he developed a much improved vacuum pump that enabled him to make his famous evacuated "Geissler tubes" for cathode-ray research. In 1858 these were used by Julius Plücker (1801–1868), Professor of Physics in Bonn, to

theoretical work on mechanics, it is probable that he was only able to devote occasional patches of time to the cathode-ray research. In whatever time he had, however, Hertz was soon able to produce an important paper on cathode rays,[145] which would turn out to be his last published paper in any field of experimental physics.

In this short but fundamental paper, Hertz shows that cathode rays pass through thin foils of gold, silver or aluminum and can be detected by the phosphorescence they produce in uranium glass after passing through the foil. He explains that this transparency to cathode rays could not be caused by macroscopic holes in the foils, since the number and size of the holes is totally inadequate to account for the amount of phosphorescence observed. This evidence would later lead J.J. Thomson (1856–1940) to the conclusion that the foils must contain holes at the atomic level, through which very small particles (such as electrons) could pass.

After Hertz made this discovery, he took Lenard into his laboratory and demonstrated his successful experiment. Lenard later described what happened:

> He [Hertz] said to me: "We ought—and I might simply do this for he was prevented—to separate two chambers with aluminum leaf, and produce the rays as usual in one of the chambers. It should then be possible to observe the rays in the other chamber more purely than has been done so far and even though the difference in air pressure between the two chambers is low because of the softness of the leaf, it might be possible to completely evacuate the observation chamber and see whether this impeded the spread of the cathode rays—in other words, find out whether the rays are phenomena in matter or phenomena in ether." He appeared to consider this last question to be the most important one.[146]

---

investigate the cathode discharge in evacuated tubes. He was the first one to show that a magnetic field shifted the position of the cathode-ray beam. Some of Geissler's tubes were still available in the physics *Kabinett* at the Bonn Institute, and this may also have sparked Hertz's renewed interest in the subject. It is worth noting that Michael Faraday had used Geissler tubes in his research, but always called them "Bonn tubes."

[145]H. Hertz, "On the Passage of Cathode Rays through Thin Metallic Layers," Paper No. 13. As FitzGerald wrote in his review of *Misc.Pprs.*, this paper "is particularly interesting as the starting-point for Lenard's work which has resulted in the discovery of the x-rays." (Paper No. 4, p. 156.)

[146]*Nobel Lectures: Physics*, vol. 1 (1967), pp. 107–108.

Introductory Biography 59

This led to Lenard's construction of what was later called a "Lenard window" for cathode-ray research. In a letter to Helmholtz on 15 December 1892 Hertz discussed Lenard's success in obtaining a beam of cathode rays outside the vacuum chamber in which they were produced, using thin foils of the kind described in Hertz's article:

> My assistant, Dr. Lenard, has made a very singular discovery here during the past weeks. He sealed off Geissler tubes with extremely thin aluminum foils, and succeeded in producing foils of a thickness that can provide completely airtight seals and yet remain thin enough to allow an appreciable part of the phosphorescence-producing cathode rays to pass through. At the same time he found that these rays, once produced, also propagate in air-filled space, to a varying degree in various gases; this opens up a whole new field of research, since the production of these rays can now be wholly separated from their observation.[147]

Lenard was to continue his work on cathode rays with such success that it would eventually win for him the Nobel Prize in Physics.

Lenard had a distinguished career as a physicist after leaving Bonn, but was less distinguished as a human being. Initially Lenard greatly admired Hertz both as a physicist and as a man, and for that reason was anxious to come to Bonn as Hertz's assistant.[148] But later Lenard claimed that he only saw Hertz once in the laboratory, and that was when Hertz showed him his cathode-ray experiment.[149] He also

---

[147]Koenigsberger (1902–1903), vol. 3, p. 64. Bopp and Gerlach (1957) have made the interesting observation that these researches of Hertz and Lenard, by enabling unambiguous experiments on cathode rays to be performed outside the Geissler tube in which they were produced, eliminated perturbations caused by static charges within the tube. It thus made up for Hertz's unsuccessful attempt to deflect cathode rays in electrostatic fields, an observation that had misled physicists for many years as to the nature of cathode rays.

[148]Shortly after Hertz's death, Lenard wrote to the astronomer Max Wolf: "For myself I must be happy that my wish has been fulfilled in that I have come to know intimately a truly great spirit, if only for a very short time." See Lenard (1957), p. 567. Another indication of Lenard's admiration for Hertz when he was with him at Bonn may be found in his brief obituary of Hertz [Lenard (1894)].

[149]Lenard (1943), p. 39. It should be noted that by this time Hertz was in increasingly bad health and not able to spend much time in the laboratory. He was also working extremely hard on his book on mechanics. Lenard, however, seemed unwilling to take these extenuating circumstances into account.

felt that Hertz had not paid sufficient attention to his work. Although a brilliant experimental physicist, Lenard was a difficult, almost paranoid, personality, who often saw rejection where none was intended.

During Hertz's final illness, he had asked Lenard to act as the editor of his *Miscellaneous Papers* and *Principles of Mechanics*, and see them through to publication.[150] Lenard fulfilled this task conscientiously until he left Bonn in fall 1894.[151] After leaving Bonn, however, Lenard claimed that his own research was greatly hampered by this editorial work on Hertz's papers and by his appointment to a post in theoretical physics (not really one of Lenard's strengths) in Breslau.[152] This is disingenuous. He probably did lose some months of research time in Bonn, but he lost a great deal more by accepting the position at Breslau (which had essentially no physics equipment), then moving to Aachen one year later, taking yet another position in theoretical physics at Heidelberg in 1896, and in 1898 moving to Kiel. Even the best experimentalist in the world could not accomplish any worthwhile research while having to set up a new research laboratory for himself at a different university every year or two.

Later in his career Lenard appears to have completely forgotten his debt to Hertz, whose work on cathode rays and the photoelectric effect had encouraged Lenard to work in these fields.[153] His research on these topics brought Lenard the Rumford Medal of the Royal Society in 1896 (jointly with Wilhelm Roentgen) and his Nobel Prize in 1905. But even this was not enough for him. He thought he should

---

[150] On this request and on the difficulties Lenard had in editing Hertz's *Mechanics*, see the letter of 27 Jan. 1894 from Lenard to Max Wolf, in Lenard (1957), p. 568.

[151] One reason Lenard left Bonn was undoubtedly that Hertz's successor as Director of the Physics Institute there was Heinrich Kayser, who had some bad experiences with Lenard in Berlin, and now found himself saddled with him as his assistant in Bonn. Kayser says that even as a doctoral student in Berlin, Lenard was terribly afraid that other students would steal his ideas, and therefore did everything possible to keep them out of his laboratory. See Kayser (1936), pp. 137, 177.

[152] *Nobel Lectures: Physics*, vol. 1 (1967), p. 114.

[153] For Lenard's research on the photoelectric effect, and the events leading up to J.J. Thomson's discovery of the electron, see Wheaton (1978).

Introductory Biography 61

have received the very first Nobel Prize in Physics, which went to Roentgen in 1901 for his discovery of x-rays.[154]

In his later years Lenard became ever more difficult. Although he had been born in Hungary, Lenard developed into an excessively nationalistic German early in his life. He became increasingly anti-Semitic and used the label "dogmatic Jewish physics" to describe theoretical physics. He even painted Hertz with the same sad brush, praising Hertz's experimental work as excellent physics, but tracing Hertz's interest in theoretical physics back to his father, whose family was Jewish.[155]

Until his death in 1894 Hertz wrongly maintained that cathode rays were some kind of unknown wave or disturbance in the ether, rather than a beam of material particles. His success in showing that electromagnetic energy propagated through space in the form of waves led many German physicists to assume that something similar was true of cathode rays. Just as they rejected a corpuscular theory for light and the rest of the electromagnetic spectrum, so they rejected a priori a corpuscular nature for cathode rays. After Hertz's death, Lenard loyally (but mistakenly) defended Hertz's erroneous position on cathode rays. At the British Association meeting in 1896 Lenard and Bjerknes, both of whom had worked with Hertz in Bonn, still argued that cathode rays were some kind of unknown ether disturbance.[156] At the same meeting all the British physicists supported the charged-

---

[154]Many physicists, including E. Goldstein in 1880 and J.J. Thomson in 1894, had come much closer to the discovery of x-rays than had Lenard, but had not grasped what they saw. Neither had Lenard. At the Kelvin jubilee at Glasgow in 1896, Quincke defended the priority of Lenard's research over that of Roentgen in the discovery of x-rays, saying that Lenard had the whole thing in his mind before Roentgen's discovery. Sir George Stokes replied to Quincke: "Lenard may have had the rays in his brain, but Roentgen got them into other people's bones." On this see Schuster (1911), pp. 77-78.

[155]On Lenard and the Third Reich, see Beyerchen (1977), especially chapter five, and Heilbron (1986), pp. 113-118 and 159-179. Lenard's gradual change in his attitude toward Hertz is discussed on p. 124 of Beyerchen.

[156]On this see Whittaker (1951), vol. 1, pp. 353-354.

particle view of cathode rays.[157] The arguments of Lenard and Bjerknes were laid to rest forever when, in 1897, J.J. Thomson performed the famous experiments in which he proved conclusively that cathode rays were indeed deflected by an electric field and must consist of beams of very small, negatively charged particles, which we now call electrons.

Hertz and Vilhelm Bjerknes

Vilhelm Bjerknes (1862–1951) was the son of the Swedish mathematician Carl Anton Bjerknes (1825–1903). The younger Bjerknes wrote to Hertz in 1889 to ask if he could be accepted as a research student in Bonn. He spent three semesters at Bonn, the summer and winter semesters of 1890 and the summer semester of 1891, during which time he published three papers in the *Annalen* describing his research on electric resonance, which had been done under Hertz's direction. In 1923 he put these papers together with four later papers on the same subject to make up a book,[158] which he dedicated to Hertz's memory. After leaving Bonn, Bjerknes switched over to applications of Helmholtz's and Kelvin's work on hydrodynamics to the motion of the Earth's atmosphere and the oceans. He became very interested in weather prediction and, as a consequence of his many contributions to the subject (including the idea of a "front"[159]), is today known as "the father of modern meteorology."[160]

Bjerknes came to Bonn expecting to find a well-equipped research laboratory and students enthusiastic about working with the discoverer of electromagnetic waves. Instead he found a laboratory that had suffered years of neglect under Clausius, and—even more surprising—that he was Hertz's only research student during his first

---

[157]As early as 1894 FitzGerald had written concerning Lenard's experiments in the 21 June 1894 issue of *Nature*: "So far the phenomena described are quite like those that would be due to moving electrified matter, and the actions are quite unlike anything we know of the properties of the ether." In FitzGerald (1902), p. 320.

[158]Bjerknes (1923).

[159]On this see Friedman (1982).

[160]For a full account of Bjerknes' remarkable career, see Friedman (1989).

semester in Bonn.[161] This provided him with an unique opportunity to develop a very close relationship with his mentor. For this reason parts of Bjerknes's twenty-six-page introduction to his book[162] provide us with an important first-hand account of Hertz's personal characteristics and his teaching and research methods.

Bjerknes refers to the "unusual friendliness, simplicity and modesty" of Hertz's demeanor.[163] According to Bjerknes, Hertz's words were always stimulating and to the point, and a student always came away with the feeling that something important had been learned. One of Hertz's great strengths was his ability to think imaginatively, using concrete pictures and examples to enlighten and clarify his ideas. Bjerknes was also impressed by Hertz's desire to help his students and to further their careers after they left Bonn. When Bjerknes was editing his first two papers for publication, Hertz insisted that all but one reference to Hertz's advice and assistance be eliminated. He apparently felt that otherwise Bjerknes would not receive enough credit for the publications. Behavior like this endeared Hertz to Bjerknes and the other Bonn students.

During Bjerknes's year in Bonn Hertz only lectured in the usual introductory physics course. Bjerknes rarely sat in on these lectures, but some thirty years later recalled "the enthusiastic presentation and delicate sense of humor that kept alive the interest of the audience."[164] Hertz emphasized the fundamentals and laid out the outline of an argument or derivation very clearly, but he saw no value in long mathematical derivations in a class or lecture. To make this point to the students in a physics colloquium, Hertz on one occasion interrupted the speaker, who had become very involved in a long derivation on the blackboard, and asked the students, one at a time: "What does m signify?"; "What does a mean?" Of course, no one could tell him, for by this time they no longer remembered the basic

---

[161] In April 1891, Lenard arrived to assume the position of assistant to Hertz and shortly thereafter J. Precht from Germany, Josef von Geitler (Hertz's cousin) from Austria, and Daniel E. Jones (the translator of the three volumes of Hertz's collected works) from Great Britain joined Bjerknes as research students in Hertz's Institute. There are four letters written by Hertz to von Geitler in 1893 in the Dibner Library of the Smithsonian Institution in Washington, D.C.

[162] Bjerknes (1923), pp. vii–xxxii.

[163] Bjerknes (1923), p. viii.

[164] Bjerknes (1923), p. xii.

concepts being discussed. He then pointed out to the students that they must be careful in their future careers not to waste time in classroom lectures on the details of mathematical derivations. The method and the results were the important thing; all the rest could be filled in later by the students for themselves.

The same approach carried over to Hertz's research. He never got bogged down in trivia. He chose important problems, and completed and published only what seemed significant for the future of physics; he would never work on unimportant problems just for the sake of publishing a paper. (This is one reason why, at certain stages of his career, he found it difficult to uncover significant problems that also appeared solvable.)

His first piece of advice to his research students[165] was the following: choose good problems; find the simplest possible apparatus and method of approach that appear likely to solve the problem; and carry the research through to completion, without stopping to improve the apparatus or to explore peripheral issues.[166] When the project is finished, look over the situation and determine whether the results warrant repeating the project with improved techniques and better equipment; then proceed accordingly. He supplemented this with some ideas about choosing good problems: separate out of the general chaos of the physical world a part that seems important and capable of being handled independently of the rest; focus your attention on this part alone and do not allow yourself to be distracted by the rest until you have completed work on the chosen part; "When that is accomplished, you will find yourself relieved and free, and with great strength for the next task." Such an approach seems to have served both Hertz and his research students well.

During Bjerknes's first semester in Bonn, Hertz visited him in his laboratory a few times each day. During these visits Hertz was clearly the teacher and Bjerknes the student. Bjerknes found Hertz outstanding (*ausgezeichnet*) as a teacher in these circumstances. He said very little, but what he said was always worthwhile, and often helped Bjerknes solve problems he could not solve for himself. On

---

[165]Bjerknes (1923), pp. xi–xii. Bjerknes himself got in trouble with Hertz by spending too much time improving methods and refining equipment for his first research project in Bonn.

[166]In his earlier researches Hertz had not always followed this advice himself. If he had, he would never have discovered the photoelectric effect!

Introductory Biography

each visit Hertz introduced only a few new ideas and allowed Bjerknes time to digest them before he returned with more. One has the impression that the same careful consideration and lively imagination that marked his dealings with Bjerknes must also have distinguished Hertz's own research and classroom teaching.

The Principles of Mechanics

Heinrich Hertz's last scientific work, *The Principles of Mechanics Presented in a New Form*,[167] was written at Bonn in the years 1891–1893. The origins of this unusual book, so different from the rest of Hertz's writings, can be traced back to three circumstances affecting his life in Bonn: his growing awareness of his poor health, which made it difficult for him to spend long hours in his beloved laboratory; his inability to find an experimental project of sufficient promise to warrant the large investment of time and physical energy it would take to complete it; and his deepening conviction that the solution to most problems in physics lay in a better understanding of the nature and properties of the ether.[168]

The first factor seemed beyond his control; it can be easily documented from a perusal of his diary for the years 1889–1893. His inability to find an attractive experimental project is clear from the many different experiments he tried and quickly abandoned during his years at Bonn. Thus, on 2 February 1891 he writes: "Took a walk, searched in vain for starting points for fresh work."[169] When he was frustrated with his efforts at experimental research at Bonn, he switched to theoretical work on the foundations of mechanics and quickly developed great enthusiasm for this research, often rising at 5:30 A.M. to devote some time to it before breakfast.

But why mechanics, and what relevance did the ether have to mechanics? Hertz firmly embraced the prevalent view of German

---

[167] Papers No. 14, 15, 16 and 17 in this book pertain to Hertz's *Principles of Mechanics*, which will be referred to in what follows simply as *Mechanics*. A good recent article on Hertz's *Mechanics* is von Reden (1988), which includes a useful summary of the contents of Hertz's book.

[168] See the last paragraph of Paper No. 10. As an example of Hertz's interest in the ether at this time (1891–1893), see the passage in Lenard's Nobel Prize Lecture cited in footnote 146 above.

[169] H. Hertz, *Erinnerungen*, p. 313.

physicists at the time that the prime task of physics was to reduce all phenomena to mechanics.[170] In addition, Hertz's research on electromagnetic waves had convinced him of the ether's reality and of its key importance in unlocking the secrets of the physical universe.

That Hertz's object in his *Mechanics* was to put mechanics on an absolutely firm foundation, so that it could later be used to reveal the structure of the ether, has been suggested by a number of historians of science, but most convincingly by Martin Klein and by Jungnickel and McCormmach.[171] This suggestion is of prime importance and makes Hertz's motivation in writing his *Mechanics* much easier to understand.

During Hertz's brief tenure in Bonn, his preoccupation with the ether is evident from his letters,[172] his theoretical papers on electromagnetism,[173] and especially his *Mechanics*. In the latter there are many references to the ether. For example, on the first page of the Author's Preface, after Hertz discusses the need to trace all phenomena back to the laws of mechanics, he goes on:

> ... it is premature to attempt to base the equations of motion of the ether on the laws of mechanics until we have obtained a perfect agreement as to what is meant by this name [i.e., by "the laws of mechanics"].[174]

Note that he does not say that it is *impossible* to base the equations of motion for the ether on the laws of mechanics; it is only *premature*, an indication that he expected that someone, perhaps he himself, would someday succeed in doing this.

In Hertz's Introduction to his *Mechanics* he seems to have the ether in mind in much of his discussion.[175] FitzGerald, in his review of the *Mechanics*, makes explicit what Hertz had only implicitly suggested:

---

[170]See, for example, the first sentence of the Author's Preface to Hertz's *Mechanics*, Paper No. 15.

[171]Klein (1972), pp. 72–73; Jungnickel and McCormmach (1986), vol. 2, pp. 141–143. See also Rosenfeld (1957), p. 1665; Klein (1970), p. 146; McCormmach (1975), p. 348; Hunt (1991), p. 179.

[172]See, for example, the letter of Hertz to Heaviside, 3 Sept. 1889; in Appleyard (1930), p. 239.

[173]See, in particular, *El.Waves*, pp. 201, 241–243, 268.

[174]H. Hertz, first page of Paper No. 15.

[175]See, e.g., footnotes 22, 37, and 39 to Paper No. 16.

Hertz sees in all actions the working of an underlying structure whose masses and motions are producing the effects on matter that we perceive, and what we call force and energy are due to the actions of these invisible structures, *which he implicitly identifies with the ether.*[176]

The suggestion of historians of science about the object of Hertz's *Mechanics* thus seems to be fully in accord with Hertz's preoccupations while he was involved in this work and is the only hypothesis consistent with all the documentary and psychological evidence available.

Hertz's *Mechanics* breaks down into a 41-page Introduction[177] and two main parts: Book I on the "Geometry and Kinematics of Material Systems"; and Book II on the "Mechanics of Material Systems," the main subject of the book.

Hertz's Introduction contains an excellent discussion of the distinctive nature of scientific knowledge. He tries to develop a moderate form of phenomenology, in which objects in nature are represented by mental "pictures" (*Bilder*), which have one requisite characteristic, that the essential relations between these pictures must agree with the actual relations between the external objects that are pictured. No other correspondence between the pictures and the real objects is required. Hertz then goes on to discuss three different systems of mechanics: the first is the conventional formulation of mechanics in terms of space, time, mass, and force as fundamental quantities; the second replaces force by energy (although Hertz points out that no adequate system of this kind had yet been formulated); the third, which is Hertz's own contribution, uses only space, time and mass as fundamental quantities, omitting both force and energy.

Book I of Hertz's *Mechanics* treats the statics and kinematics of material systems as a branch of mathematics, and contains little that is new or different. Book II, which is the heart of the *Mechanics*, uses the three fundamental quantities—space, time and mass—together with the following Fundamental Law, to put mechanics in so perfect a form "that there should no longer be any possibility of doubting it."[178]

---

[176]Paper No. 17, p. 371 [italics not in original]. FitzGerald's review contains a fine summary of the *Mechanics*.
[177]Paper No. 16.
[178]See footnote 15 to Paper No. 16.

Fundamental Law: Every free system persists in its state of rest or of uniform motion in one of its straightest paths.[179]

For Hertz this law, which combines Newton's law of inertia with Gauss's Principle of Least Constraint, "... satisfies the requirement that the whole of mechanics can be developed from it by purely deductive reasoning without any further appeal to experience."[180] Starting from this law, his treatment proceeds in a purely logico-deductive fashion, with a long series of definitions, propositions, and correlaries that remind one of Euclid's *Elements.* At the end Hertz does not claim to have discovered anything new in mechanics; all he has done is improve the order and arrangement of the subject. His interest clearly was more in the logical and philosophical aspects of mechanics than in applications.

Hertz wrote a first draft of his *Mechanics* in one year and then spent two years in polishing and improving it. He rewrote the whole manuscript three or four times in an effort to achieve "absolute clarity." His discussion of the fundamental concepts of mechanics helped rid mechanics of some of the vagueness and lack of rigor that marked much of nineteenth-century physics, but his attempt to eliminate force as a fundamental quantity introduced new problems. He had to replace "force" by an elaborate scheme of "concealed masses" (*verborgene Massen*) and "concealed motions" (*verborgene Bewegungen*), which acted in such a way as to simulate the forces we seem to observe in nature.[181] This bought for Hertz some relief from the need to understand precisely the nature of "force" (a concept he had always found confusing).[182] However, it did so at the expense of

---

[179]H. Hertz, *Mechanics,* p. 144. Here by a "straightest path" Hertz means the path of least curvature for the whole system. The statement of this Law in Hertz's Introduction (Paper No. 16, p. 350) differs from this one in wording, but the meaning is the same. This fundamental law was referred to at the turn of the century in Germany as the *Hertzsche Prinzip.*

[180]H. Hertz, Paper No. 16, p. 326.

[181]Martin Klein calls this proposal of Hertz "an explanation in the Cartesian tradition, a mechanics from which dynamics would be eliminated and which would consist exclusively of kinematics." Klein (1972), p. 74. For this reason Emil Wiechert (1861–1928) gave the name *Kinetische Mechanik* to Hertz's *Mechanics.*

[182]See, for example, Hertz's letter of 13 Jan. 1878 to his parents, *Erinnerungen,* pp. 75–77.

Introductory Biography

requiring the understanding of unobservable motions of unobservable masses, which were linked together, and to the observable masses of the system, by rigid connections. This was Hertz's way of eliminating all action-at-a-distance from physics, but he did this by introducing something looking very much like a model of an ether that could transmit electric and magnetic forces through space.

Most physicists have found Hertz's *Mechanics* a beautiful, logical system, but of little practical value. They have not been sympathetic to Hertz's desire to eliminate force as a fundamental quantity and have been uneasy with his hidden masses and hidden motions. Ludwig Boltzmann commented on the sad fact that Hertz died before his *Mechanics* was published and so was unable to respond "to the thousand requests for clarification that are certainly not on the tip of my tongue alone."[183] Ernst Mach wrote: "As an ideal program Hertz's mechanics is simpler and more beautiful, but for practical purposes our present system of mechanics is preferable, as Hertz himself, with his characteristic candor, admits."[184] Even Helmholtz in his Preface to Hertz's *Mechanics* has very little of substance to say about Hertz's book and certainly displays little enthusiasm for it. He was unhappy that Hertz did not provide examples of how these "hypothetical mechanisms" were supposed to replace the forces of the conventional mechanics.[185]

For the contemporary physicist, the *Mechanics*—which Hertz considered important because of its relevance to the enigma of the ether—has been rendered irrelevant by the ether's demise when relativity and quantum mechanics came into being at the beginning of this century. At present, the *Mechanics* seems overtaken by the progress of physics, but possibly Hertz's three years of intense work

---

[183] Boltzmann (1974), p. 90.

[184] Mach (1960), p. 324. Mach, who was twenty years older than Hertz, only met Hertz once, at the 1891 meeting of the Gesellschaft Deutscher Naturforscher und Ärzte in Halle. Before that meeting they exchanged letters. In his letter to Hertz, Mach expressed his desire to meet the author of Hertz's 1884 paper (Paper No. 3) and of Hertz's two theoretical papers on electromagnetism published at Bonn in 1890. The latter papers Mach described as approaching "the ideal of a physics free of mythology." Mach did not have the same high opinion of Hertz's *Mechanics*, however, since he thought that Hertz's concealed masses and concealed motions were reintroducing mythology into physics. For this interchange of letters between Hertz and Mach, see Thiele (1968).

[185] Paper No. 14, pp. 317–318.

may one day prove to have more relevance to the world of physics than we are willing to allow it today.

Philosophers of science, on the other hand, have always been enthusiastic about Hertz's *Mechanics*, especially his Introduction, because it contains an impressive discussion of the epistemology of physics. Braithwaite calls Hertz "the most philosophically profound of the great nineteenth-century physicists who wrote on the philosophy of science."[186] Ludwig Wittgenstein read Hertz's *Mechanics* while an engineering student in Linz and was captivated by it. Hertz continued to exert a strong influence on Wittgenstein's thought throughout his life.[187] Wittgenstein considered the Introduction to Hertz's *Mechanics* an authoritative analysis of fundamental ideas about the physical world and was strongly influenced by it in writing his *Tractatus*. And many other philosophers of science have expressed similar enthusiasm for Hertz's excursion into philosophy.[188]

Helmholtz and Hertz's Mechanics

Hertz's *Mechanics* is a clear example of the philosophical impulse that motivated Hertz, as it had Helmholtz throughout his long career. From 1883 to his death in 1894, Helmholtz had endeavored to find a grand unifying principle of physics.[189] He thought for a while that he had found it in the Principle of Least Action, but his efforts to derive all of physics from this principle were doomed to the same frustration Einstein experienced in his search for a unified field theory.

Hertz's diary for 12 June 1892 indicates that he had completely rewritten his "paper" on mechanics, and that it had begun to take palpable shape. Soon thereafter an illness kept him from any serious research and writing until near Christmas. It was only on 15 December

---

[186] Braithwaite (1953), p. 90.

[187] On Hertz and Wittgenstein, see Janik and Toulmin (1973), pp. 139–146, 179–191. On page 200 the same authors write: "For it was during that war (World War I) that these thoughts—drawn from Kraus and Loos, Hertz and Frege, Schopenhauer, Kierkegaard and Tolstoy—coalesced into the unity which was the *man* Ludwig Wittgenstein."

[188] Among others, Cassirer (1950), Cohen (1956), Cooke (1974), and Barker (1980). For a further discussion of the relationship of Hertz's ideas on mechanics to those of Ernst Mach, see footnote 9 to Paper No. 15.

[189] See Koenigsberger (1902–1903), vol. 2, pp. 315–335.

1892 (in the same letter in which he reported Lenard's breakthrough on cathode rays) that Hertz wrote to Helmholtz:

> Of late I have been devoting myself entirely to theoretical work on topics suggested to me by the study of your papers on the Principle of Least Action.[190]

It is no great surprise, then, that when his *Mechanics* was published after Hertz's death in 1894, the first reference in the Author's Preface is to Helmholtz's papers on the Principle of Least Action and on cyclical systems. Hertz there makes explicit his dependence on Helmholtz's work:

> Both in its broad features and in its details my own investigation owes much to the above-mentioned papers [of Helmholtz]; the chapter on cyclical systems is taken almost directly from them.[191]

It is also clear that the hidden masses and hidden motions introduced by Hertz in his *Mechanics* can be traced back to these same papers of Helmholtz.

Given the admitted dependence of Hertz's *Mechanics* on Helmholtz's earlier work, an event that occurred when Philipp Lenard was preparing Hertz's book for publication seems at first hard to understand. On 28 April 1894 Lenard wrote to Helmholtz seeking help with two passages in the Second Part of the *Mechanics*, which Hertz had sent to the printer reluctantly and only because his health would not allow him to improve it further. Lenard expressed great hesitation about making any changes in Hertz's text without advice from an expert like Helmholtz. On May 21 Helmholtz replied that he could not correct Hertz's text or approve of its correction, since he did not have the time for the "tranquil consideration and thorough understanding" required. He goes on:

> I can only say that I am just beginning to see what his aim is, and this merely since I received the last set of proofsheets a few days ago. *Till then I had not the slightest inkling of what he was driving at.*[192]

---

[190]Koenigsberger (1902–1903), vol. 3, pp. 63–64. One of the papers Hertz refers to here is Helmholtz (1892).

[191]Paper No. 15, p. 320.

[192]Koenigsberger (1902–1903), vol. 3, pp. 104–105 [italics not in original]. Helmholtz had suffered a severe fall aboard ship on his way back to

It seems clear then that, even though some of the elements of Hertz's *Mechanics* derive from Helmholtz, Hertz had assembled them into a logical system that Helmholtz found strange. Helmholtz openly expressed his doubts about Hertz's approach in his Preface to Hertz's *Mechanics*[193] and seemed to fear that Hertz was following Maxwell and Lord Kelvin in seeking another mechanical model for the ether.

While working on his *Mechanics* during 1891, Hertz found time to devote to events connected with Helmholtz's seventieth birthday celebration. During August he worked on an article, which appeared in the *Münchener Allgemeine Zeitung* on 31 August 1891, to commemorate Helmholtz's birthday.[194] Hertz also carried on an extensive correspondence with Professor F. Neesen in Berlin in an attempt to organize a tribute to Helmholtz from his former assistants. The decision was made to present to Helmholtz an album of photographs significant in his life: family, students, assistants, colleagues, laboratories, institutes, libraries, homes, etc.[195] On 1 November 1891 Hertz travelled to Berlin and at the banquet in the Kaiser's Palace the next day made the presentation to Helmholtz on behalf of all his assistants. What Hertz said on that happy occasion has not been preserved, but it is safe to assume that it was a variation on the theme of his newspaper article in August of the same year. In that piece he wrote about Helmholtz and the Principle of Least Action:

> An investigator of this stamp treads a lonely path: years pass before even a single disciple is able to follow in his steps.[196]

Helmholtz was fortunate to have had such a dedicated disciple in Hertz, but even in this case Hertz had grown increasingly independent of his mentor's ideas. Hertz himself never found a disciple with the

---

Germany from the United States on 14 Oct. 1893, from which he never fully recovered. This may be a partial explanation of his quoted statement.

[193]See the last few pages of Paper No. 14. As D'Agostino has perceptively remarked, "In this sober but firm statement [of Helmholtz] Hertz is considered as the third representative of the school of [William] Thomson and Maxwell, rather than the intellectual pupil of Helmholtz himself." See D'Agostino (1971), p. 645.

[194]This is Paper No. 2.

[195]Letter from H. Hertz to F. Neesen, 26 Aug. 1891; Deutsches Museum, Munich: Hertz Archives, No. 3217. Other letters on the same subject are Nos. 3216, 3218, 3219, 3220.

[196]Paper No. 2, p. 120.

Introductory Biography

understanding of his *Mechanics* needed to advance his work further.[197] As a consequence, while his *Electric Waves* is still an important landmark in the history of physics, the *Principles of Mechanics* is all but forgotten by physicists today.[198]

Final Illness and Death

By the time Hertz moved to Bonn in April 1889, his experiments with electromagnetic waves had brought him fame throughout the scientific world.[199] He was still only thirty-two and must have looked forward to many additional years of success and happiness. But some foreboding clouds began to appear in an otherwise bright sky. Hertz had trouble with his teeth and his eyes before leaving Karlsruhe. During June and July 1889, after the exertions of the move to Bonn, he was exhausted and depressed. During September and October he was bothered by a foot infection; the foot had to be lanced and, even then, healed very slowly. He was upset by all this concern with his body and confided to his parents in November that he felt "plagued by one minor affliction after another."[200]

The next year and a half brought some improvement in his health. During the summer of 1890 his two months of military duty as a reserve officer, with plenty of fresh air and warm sunshine, restored his

---

[197] Paul Ehrenfest (1880–1933) and Vilhelm Bjerknes did apply the insights of Hertz's *Mechanics* to practical problems, but the results obtained could have been obtained by more traditional methods. On Ehrenfest's use of Hertz's ideas in hydrodynamics, see Klein (1970), pp. 53–74; on Bjerknes' applications to meteorology see Friedman (1989), chapters 1 and 2.

[198] One reason for this may have been that Hertz had never clarified and refined his ideas on mechanics in discussions with his colleagues. In a letter to his parents on 19 Nov. 1893, Hertz wrote with reference to the *Mechanics*: "It frightens me to come out with something that I have never talked over with any human being." (*Erinnerungen*, p. 343).

[199] Hendrik Lorentz (1853–1928), in his Nobel Prize address in 1902, remarked about Hertz: "Immediately after Maxwell I named [as one of the founders of Maxwell's electromagnetic theory] Hertz, that great German physicist, who, if he had not been snatched from us too soon, would certainly have been among the very first of those whom your academy would have considered in fulfilling your annual task." In *Nobel Lectures: Physics*, vol. 1 (1967), p. 16.

[200] H. Hertz, *Erinnerungen*, p. 299.

health and good spirits, and he was in fine shape for his trip to England in December 1890. In early 1892 he made a trip to Switzerland to visit Sarasin[201] in Geneva, greatly enjoyed the experience and ended the trip with a few days vacation at Baden-Baden, the beautiful spa not too far from Karlsruhe.

In July 1892 Hertz caught a bad head cold, which gradually developed into severe nose and ear problems. Major operations were performed to remove a tumor on the pharynx and to clear an ear infection. This led to some improvement, but to Hertz his convalescence seemed "endlessly long." On 30 November 1892 he wrote to his friend and former Karlsruhe colleague, the historian Aloys Schulte: "I am still a veritable Lazarus, but we prefer to be silent about this."[202] In December he wrote a sad letter to Emil Cohn, a physics colleague at Strassburg, in which he gave some further details of his illness. The healing process was so slow that he was forced to cancel all lectures and other work at the Institute.[203] He goes on to say: "One must be sick oneself, to understand what sickness really is."[204] By 3 January 1893 his spirits were even lower, and he wrote a letter to his parents that indicated his near despair:

> ... Nothing is harder than a struggle fought no longer for victory, but merely so as not to give up without making a decent stand.[205]

These words were written one year before Hertz's death. That final year was marked by vacation trips with his wife to Milan, to Santa

---

[201] Éduard Sarasin (1843–1917), Swiss physicist, with whom Hertz had corresponded about electromagnetic waves when he was in Karlsruhe. Hertz had also published a summary of his electromagnetic-wave research in the Swiss journal, *Archives des Sciences physiques et naturelles*, which Sarasin edited. On Hertz's previous interactions with Sarasin, see footnote 26 to Paper No. 5.

[202] Letter from the Schulte Nachlass in the Bonn University archives; reprinted in Bonn University (1958), p. 46.

[203] Letters from Hertz to the Rector of the University asking for dispensations from his faculty duties are reprinted in Bonn University (1958), pp. 10–11.

[204] "... man muss eben selbst krank sein, um Verständnis dafür zu haben." Letter of 9 Dec. 1892 from H. Hertz to E. Cohn; Deutsches Museum, Hertz Archives, No. 3207.

[205] H. Hertz, *Erinnerungen*, pp. 333.

Introductory Biography 75

Margherita on the Riviera and to Kirchberg bei Reichenhall, a German health resort near Salzburg, in search of some improvement in his condition. While on vacation he found that he still had a fair amount of strength and endurance (even enough to climb a mile-high mountain), but his head maladies did not desist. Despite this, he was able to resume his lectures in May and was surprised at how well they went.

In September Hertz was advised by his doctors to have another operation to clear up a very old and stubborn tooth abscess, which threatened to cause a more general and serious blood poisoning. After the operation a letter to his parents indicates that, despite his suffering, his sense of humor was still firmly in place: "The operation did not harm me very much, nor does it seem to have harmed my illness either."[206]

Hertz's lectures during the winter semester began reasonably well, but in early December he experienced a severe pain, which he attributed to rheumatism. He pushed ahead with his class lectures anyway, but the pain grew so intense that his lecture on December 7 proved to be the final one of his academic career. On 9 December 1893 he sent his last, very sad letter to his parents; it reads in part:

> If anything should really befall me, you are not to mourn; rather you must be a little proud and consider that I am among the specially elect destined to live for only a short time and yet to live enough. I did not desire or choose this fate, but since it has overtaken me, I must be content; and, if the choice had been left to me, perhaps I would have chosen it myself.[207]

Shortly after he completed this letter, blood poisoning spread rapidly through his system, and for three weeks he was in great pain.[208] His wife and his mother were at his bedside; he was conscious and in

---

[206]H. Hertz, *Erinnerungen*, p. 341.
[207]H. Hertz, *Erinnerungen*, p. 345.
[208]In a memorial address on Hertz, Hermann Ebert said: "According to the testimony of those who were near him, and particularly of his wife, who nursed him faithfully to the last, his sufferings were indescribable." See Ebert (1894), English translation, p. 273.

full possession of his mental powers to the end.[209] He died peacefully on 1 January 1894 in the thirty-sixth year of his life, only sixteen years after he had begun his university studies in Munich. His doctor gave the cause of death as *septicemia*, blood poisoning; no autopsy was performed.

Hertz was buried in the cemetery of Hamburg-Ohlsdorf under a simple tombstone that merely gives his full name, the fact that he was a professor of physics, and the dates of his birth and death.[210] At the funeral in Hamburg, Bonn Professor Hubert Ludwig praised Hertz's "noble simplicity and genuine modesty." This sums up his character as well as any short phrase can, especially for a man as complicated beneath the surface as was Hertz.[211] In his eulogy before the Berlin Physical Society on 16 February 1894,[212] Max Planck emphasized Hertz's great sense of duty and rare gift of humor, even amid adversity. Perhaps these were the qualities that endeared him to the gods for, as his daughter, Mathilde, later wrote, applying to her father the words of Goethe: "The gods give their favorites everything in full, all the limitless joys, all the limitless sorrows."[213]

When first informed that Hertz died at age thirty-six, the immediate reaction of most people is: "Why did he die so young?" The early death of Hertz seems inexplicable to us today, accustomed as we are to highly competent medical care and wonder drugs. But a number of factors may help explain the terribly premature death of this great physicist. First of all, according to his mother, when Heinrich was born their excellent family physician "was barely able to save the half-dead child for us and for the world."[214] His mother nourished him slowly back to health, but he was weak and prone to disease throughout his

---

[209]These details are taken from Bonfort (1894). Bonfort's sketch of Hertz reads as if she had personal knowledge of Hertz, but it is lifted—in many spots, word-for-word—from Daniel Jones's obituary of Hertz in *Nature* for 18 January 1894. Jones did have personal knowledge of Hertz, having studied under his direction in Bonn in 1891–1892.

[210]A photograph of Hertz's tombstone appears as the frontispiece of this book.

[211]There is a good discussion of Hertz's character and personality in Zenneck (1929), pp. 29–33.

[212]Paper No. 18.

[213]Preface to the second edition of H. Hertz, *Erinnerungen*, p. xv.

[214]In H. Hertz, *Erinnerungen*, p. 1.

Introductory Biography

life. Thus in 1881, when he was only twenty-four and should have been at the peak of his powers, he complained in a letter from Berlin to his parents that he was always so tired in the evening. A few years later, while still in Berlin, he had worrisome stomach problems. A month of summer military duty improved his health and he confided to his parents that he felt completely well at the time—an intimation that this was often not the case.

But his main bout with illness began near the end of his stay in Karlsruhe, with his eye and tooth problems, and his health deteriorated noticeably when he moved to Bonn in 1889. One possible reason has been suggested by Heinrich Kayser, who succeeded Hertz as director of the Bonn Physics Institute. Kayser writes about the Bonn Physics Institute as follows:

> In order to make my description of the health conditions complete, it is necessary to add that the ground floor rooms before their use as the Physics Institute had served for many years as the clinic for contagious diseases. On my arrival in Bonn I was made aware of this by many of my medical colleagues who emphatically warned me about the dangerous situation. The doctor who had attended Hertz maintained with complete certitude, that Hertz was killed "at the Institute" [*am Institut*]. Hertz was a healthy man when he came to Bonn, but after a while he suffered from suppurating inflammations of the head cavities; he underwent many difficult operations; after long vacations he appeared to recover, but as soon as he went back to work at the Institute the malady returned, until it finally carried him off.[215]

Kayser goes on to say that an assistant (Hagenbach) after one winter at the Institute "developed some very suspicious symptoms," and Kayser himself became ill from a serious sinus inflammation after a brief time there. As a result a hygienist was sent from Berlin and spent months boring holes in the walls of the Institute searching for pathogens. He found none, but concluded that the dampness of the Institute was so extreme that it could easily be a menace to health.[216]

---

[215] Kayser (1936), pp. 181–182. Since Kayser was eighty-four when he wrote this, his account may not be completely accurate (e.g., Hertz was not a really healthy man when he came to Bonn or even before that, as we have seen). But Kayser's account certainly points up the fact that, for whatever reason, the Bonn Physics Institute was not a healthy place.

[216] The southern wing of the university originally consisted of a two-story structure surrounding a central courtyard or garden. A prince-elector who

This seems a more likely explanation of Hertz's illness at Bonn and could also explain the symptoms displayed by Hagenbach and Kayser.

Now Heinrich Hertz was a bright, alert scientist who had been informed of the possible contamination of part of his Institute and must have observed, and even felt in his bones, the dampness of his surroundings.[217] If, as Kayser states and as Hertz's diary supports, Hertz's health improved when he was away from the Institute, and deteriorated when he returned to it, why did he not do something about this destructive situation? Was it an excessive sense of duty that kept him at the Institute even when it was undermining his health? Was he reluctant to confront Althoff and the Education Minister with the health hazards of his Institute? We will probably never know the answers to these questions. But what does seem clear is that Hertz's low resistance to infection, and the excessive dampness—even possible contamination—of his Institute's rooms, conspired to deprive the world of one of its best scientists and finest men at a lamentably early age.

VI. Heinrich Hertz's Importance in the History of Physics

Since Hertz's most important research papers are contained in the present collection and have been discussed earlier in this biography, I

---

at one time resided in the second-floor rooms desired direct access to the garden from his rooms. The entire courtyard was therefore filled with soil to the second-floor level and a garden planted, including trees that were quite tall when Kayser first saw them. This garden collected large amounts of rain-water, which tended to seep through the walls of the first floor. These walls became saturated with water and never dried out, since they had no access to outside air or sunlight. This made all the rooms on the first floor excessively damp. In 1884 Clausius, who had been seeking more space for physics, agreed to take over the first-floor rooms as part of an expanded Physics Institute. Kayser, who never minced words, refers to the "folly" of Clausius' actions. On this, see Kayser (1936), pp. 180–184.

[217]Hertz had chosen for his own office a small inside room at the front corner of the building, and used the long corridor that connected the inner rooms for his electromagnetic-wave research. On this see Konen (1933), p. 350.

shall limit myself here to a summary and brief evaluation of his contributions to physics.[218]

In retrospect, the most impressive traits marking Hertz's brief research career were a sense for the truly important problems in physics, and the felicitous ability to solve these problems—either at his desk or at his laboratory bench—as the particular project required. As Lampariello[219] has pointed out, neither Faraday's experimental ability nor Maxwell's theoretical insight sufficed to solve the mystery of electromagnetic waves. The discovery of such waves required the happy combination of theory and experiment that Hertz brought to the problem; this is his claim to fame. The scope of Hertz's abilities becomes clear if one tries to read his *Principles of Mechanics* and his *Electric Waves* in quick succession. The author of the former appears to be an abstract, mathematically inclined physicist, with strong philosophical interests; the author of the latter is the most practical kind of experimentalist, fully capable of designing and building the equipment he needs for a research project and enjoying even the arduous labor often demanded by experimental research. And yet both books were written by the same man—Heinrich Hertz!

We have seen that Hertz pursued three main lines of research in physics: electromagnetism, what we would today call electron physics (cathode rays and the photoelectric effect) and theoretical mechanics.

It was his combined theoretical and experimental research on electromagnetism that first brought Hertz his deserved fame. These contributions produced a true revolution in physics, leading directly to the development of modern field theory. In the words of Max Planck, always a great admirer of Hertz,

> Thus gradually the universal significance of Maxwell's ideas became more and more recognized on all sides, till at last the crucial experiments of Heinrich Hertz with very rapid electrical oscillations were crowned with an unexampled success, by the production of electrical waves of a few centimeters wavelength. Through this discovery, which produced the greatest sensation in all the scientific world, the speculations of Maxwell were

---

[218]One of the best summaries of Hertz's contributions to physics is Max von Laue's account in *Die grossen Deutschen* (Propyläen-Verlag, Berlin, 1957), vol. 4, pp. 103-112. This is reprinted in both German and English in H. Hertz, *Erinnerungen*, pp. xviii-xxxvii. There are also excellent accounts in McCormmach (1975) and Zenneck (1946).

[219]Lampariello (1955), p. 8.

translated into fact and a new epoch of experimental and theoretical physics was begun.[220]

Along more practical lines, one of Hertz's papers, "The Forces of Electrical Oscillations, handled according to Maxwell's Theory,"[221] laid the foundations for the development of wireless telegraphy by Popov, Marconi, Braun, Lodge, and their successors, whose contributions led to our present worldwide system of radio, television, and microwave communications.

Hertz had, by his experiments, helped replace a mechanistic world picture by an electromagnetic one. This latter world view did not last very long, but it was an essential prerequisite for the development of relativity and quantum theory. Here too Hertz played a major role.

The consequence of Hertz's research on electromagnetic waves was the elimination of all action-at-a-distance theories from electromagnetism and their replacement by the Faraday-Maxwell field theory. This left only gravitation as an action-at-a-distance phenomenon, and Hertz was suspicious of even this lone exception. His great success in proving the superiority of Maxwell's field theory over all competing theories caused all action-at-a-distance theories gradually to fall out of favor and prepared the intellectual climate for Einstein's theory of gravitation, in which gravitational effects, like electromagnetic waves, propagate at the speed of light.

The first volume of Einstein's *Collected Papers* contains a remarkable letter written by Einstein in August 1899 to his fiancée, Mileva Marić, in which he first reveals his interest in the electrodynamics of moving bodies.[222] He says that this interest was stimulated by a rereading of Hertz's *Electric Waves*, particularly by Hertz's 1890 paper, "On the Fundamental Equations of Electromagnetics for Bodies in Motion." Einstein criticizes Hertz's retention of the ether, and especially his assumption that the ether imbedded in an object is carried along by the object at the same speed as the object itself. But Hertz's paper led Einstein to a more serious study of the electrodynamics of moving bodies, and ultimately to his special theory of relativity. It is for this reason that Max Born devotes a section of his

---

[220]Max Planck, in J.J. Thomson (1931), p. 62. See also Harman (1982), p. 107.

[221]H. Hertz, *El.Waves*, pp. 137–159.

[222]Einstein (1987), vol. 1, p. 226. See also Stachel (1987), pp. 45–47.

Introductory Biography 81

book on Einstein's relativity theory to a discussion of this paper of Hertz.[223]

In Einstein's first paper on the special theory of relativity he mentions only three physicists, Maxwell, Lorentz and Hertz. In 1896, when Einstein was seventeen, he entered the Polytechnic in Zürich to study mathematics and physics. He initially spent most of his time in the physics laboratory, but studied the works of "Kirchhoff, Helmholtz, Hertz, etc." at home.[224] As Einstein's brief 1949 autobiography[225] indicates, there is little doubt that Einstein knew Hertz's work very well and that he profited from many of his ideas.

Some writers have even gone so far as to suggest that Hertz's idea of eliminating all forces from mechanics (and thus reducing dynamics to kinematics) was a foreshadowing of Einstein's general theory of relativity, in which objects move along paths determined not by gravitational forces exerted by other objects, but by the properties of the space through which they move.[226]

Hertz also had a part in the development of quantum physics. In Planck's theory of black-body radiation, he chose to consider an array of dipole oscillators in the walls of his *Hohlraum* as the source of the radiation filling the cavity. Planck did this because Hertz had worked out the equations for the emission and absorption of energy by such oscillators. This was a major ingredient in Planck's development of his final black-body equation, as he related in his Nobel Prize address in 1919.[227] In the same address he refers to his use of the work of V. Bjerknes on the damping of high-frequency electric oscillations, work that had been carried out at the Physics Institute in Bonn under Hertz's direction.

---

[223]Born (1962), pp. 189-199.

[224]Einstein (1979), p. 15. Since Hertz's *El.Waves* had been published in 1892, his *Mechanics* in 1894, and *Misc.Pprs.* in 1895, it is clear that there must have been great interest in Hertz's papers when Einstein began his studies at the Zürich *Polytechnic* in 1896.

[225]Einstein (1979). See, e.g., the reference on p. 33 to Hertz's 1890 paper mentioned above.

[226]Unsöld (1970), pp. 341-342; von Reden (1988), p. 92. The latter author writes: "In this light the Hertzian construct can be interpreted as a foreglimpse of the Einsteinian world."

[227]*Nobel Lectures: Physics*, vol. 1 (1967), p. 408.

Hertz's experiments on cathode rays also led to a series of important contributions to modern physics. Lenard, J.J. Thomson, and others improved on Hertz's work in experiments that resulted in Lenard's Nobel Prize in 1905 "for his work on cathode rays" and in Thomson's Nobel Prize in 1906 "for his theoretical and experimental investigation of the conduction of electricity by gases." The work of Hertz and Lenard was also important for Roentgen's discovery of x-rays in late 1895. Hertz and Lenard had clearly produced x-rays in some of their earlier experiments, but were never able to distinguish them from cathode rays because they had a mistaken idea of what cathode rays really were. In his fundamental 1896 paper on x-rays Roentgen refers to the research of Hertz and Lenard and indicates how helpful it was to him in his own great discovery, which ushered in the era of modern physics.[228]

Hertz's work on the photoelectric effect, although necessarily incomplete (the existence of electrons was unknown at the time) provided a firm experimental foundation for the work of Hallwachs, Elster and Geitel, Lenard and Millikan. The experimental data of these researchers, and especially of Millikan, led, in turn, to the verification of Einstein's quantum equation for the photoelectric effect.

Hertz's *Principles of Mechanics* had little impact on the development of classical mechanics, for it was hard to apply his ideas to practical problems. The ideas expressed in the Introduction to his *Mechanics*, however, have had a potent influence on the development of physics as we know it today. Hertz's Introduction contained illuminating suggestions on the proper approach to be taken toward the physical universe by theoretical physicists. It expressed the view that it was not the physicist's task to obtain from nature its deepest truths; rather it was to create models or pictures (*Bilder*) of reality that could be used to derive mathematically results that could then be compared with experiment. As Hertz stressed, all that could be asked of any model was that it predict or reproduce experimental results. Two very different models could lead to the same correct results, in which case a choice could be made between them on the basis of criteria spelled out by Hertz. This approach reflects the true spirit of

---

[228]W. Roentgen, "On a New Kind of Rays," translated in Boorse and Motz (1966), pp. 389–401; on p. 395.

modern physics.[229] In this sense Hertz's contributions to physics in his *Mechanics* may be deserving of the same enthusiastic admiration that physicists have always directed to his work on electromagnetism.

These varied contributions of Hertz show him to have been the last great classical physicist and the precursor of a new generation of modern physicists like Einstein and Planck. It is remarkable how highly physicists at the turn of the century rated the importance of Hertz's contributions to physics. Whether it was in Nobel Prize Lectures,[230] inaugural addresses by new members of the Berlin Academy of Sciences,[231] or correspondence and published papers, physicists raised their voices to praise his work. The leading German physicists of the day had profound respect for him both as a physicist and as a man and had expected him to take Helmholtz's place as the leader of German physics. But the year 1894, the year Max Planck aptly named the "Black Year in German Physics," carried off both its leader, Helmholtz, and his heir-apparent, Hertz. With their passing, physics in Germany seemed in danger of losing its way. Fortunately, however, Hertz's approach continued to direct German physics for a sufficient time after his death to allow new pathfinders in the person of Planck and Einstein to assume the lead. As Ludwig Boltzmann wrote to Helmholtz five days after Hertz's death:

> One should emphasize the extraordinary import of Hertz's discoveries in relation to our whole concept of Nature, and the fact that beyond a doubt they have pointed out the only true direction that research can take for many years to come.[232]

This appreciation of Hertz's research contributions was not confined to Germany. Similar tributes came from Lorentz in Leyden, FitzGerald in Dublin, Poincaré in Paris, J.J. Thomson in Cambridge,

---

[229] On this see Blackmore (1972), p. 119, and the papers of Boltzmann and Poincaré cited in his note 15 on page 337.

[230] *Nobel Lectures: Physics*, vol. 1 (1967). See, in particular, the lectures of Lorentz, Lenard, J.J. Thomson and Planck. For example, J.J. Thomson comments on Hertz's discovery that cathode rays pass through metal foils: "... this led me to investigate more closely the nature of the particles which form the cathode rays." (p. 147)

[231] Kirsten and Körber (1979), *passim*. For instance, see Heinrich Rubens' inaugural address, pp. 230–232.

[232] Letter of 6 Jan. 1894 from L. Boltzmann to H. von Helmholtz; in Koenigsberger (1902–1903), vol. 3, p. 100.

and from scientific societies in London, Manchester, Vienna, Rome, Geneva and Turin, among others. Even in the United States, where physics was still in an embryonic state in 1894, the very first volume of the *Physical Review* devoted four pages to an account of Hertz's death and his contributions to physics.[233]

One of Helmholtz's last acts before he suffered a stroke on 12 July 1894, which led to his death on 8 September, was to urge the president of the Peter-Wilhelm-Mueller Institute of Frankfurt am Main to give a posthumous prize of 15,000 M to Hertz, to be used for the support of his wife and two daughters, one age six and the other only three. Helmholtz pointed out that such a gesture would honor Hertz's great scientific achievements, and at the same time remove a reproach from the German nation, "inasmuch as during his lifetime Hertz was much less honored by German citizens than by those of other countries."[234] Legalities unfortunately prevented the award of such a posthumous prize, but Helmholtz's words show that at the time of his death Hertz's fame had indeed spread throughout the western world.

During the one hundred years since his death, physicists have retained their admiration for Heinrich Hertz as a model for their scientific work. In 1956 Robert Cohen gave eloquent expression to their views:

> He was almost the ideal scientist, an extraordinary experimenter and a master of conceptual thinking. In his work, theory and experimental practice were of mutual influence, his theoretical discussions leading to experiments and his experimental results leading to his acute theoretical discussions. His like is rare enough within science—perhaps Enrico Fermi is the most distinguished example in our own day—but his fusion of theory and experiment with a creative interest in philosophical and logical foundations is nearly unique. Only Helmholtz comes readily to mind, and one can only wonder, regretfully, how far Hertz might have gone beyond his master and friend.[235]

---

[233]Nichols (1894)
[234]Koenigsberger (1902–1903), vol. 3, p. 121.
[235]Cohen (1956), Section 5.

PART I

Hertz's Childhood and Young Manhood (1857–1878): Hamburg, Frankfurt, and Munich

> I am burning with impatience to reach the frontiers of what is already known and to go on exploring into unknown territory.
>
> *Heinrich Hertz (1877)*

Heinrich Hertz with his mother and two of his brothers. Heinrich is the second from the left. His brother Gustav is beside him, and Rudi is in his mother's arms. The photo was probably taken in 1863. Courtesy of Deutsches Museum, Munich.

Introduction

The one paper that we have chosen to illuminate Hertz's childhood and young manhood is Philipp Lenard's "Introduction" to Hertz's *Miscellaneous Papers* (Leipzig: J.A. Barth, 1895). Lenard was an assistant to Hertz in Bonn and, after Hertz's death on 1 January 1894, oversaw the publication of both Hertz's *Miscellaneous Papers* and his *Principles of Mechanics* (Leipzig: J.A. Barth, 1894). Lenard did a competent and careful job of seeing these works of Hertz through the press.

In Lenard's "Introduction" he confines himself in great part to Hertz's early years as a physics student and physicist in training, first in Munich (1877–1878) and then in Berlin (1878–1883). Hertz's early life in Hamburg from 1857 to 1875 is discussed in a charming manner by his mother, Anna Elisabeth Pfefferkorn Hertz, at the beginning of Hertz's *Erinnerungen* (pp. 1–21). The events of the years 1875–1877, when Hertz was preparing to be an engineer, are described in letters to his parents also contained in his *Erinnerungen* (pp. 23–59). During that period he was restless and uncertain about what he wanted to do with his future.

The history of how Hertz finally made his decision to turn from engineering to pure science is very well related by Lenard in the first few pages of this introduction. Hertz's letter of 1 November 1877 to his parents is particularly revealing in this regard.

PAPER NO. 1

Introduction to Heinrich Hertz's *Miscellaneous Papers* (1895) by Philipp Lenard

(Heinrich Hertz, *Miscellaneous Papers*, pp. ix–xxvi)[1]

In October 1877, at the age of twenty, Heinrich Hertz went to Munich to continue his engineering studies. He had chosen this as his profession, and had already made some progress in it; for in addition to completing the usual year of practical work, he had thoroughly grounded himself in the basic mathematics and science needed for his further studies. He had now to apply himself to engineering work proper, to the technical details of his profession. At this point he began to doubt whether his natural inclinations lay in the direction of this work—whether he would find engineering as satisfactory as the studies that led up to it. The study of natural science had been a delight to him; now he feared lest his lifework should prove a burden. He stood at the parting of the ways. In the following letter he consults his parents in the matter.

Munich, 1 November 1877

My dear Parents,

    No doubt you will wonder why this letter follows so quickly after my previous one. I had no intention of writing so soon again, but this time it is about an important matter that will not brook any long delay.

    I really feel ashamed to say it, but I must: now at the last moment I want to change all my plans and return to the study of natural science. I feel that the time has come for me to decide either to devote myself to this entirely or else to say good-bye to it; for if I give up too much time to science in the future it will end in my neglecting my professional studies and becoming a second-rate engineer. Only recently, in arranging my plan of studies, have I clearly seen this—so clearly that I can no longer feel any doubt about it; and my first impulse was to renounce all unnecessary dealings with mathematics and natural science. But then, all at once, I saw clearly that I could not bring myself to do

---

[1] [In the following footnotes, the editor's are printed in square brackets; those in the original article have no brackets.]

this; that these had been my real occupation up to now, and were still my chief joy. All else seemed hollow and unsatisfying. The conviction came upon me quite suddenly, and I felt inclined to sit down and write to you at once. Although I have restrained myself for a day or two, so as to consider the matter thoroughly, I can come to no other conclusion. I cannot understand why all this was not clear to me before; for I came here filled with the idea of working at mathematics and natural science, whereas I had never given a thought to the essentials of my professional training—surveying, building construction, builders' materials, and such like.

I have not forgotten what I often used to say to myself, that I would rather be a great scientific investigator than a great engineer, but would rather be a second-rate engineer than a second-rate investigator. But now when I am in doubt, I think how true is Schiller's saying, "*Und setzet Ihr nicht das Leben ein, nie wird Euch das Leben gewonnen sein,*"[2] and that excessive caution would be folly. Nor do I conceal from myself that by becoming an engineer I would be more certain of earning my own livelihood, and I regret that in adopting the other course I shall probably have to rely upon you, my dear father, all the longer for support. But against all this there is the feeling that I could devote myself wholly and enthusiastically to natural science, and that this pursuit would satisfy me; whereas I now see that engineering science would not satisfy me, and would always leave me hankering after something else. I hope that I am not deceiving myself in this, for it would be a great and woeful piece of self-deception. But of this I feel positive, that if the decision is in favor of natural science, I shall never look back with regret towards engineering science, whereas if I become an engineer I shall always be longing for the other; and I cannot bear the idea of being able to work at natural science only for the purpose of passing an examination.

When I think of it, it seems to be that I used to be much more frequently encouraged to go on with natural science than to become an engineer. I may be better grounded in mathematics than many, but I doubt whether this would be much of an advantage in engineering; so much more seems to depend, at any rate in the first ten years of practice, upon business capacity, experience, and knowledge of data and formulae, which do not happen to interest me. This and much else I have carefully considered (and shall continue to think over until I receive your reply), but when all is said and done, even admitting that there are many sound practical reasons in favor

---

[2] ["Unless you are willing to risk your life, you can never hope to save it."]

of becoming an engineer, I still feel that this would involve a sense of failure and disloyalty to myself, to which I would not willingly submit if it could be avoided.

And so I ask you, dear father, for your decision rather than for your advice; for it isn't advice that I need, and there is scarcely time for it now. If you will allow me to study natural science I shall take it as a great kindness on your part, and whatever diligence and love can do in the matter, that they shall do. I believe this will be your decision, for you have never put a stone in my path, and I think you have often looked with pleasure on my scientific studies. But if you consider it best for me to continue in the path on which I have started (which I now doubt), I will carry out your wish, and do so fully and freely; for by this time I am sick of doubt and delay, and if I remain in the state I have been in lately I shall never make a start.... So I hope to have an early answer, and until it comes I shall continue to think the matter over. Meanwhile I send my love to you all, and remain your affectionate son,

Heinrich

Matters were decided as he had hoped, and, full of joy at being able to carry out his wishes, Hertz now proceeded to arrange his plan of studies. He remained altogether a year at Munich. He devoted the winter semester of 1877–78 in all seclusion to the study of mathematics and mechanics, using for the most part original treatises such as those of Laplace and Lagrange. Most of the following summer semester he spent at practical work in the physics laboratory. By attending the elementary courses in practical physics at the University (under Jolly) and at the same time in the Polytechnic (under Bezold), he was able to supplement what he had already learned by means of his own home-made apparatus.[3]

Thus prepared he proceeded in October 1878 to Berlin, eager to become a pupil of Helmholtz and Kirchhoff. When he had arrived there, in looking at the notices on the black notice-board of the University his eye fell on an announcement of a prize offered by the Philosophical Faculty for the solution of a problem in physics. It referred to the question of electric inertia. To him it did not seem as hopelessly difficult as it might have appeared to many of his contemporaries, and he decided to have a try at it.

---

[3][The two professors mentioned here are Philipp von Jolly (1809–1884) and Wilhelm von Bezold (1837–1907).]

This brings us to the beginning of his first independent research.[4] We cannot read without astonishment the letters in which this student of twenty-one reports to his parents the starting of an investigation which might well be taken for the work of an experienced investigator.

Berlin, 31 October 1878

I have been attending lectures—Kirchhoff's—since Monday; another course only begins on Wednesday next. Besides this I have also started practical work; one of the prize problems for this year falls more or less in my line, and I am going to work on it. This was not what I intended at first, for a course of lectures on mineralogy, which I wished to attend, clashed with it; but I have now decided to let these wait until the next semester. I have already discussed the matter with Professor Helmholtz, who was good enough to put me on the track of some of the literature.

A week later we find him already at his experiments.

6 November 1878

Since yesterday I have been working in the laboratory. The prize problem runs as follows: If electricity moves with inertia in bodies, then this must, under certain circumstances, manifest itself in the magnitude of the extra-current (i.e. of the secondary current which is produced when an electric current starts or stops). Experiments on the magnitude of the extra-current have to be made such that a conclusion can be drawn from them as to the inertia of the electricity in motion. The work has to be finished by 4 May; it was given out as early as 3 August, and I am sorry that I did not know of it before. I ought, however, to say that at present I am only trying to work out the problem, and I may not succeed in solving it satisfactorily; so I would not readily have spoken of it at all, if it were not necessary by way of explanation. Anyhow, I find it very pleasant to be able to attack such an investigation. So yesterday I informed Professor Helmholtz that I had considered the matter and would like to start work. He then took me to the demonstrators and very kindly remained some twenty minutes longer, talking with me

---

[4] [The research mentioned led to the article: "Experiments to Determine an Upper Limit to the Kinetic Energy of an Electric Current," *Misc. Pprs.*, pp. 1–34.]

about it, as to how I had better begin and what instruments I should require.

So yesterday and today I have begun to make my arrangements. I have a room all to myself as large as our morning room,[5] but nearly twice as high. I can come and go as I like, and you will easily see that I have room enough. Everything else is capitally arranged.... Nothing could be more convenient, and I can only hope now that my work will come up to its environment. Of course at present I am only getting things ready, but I feel how pleasant it will be to have the resources of a good laboratory at my back. My galvanometer, which at home stood upon the lathe,[6] now stands upon an iron bracket set into the wall. The reading telescope can be adjusted in all directions by screws, which is certainly more convenient than propping it up on books....

Every morning I hear an interesting lecture, and then go to the laboratory, where I remain, barring a short interval, until four o'clock. After that I work in the library or in my rooms; up till now there has been plenty to do in hunting up the literature on extra-currents. (It seems that there is a paper, of which, however, I have only seen an excerpt, in which some one shows that no such current exists; it is to be hoped that the man is quite wrong.)

17 November 1878

My work goes on steadily. The first thing I found out was that a bracket in the Dorotheenstrasse is much more shaky than an ordinary table in the Magdalenenstrasse.[7] At my request I have been shifted into another room, in which there is a brick pillar.... Helmholtz is very kind; he comes in every day for a few minutes, and has a look at how things are getting on. The task upon which I am engaged is not particularly rewarding, for in all likelihood the result will be negative: i.e. certain things will not happen, and on the whole this is less exciting than when something does happen; but it can't be helped in this case.

---

[5]A large room in his parents' house.

[6]He made good use of this lathe, which he had obtained at age sixteen.

[7][Helmholtz's laboratory was on the Dorotheenstrasse in Berlin; the Hertz family lived on Magdalenenstrasse in Hamburg all during Hertz's life.]

24 November 1878

I am now thoroughly happy, and could not wish things better. I spend the greater part of the day working in the laboratory, and unfortunately the days are so short that when the greater part is gone, scarcely anything is left. Most of this greater part is spent upon things that are very useless, or at any rate don't teach one much, such as cutting corks and filing wires, and the observations themselves are naturally not very delightful. Possibly it may be doubtful whether it is quite right for me to spend so much time at these things when I still have so much to learn.

And yet I feel that it is right; to obtain information for myself and for others directly from nature gives me so much more satisfaction than to be always learning it from others and for myself alone—so much more that I can scarcely express it. When I am only studying books I am never free of the feeling that I am a perfectly useless member of society.

It is odd to think that I am now working on a rather specialized research project in electricity, whereas only about half a year ago I scarcely knew any more about it than what still remained in my memory since the time I was with Dr. Lange.[8] I hope my work won't suffer from this. At present it looks promising. I have already surmounted the difficulties which Helmholtz pointed out to me at the start as being the principal ones; and in a fortnight, if all goes well, I shall be ready with a scanty kind of solution, and shall still have time left to work it up properly.

He asks his parents to send on a tangent galvanometer which he had made during the last holidays at home, without having any suspicion that it would so soon be used in this way.[9]

A week later, in writing to report progress, he is not so cheerful. "When one difficulty is overcome, a bigger one turns up in its place." These were the difficulties mentioned on pp. 5–6 [of *Misc.Pprs.*] The Christmas holidays were now at hand, and while at home in Hamburg he made the commutator, on which he later reports.

---

[8]The Head-Master of the *Bürgerschule*, which Hertz attended up to his sixteenth year. [Lenard is wrong about this: Hertz attended a private school, of which Lange was Head-Master.]

[9]This is the galvanometer referred to on p. 12 [of *Misc.Pprs.*]—a simple wooden disc turned upon the lathe and wound with copper wire, with a hole in the center for the magnet. It is still in good order [in 1895].

> 12 January 1879
>
> The apparatus which I made works very well, even better than I had expected; so that within the last three days I have been able to make all my measurements over again, and more accurately than before.

Within three months after he had first turned his attention to this investigation he is able to report the conclusion of the first part of it.

> 21 January 1879
>
> It has delighted me greatly to find that my observations are in accordance with the theory, and all the more because the agreement is better than I had expected. At first my calculations gave a value which was much greater than the observed value. Then I happened to notice that it was just twice as great. After a long search among the calculations I came upon a 2 which had been forgotten, and then both agreed better than I could have expected. I have now set about making more accurate observations; the first attempt has turned out badly, as generally happens, but I hope in due course to pull things into shape. The apparatus which I have made at home really works well, so well that I wouldn't exchange it for one made out of gold and ivory in the best workshop. (Mother might like to hear this, and if I find that it pleases her I will try it again.)

Ten days later the experiments with rectilinear wires were completed.

> 31 January 1879
>
> I have now quite finished my research, much more quickly than I had expected. This is chiefly because the more accurate set of experiments has led to a very satisfactory, although negative, result: i.e. I find that, to the greatest degree of accuracy I can obtain, the theory is confirmed. I should much have preferred some positive result; but as there is nothing of the kind here I must be satisfied. My experiments agree as well as I could wish with the current theory, and I do not think that I can push matters any further with the means now at my disposal. So I have finished the experiments, and hope the Commission will be satisfied; as far as I can see, any further experiments would only lead to the same result. I shall begin writing my paper in a few days; just at present I don't feel in the mood for it.

The paper was written during a period of military service at Freiburg.

In these successive reports on his work we nowhere find signs of his having encountered difficulties in developing the theory for it; and this is all the more surprising because at this time he could scarcely have made any general survey of what was already known. But it is clear that even at this early stage he was able to find his own way through regions yet unknown to him, and to do this without first searching anxiously for the footprints of other explorers. Thus just about this time he writes as follows:

9 February 1879

Kirchhoff has now come to magnetism in his lectures, and a great part of what he tells us coincides with what I had worked out for myself at home last autumn. Now it is by no means pleasant to hear that all this has long since been well known; still it makes the lecture all the more interesting. I hope my knowledge will soon grow more extensive, so that I may know what has already been done, instead of having to take the trouble of finding it out again for myself. But it is some satisfaction to find gradually that things which are new to me make their appearance less frequently; at any rate that is my experience in the special field in which I have done my experiments.

His research gained the prize.

4 August 1879

Happily I have not only obtained the prize, but the decision of the Faculty has been expressed in terms of such commendation that I feel twice as proud of it.... I had gone with Dr. Katz and Levy [to hear the public announcement of the decision] without having said anything to anybody, and fully determined not to show any disappointment if the result was unfavorable.

11 August 1879

I have chosen the medal, in accordance with your wish, for the prize. It is a beautiful gold medal, quite a large one, but by a piece of incredible stupidity it has no inscription whatever on it, nothing even to show that it is a University prize.

This prize research was Hertz's first investigation, and it is to this he refers in the Introduction[10] to his *Electric Waves*, as being engaged upon it when von Helmholtz invited him to attack the problem[11] proposed for the prize of the Berlin Academy. For reasons now known to us, he gave up the idea of working at the problem. He preferred to apply himself to other work, which was perhaps of a more modest nature, but promised to yield some tangible result.[12] So he turned his attention to the theoretical investigation "On Induction in Rotating Spheres."[13] This extensive investigation was completed in an astonishingly short time. The first sketch of it, which still exists, is dated from time to time in Hertz's handwriting, and one sees with surprise what rapid progress he made from day to day. He had made preliminary studies at home during the autumn vacation of 1878, and the results of these are partly contained in the paper "On the Distribution of Electricity over the Surface of Moving Conductors,"[14] which was first published two years later. In November 1879, he had begun to work on induction [in rotating spheres], and no later than the following January this investigation was submitted to the Philosophical Faculty as his research dissertation for the doctoral degree. We hear of this rapid progress in the letters to his parents:

27 November 1879

I secured a place in the laboratory and started working there at the beginning of term, but do not feel much drawn in that direction just now. I am busy with a theoretical investigation which gives me great pleasure,[15] so I work at this in

---

[10][Paper No. 5 in this collection.]

[11]This latter seems to be the problem in electromagnetics to which Helmholtz refers in his Preface to Hertz's *Principles of Mechanics* [Paper No. 14 in this collection] as having been proposed by himself in the belief that it was one in which his pupil [Hertz] would feel an interest.

[12][A relevant unpublished manuscript containing some calculations on this prize problem has been discovered by O'Hara and Pricha. On this see footnote 22 in the Introductory Biography in this book.]

[13][*Misc. Pprs.*, pp. 35–126.]

[14][*Misc. Pprs.*, pp. 127–136.]

[15][This ability to switch from experimental work to theory, and back again to experiment, was remarkable in Hertz. He continued to switch back and forth throughout his brief career, and with equal success in both theory and experiment.]

my rooms instead of going to the laboratory; indeed I wish that I had made no arrangements for practical work. The investigation which I now have in hand is closely connected with what I did at home. Unless I discover (which would be very disagreeable) that this particular problem has already been solved by some one else, it will become my dissertation for the doctorate.

13 December 1879

There is little news to send about myself. I have been working away, with scarcely time to look about me, at the research which I have undertaken. It is getting on as well and as pleasantly as I could wish.

17 January 1880

As soon as I got here [from Hamburg, after the Christmas holidays] I settled down to my research, and by the end of the week had it ready: I had to keep working hard at it, for it became much more extensive than I had expected.

In its extent this second research differs from all of Hertz's other publications [in Berlin]; he had clearly decided to follow the usual custom with respect to doctoral dissertations. Although long, it will be found to repay the most careful study. The decision of the Berlin Philosophical Faculty (drawn up by Helmholtz) was *Acuminis et doctrinae specimen laudabile*.[16] Together with a brilliant examination it gained for him the title of Doctor, with the added designation *magna cum laude*, which is but rarely given in the University of Berlin.

In the following summer of 1880 Hertz was again engaged upon an experimental investigation on the formation of residual charge in insulators. He did not seem well satisfied with the result; at any rate he did not consider it worth writing up. It was only by Helmholtz's special request that he was subsequently induced to give an account of this research at a meeting of the Physical Society of Berlin on 27 May 1881. It did not appear in Wiedemann's *Annalen* until three years later, after the quantitative data had been recovered by a repetition of the experiments made for this purpose at Hertz's suggestion.[17]

---

[16]["A praiseworthy example of insight and knowledge."]
[17][*Misc.Pprs.*, pp. 255–260.]

Soon afterwards, in October 1880, Hertz became assistant to Helmholtz. He now revelled in the enjoyment of the resources of the Berlin Institute. He was soon engaged, in addition to the duties of his office, upon many problems both experimental and theoretical; he expressed his regret at not being able to use all the resources at his disposal, and not being able to solve all the problems at one fell swoop. At this time he sowed the seeds which during his three years' term as assistant developed one after the other into the investigations which appear in *Miscellaneous Papers*.[18]

He was first attracted by a theoretical investigation "On the Contact of Elastic Solids." During the frequent discussions on Newton's rings in the Physical Society of Berlin it had occurred to Hertz that, although much was known in detail as to the optical phenomena which takes place between the two glasses, very little was known as to the changes of form which they undergo at their point of contact when pressed together. So he tried to solve this problem and succeeded in doing so. Most of the investigation was carried out during the Christmas vacation of 1880. Its publication, at first in the form of a lecture to the Physical Society (on 21 January 1881), was at once greeted with much interest. A new light had been thrown on the phenomena of contact and pressure, and it was recognized that this had an important and direct bearing upon the conduct of all delicate measurements. For example, determinations of a base-line for the great European measurement of a degree were just then being calculated at Berlin. The steel measuring-rods used in these determinations were lightly pressed against each other with a glass sphere interposed between them. This elastic contact necessarily introduced an element of uncertainty depending upon the pressure exerted; a method of ascertaining its magnitude with certainty was wanting. Now the question could be answered definitively and immediately. In technical circles equal interest was exhibited, and this induced Hertz to extend the investigation further and to allow it to be published not only in Borchardt's *Journal*, but also in a technical

---

[18][These researches include a great variety of topics, including elasticity, properties of liquids, meteorology, electricity, and the electric discharge from a cathode, and are to be found in *Misc.Pprs.*, pp. 137–272. None of these lesser papers of Hertz is included in the present volume.]

journal, with a supplement on "Hardness."[19] About this he writes to his parents as follows:

9 May 1882

I have been writing a great deal lately; for I have rewritten the investigation once more for a technical journal in compliance with suggestions which reached me from various directions.... I have also added a chapter on the hardness of bodies, and hope to lecture on this to the Physical Society on Friday. I have had some fun out of this too. For hardness is a property of bodies of which scientific men have as clear, i.e. as vague, a conception as the man in the street. Now as I went on working it became quite clear to me what hardness really was. But I felt that it was not in itself a property of sufficient importance to make it worthwhile writing about it in particular; nor was such a subject, which would necessarily have to be treated at some length, quite suitable for a purely mathematical journal. In a technical journal, however, I thought I might well write something about the matter. So I went to look around the library of the *Gewerbeakademie*, to see what was known about hardness. And I found that there really was a book written on it in 1867 by a Frenchman. It contained a full account of earlier attempts to define hardness clearly, and to measure it in a rational way, and of many experiments made by the writer himself with the same object, interspersed with assurances on the importance of the subject. Altogether it must have involved a considerable amount of work, which was labor lost—so I think, and he partly admits it—because there was no right understanding at the bottom of it, and the measurements were made without knowing what had to be measured. So I concluded that now I might with a quiet conscience make my paper a few pages longer; and as a consequence I have naturally had much more pleasure in writing the paper than I would have had otherwise.

While these problems on elasticity were engaging his attention, Hertz was also busy with some research on evaporation[20] and with further investigation of the "Kinetic Energy of Electricity in Motion."[21]

---

[19][The first article mentioned by Lenard appeared in Borchardt's *Journal für die reine und angewandte Mathematik* 92, 156–171 (1881); *Misc.Pprs.*, pp. 146–162; the second appeared in *Verhandlungen des Vereins zur Beförderung des Gewerbefleisses* (November 1882); *Misc.Pprs.*, pp. 163–183.]

[20][*Misc.Pprs.*, pp. 186–206.]

[21][*Misc.Pprs.*, pp. 137–145.]

Lenard's Introduction to Hertz's Miscellaneous Papers

Both of these had been commenced in the summer of 1881. In order to push ahead with this three-fold task to his own satisfaction, he devoted the greater part of the autumn vacation to it. Thus the investigation on electric inertia was soon finished; on the other hand, the evaporation problems took up much more time without giving much satisfaction.

15 October 1881

I am now devoting myself entirely to the research on evaporation, which I began thinking of in the spring, and of which I have now some hope.

10 March 1882

The present research is going on anything but satisfactorily. Fresh experiments have shown me that much, if not all, of my labor has been misapplied; that sources of error were present which could scarcely have been foreseen, so that the beautiful positive result which I thought I had obtained, turns out to be nothing but a negative one. At first I was quite upset, but have plucked up courage again; I feel as fit as ever now, only I do regret the valuable time that cannot be recovered.

13 June 1882

I am writing out my paper on evaporation, i.e. as much of the work as turns out to be correct; I am far from being pleased with it, and feel rather glad that I am not obliged to work it out completely, as I originally intended.

In the midst of this period of strenuous exertion comes the somewhat refreshing episode of the invention of the hygrometer.[22] In sending a charming description of this little instrument, "so simple that there is scarcely anything in it," Hertz explains to his parents how the air in the room of a house should be kept moist in winter. There can be no harm in including his explanation here.

2 February 1882

I may here give a little calculation which will show father how the air in the morning-room should be kept moist. On an average the atmosphere contains half as much water vapor as is required to saturate it; in other words, the average relative humidity is 50 percent. Assume then that this proportion is

---

[22][*Misc.Pprs.*, pp. 184–185.]

suitable for me, that it is the happy—or healthy—mean. In a cubic meter of air there should then be definite quantities of water, which are different for different temperatures—2.45 grams at 0°C, 4.70 grams at 10°C, and 8.70 grams at 20° C, for these amounts would give the air a relative humidity of 50 percent. Now let us assume that the temperature is 0°C out of doors, and 20°C in the (heated room). Then in the room there would be (since the air comes ultimately from the outside) only 2.45 g of water in each cubic meter of air. In order to get the correct proportion there should be 8.70 g of water. Hence the air is relatively very dry and needs 6.25 g of water more per cubic meter. Since the room is about 7 meters long, 7 meters wide, and 4 meters high, it contains $7 \times 7 \times 4$ cubic meters, and the additional amount of water required in the room is $7 \times 7 \times 4 \times 6.25$ grams, or nearly 1.25 liters. Thus if the room were hermetically closed, 1.25 liters of water would have to be sprinkled about in order to secure the proper degree of humidity.

Now the room is not hermetically closed. Let us assume that all the air in it is completely changed in n hours; then every n hours 1.25 liters of water would have to be sprinkled about or evaporated into it. I think we may assume that through window-apertures, opening of doors, etc., the air is completely changed every two or three hours; hence from $5/8$ to $5/12$ of a liter of water, or a big glassful, would have to be evaporated per hour. All this would roughly hold good whenever rooms are artificially heated, and the external temperature is below 10°C. If you were to set up a hygrometer and compare the humidity when water is sprinkled and when it is not, you could from this find within what time the air in the room is completely changed....

This has become quite a long lecture, and the postage of the letter will ruin me; but what wouldn't a man do to keep his dear parents and brothers and sister from complete desiccation?

As soon as the research on evaporation was finished Hertz turned his attention to another subject, in which he had always felt great interest—that of electric discharges in gases. He had only been engaged a month upon this when he succeeded in discovering a phenomenon accompanying the spark discharge which had previously remained unnoticed.[23] But he was too keen to allow this to detain him long; he at once made plans for constructing a large secondary battery, which seemed to him to be the most suitable means for obtaining

---

[23] [Misc.Pprs., pp. 216–223.]

information of more importance. His letters tell us how he attacked the problem.

> 29 June 1882
>
> I am busy from morning to night with optical phenomena in rarefied gases, in so-called Geissler tubes—only the tubes I mean are very different from the ones you see displayed in public exhibitions. For once I feel an inclination to take up a somewhat more experimental subject and to put the exact measurements aside for a while. The subject I have in mind is involved in much obscurity, and little has been done on it; its investigation would probably be of great theoretical interest.[24] So I should like to find in it material for some new research; meanwhile I keep rushing about without any fixed plan, finding out what is already known about it, repeating experiments and setting up others as they occur to me; all of which is very enjoyable, inasmuch as the phenomena are in general exceedingly beautiful and varied. But it involves a lot of glass-blowing; my impatience will not allow me to order from the glass blower today a tube which would not be ready until several days later, so I prefer to restrict myself to what can be achieved by my own slight skill in the art. In point of expense this is an advantage. But in a day one can only prepare a single tube, or perhaps two, and make observations with these under varied conditions, so that naturally it is laborious work. At present, as already stated, I am simply roaming about in the hope that one or other of the hundred remarkable phenomena which are exhibited will throw some light upon the path.
>
> 31 July 1882
>
> I have made some preliminary attempts in the way of building up a battery of 1000 cells. This will cost some money and a good deal of trouble; but I believe it will prove a very efficient means of pushing ahead with the investigation, and will amply repay its cost.

After devoting the first half of the ensuing autumn vacation to recreation, he begins the construction of the battery.

---

[24][Hertz retained this great interest in "cathode-rays" throughout his life. He seems to have had the insight that they might reveal something important about the structure of matter and radiation—as indeed they did, since they ultimately led to the discovery of the electron and of x-rays. As an example of Hertz's later work on cathode-rays, see Paper No. 13 in this book.]

6 September 1882

I am now back again, after having had a good rest, and as there is nothing to disturb me here I have at once started fitting up the battery. So I am working away just like a mechanic. Every turn and twist has to be repeated a thousand times; so that for hours I do nothing but bore one hole and then another, bend one strip of lead after the other, and then again spend hours in varnishing them one by one. I have already finished 250 cells, and the remaining 750 are to be made forthwith; I expect to have the lot ready in a week. I don't like to interrupt the work, and that is why I haven't written to you before. For a while I feel quite fond of this monotonous mechanical occupation.

20 September 1882

The battery has practically been ready since the middle of last week; since last Sunday night it has begun to spit fire and light up electric tubes. Today for the first time I have made experiments with it—ones which I could not have carried out without it.

7 October 1882

I managed to get the battery to work satisfactorily, and a week ago succeeded in solving, to the best of my knowledge, the first problem which I had proposed to myself (a problem solved, when it really is solved, is a good deal!). But even this first stage was only attained with much trouble, for the battery turned sick, and its sickness has proved to be a very dangerous one.

By preventive measures the battery was kept going for yet a little while, and later on he reports, "Battery doing well." How the battery finally came to grief is explained in the account of the research.[25] By its aid he was able, in six weeks of vigorous exertion, to bring to a successful issue most of the experiments which he had planned. The investigation was first published in April 1883, at Kiel, in connection with Hertz's assuming the position of *Privatdozent* there.[26] It brought

---

[25][*Misc.Pprs.*, pp. 224–254.]

[26][Hertz used this cathode-discharge research as his *Habilitationsschrift* (sort of a "super dissertation") required by German universities of anyone desirous of teaching in a university. The submission of this paper and the completion of other requirements at Kiel led to his appointment as *Privatdozent* there. This gave him the right to teach. Unlike many *Privatdozenten*, who received no fixed salary from the state, he was to receive a

him recognition from a professor who rarely bestowed such tokens of esteem, and whose opinion he valued greatly. Hertz treasured as precious mementoes two letters from Helmholtz. One of these was his letter of appointment as assistant to Helmholtz in Berlin; the other is the following:

Berlin, 29 July 1883

Geehrter Herr Doktor!

I have read with the greatest interest your investigation on the cathode discharge, and cannot refrain from writing to say "Bravo!" The subject seems to me to be one of very wide importance. For some time I have been thinking whether the cathode rays may not be a mode of propagation of a sudden impact upon the Maxwellian electromagnetic ether, in which the surface of the electrode forms the first wave-surface. For, as far as I can see, such a wave should be propagated just as these rays are. In this case deviation of the rays through a magnetization of the medium would also be possible. Longitudinal waves could be more easily conceived; and these could exist if the constant k in my electromagnetic researches were not zero. But transverse waves could also be produced.[27]

You seem to have similar thoughts in your own mind. However that may be, I should like you to feel free to make any use of what I have mentioned above, for I have no time at present to work at the subject. These ideas suggest themselves so readily in reading your investigation that they must soon occur to you if they have not already done so. . . .

With kindest regards, yours, H. Helmholtz

While still busily engaged in completing this investigation on the [cathode] discharge, Hertz began to reflect upon another problem which seems to have been suggested to him by sheets of ice floating upon water during the winter.

---

modest yearly stipend of 500 Thaler ($375). He was expected to devote himself primarily to his own research, while offering a few lecture courses, for which he received additional student fees.]

[27][Here Helmholtz was wrong about the nature of cathode rays, and may have influenced Hertz, who maintained up till his death that they were some kind of unusual phenomenon in the ether. Oddly enough, just a few years earlier Helmholtz had suggested to Eugen Goldstein that cathode rays might be charged particles, but Goldstein rejected this idea, which turned out to be the correct one. On this see Schuster (1911), p. 55.]

Berlin, 24 February 1883

My researches are going on all right. From the date of my last letter until today I have been wholly absorbed in a problem which I cannot keep out of my head, namely the equilibrium of a floating sheet of ice upon which a man stands. Naturally the sheet of ice will become somewhat bent, but what form will it take, what will be the exact amount of the depression, etc.? One arrives at quite paradoxical results. In the first place a depression will certainly be produced underneath the man; but at a certain distance there will be a circular elevation of the ice; after this there follows another depression, and so on. As a matter of fact the elevations and depressions decrease so rapidly that they can never be seen; but to the eye of the mind their endless series is perceptible.

Even more paradoxical is the following result. Under certain circumstances a disc heavier than water, and which would therefore sink when laid upon water, can be made to float by putting a weight on it; and as soon as the weight is taken away it sinks. The explanation is that when the weight is put on, the disc takes the form of a boat, and thus supports the weight and itself. If the load is gradually removed, the disc becomes flatter and flatter; and finally there comes an instant when the boat becomes too shallow and so sinks, together with what is left of the load. This is the theoretical result and the way I explain it to myself, but meanwhile there may be errors in the calculation.

Such a subject has a peculiar effect on me. For a whole week I have been struggling to be finished with it, because it is of no great importance, and I have other things to do, e.g. I ought to be writing up the research which is to serve for my *Habilitation* at Kiel, which is already in my mind but without a bit of it on paper. Still it seems impossible to finish it off properly; there always remains some contradiction or improbability, and so long as anything of that sort is left I can scarcely take my mind away from it. Then, too, the formulae which I have deduced for the accurate solution are so complicated that it takes a lot of time and trouble to make out clearly their meaning. But if I take up a book or try to do anything else, my thoughts continually hark back to it. Shouldn't things happen in this way or that? Isn't there still some contradiction here? All this is a perfect plague when one doesn't attach much importance to the results.

Soon afterwards Hertz had to move to Kiel. This move, his *Habilitation*, and his lectures there took up much of his time, so that his investigation on floating plates was not published until a year later. Its place was taken by the investigation of the fundamental equations

Lenard's Introduction to Hertz's Miscellaneous Papers 107

of electromagnetics.[28] At this time he kept a diary, from which it appears that in May 1884 he was alternately working at his lectures, at electromagnetics, and at microscopic observations taken up by way of change. On six successive days there are brief but expressive entries—"Hard at Maxwellian electromagnetics in the evening," "Nothing but electromagnetics"; and then follows on the next day, the nineteenth of May—"Hit upon the solution of the electromagnetic problem this morning." This will remind the reader of Helmholtz's remark that the solution of difficult problems came to him suddenly, and then often unexpectedly, when a period of vigorous battling with the difficulties had been followed by one of complete rest.

In close connection with this subject, and immediately following it in order of time, came the paper "On the Dimensions of Magnetic Pole in Different Systems of Units."[29] Directly after this came the meteorological paper "On the Adiabatic Changes of Moist Air."[30] A diagram illustrating the latter is reproduced here from the original;[31] the drawing of this, as a recreation after other work, seems to have given Hertz great pleasure.

We may complete our account of Hertz's scientific work during his two years at Kiel by adding that at this time he repeatedly, although unsuccessfully, attacked certain hydrodynamic problems, and that his thoughts already turned frequently towards that field [electrodynamics] in which he was afterwards to reap such a rich harvest. Nearly five years before he had carried out his investigation "On Electric Radiation" [1888] we find in his diary the notable remark—"27 Jan. 1884. Thought about electromagnetic rays," and again, "Reflected on the electromagnetic theory of light." He was always full of schemes for investigations, and never liked to be without some experimental work. So he did his best to fit up in his house a small laboratory with home-made apparatus, thus transporting himself back to the times when chemists worked with the modest spirit lamp. But before his experiments were concluded or any of his schemes carried out he was called to Karlsruhe, and his move there [from Kiel] relieved him from

---

[28][Paper No. 3 in this book.]

[29][*Misc. Pprs.*, pp. 291–295.]

[30][*Misc. Pprs.*, pp. 296–312.]

[31][This diagram is included at the back of Hertz's *Miscellaneous Papers*, but is not included in the present volume.]

much unprofitable exertion caused by the lack of proper experimental facilities.

This brings us to the end of the series of papers around which we have grouped the events of the author's life. After this follow the great electrical investigations which now form the second volume of his collected works [*Electric Waves*]. At this point we have introduced the lecture which Hertz gave at Heidelberg on these discoveries,[32] and which will still be fresh in the remembrance of many who heard it.

After this follows the last experimental investigation that Hertz made. While his colleagues, and in Bonn his pupils as well, were eagerly pushing forward into the territory which he had opened up, he returned to the study of electric discharges in gases, which had interested him before [in Berlin]. Again he was rewarded by an immediate and unexpected discovery. Early in the summer semester of 1891 he found that cathode rays could pass through metals. The investigation was soon interrupted, but was published early in the ensuing year;[33] from now on the subject matter of his last work, the *Principles of Mechanics*, wholly absorbed his attention.

---

[32] ["On the Relations between Light and Electricity"; Paper No. 10.]

[33] ["On the Passage of Cathode Rays through Thin Metallic Layers"; Paper No. 13.]

PART II

Berlin (1878–1883): Advanced Studies in Physics, Friedrich-Wilhelm University

> Everyone who has had the good fortune to work even for a brief period under Helmholtz's guidance feels that in this sense he is above all things his pupil, and remembers with gratitude the consideration, the patience, and the good-will shown to him. Of the many pupils now scattered over the world there is not one who will not today think of his master with love as well as admiration....
>
> *Heinrich Hertz (1891)*

Hermann von Helmholtz (1821–1894), the leading German physicist in the last half of the nineteenth century. This engraving of Helmholtz by T. Johnson is based on a photograph taken by the famous American photographer Matthew Brady in September 1893, when Helmholtz was visiting the United States. Courtesy of Burndy Library.

Introduction

While an advanced student and assistant to Helmholtz in Berlin from 1878 to 1883, Hertz completed the research for fifteen published papers on a great variety of topics in physics. Many of these are discussed in Lenard's "Introduction" to *Miscellaneous Papers* (Paper No. 1), and also in a lengthy review in *Nature* of the same collection of articles by George Francis FitzGerald (reprinted here as Paper No. 4).

Although all Hertz's early articles are well-conceived and executed, most are on subjects of little relevance to present-day physics. The major exceptions are the papers, "On a Phenomenon Which Accompanies the Electric Discharge," and "Experiments on the Cathode Discharge," both published in 1883 (*Miscellaneous Papers*, pp. 216–254). These are important because they reveal Hertz's early involvement with cathode rays, a topic in which he remained interested all his life and to which he returned for his last experimental research in 1892. But these papers are quite qualitative, and the second is marred by Hertz's conclusion that cathode rays could not be charged particles because he could not deflect them with an electrostatic field nor observe any magnetic effect of the cathode beam outside his Geissler tube. This led Hertz to surmise that cathode rays must be some new sort of electromagnetic phenomenon in the ether. In his review (Paper No. 4) FitzGerald was able to explain the experimental problems that misled Hertz in this matter. Because of these complications we have not included these early cathode-ray papers in this collection.

Instead we have chosen a single paper that goes to the heart of Hertz's experience in Berlin from 1878 to 1883. These were the years he studied and developed as a physicist under the direction and mentorship of Hermann von Helmholtz (1821–1894). During these five years Hertz developed from a student and assistant of Helmholtz to his colleague and personal friend. No man, except perhaps his father, had a more positive influence on Hertz, who always retained an attitude of near hero-worship for Helmholtz, even when he disagreed with him on particular points in physics. Hertz's approach to physics, his understanding of what the truly important problems were, and his ability to devise ways of solving these problems trace themselves back to Helmholtz. For this reason, in his later years as a physicist, Hertz still rejoiced when Helmholtz took the time to show his approval of what his former student had achieved. (As an example, see Helmholtz's letter of 29 July 1883 to Hertz on his cathode-ray work, which is reprinted in Paper No. 1.)

For this reason we include in this section a popular piece written by Hertz for Helmholtz's seventieth birthday and published in the supplement to the *Münchener Allgemeine Zeitung* for 31 August 1891. It shows Hertz's great appreciation of Helmholtz as a physicist and his love for him as a man; at the same time it reveals to us a great deal about Hertz himself, especially his considerable literary ability.

PAPER NO. 2

Hermann von Helmholtz: On 31 August 1891

(Heinrich Hertz, Supplement to the *Münchener Allgemeine Zeitung* for 31 August 1891; *Miscellaneous Papers*, pp. 332–340.)

In Germany the men who now stand upon the threshold of old age have inaugurated and lived through a period of rare felicity and success. They have seen aims attained and desires realized, and this not only in matters political: they have seen mighty developments in the arts of peace; they have seen our Fatherland take its place in the front rank of nations, not only in our own estimation but in that of others. Even in the beginning of this century the natural sciences were far from being neglected in Germany: the labors of a Humboldt, the undying fame of a Gauss, were sufficient to keep alive respect for German research. But side by side with the wheat of true effort there sprang up the tares of a false philosophy[1] which flourished so luxuriantly as to hinder the full growth of the crop. Up to the middle of the century sober progress along the path of experimental investigation lacked the glory which accompanies international success; and the successes of a fictitious natural philosophy were very properly not greeted with the same exultation abroad as in Germany. Germans followed eagerly and diligently the discoveries made in other lands; but they always expected the great discoveries and successes to come from Paris and London. To these places young investigators travelled to see famous scientific men and to learn how great investigations were carried out; from these sources they obtained the materials for their own researches; there alone could new discoveries be properly and authentically published. They found it hard to believe that things could ever be otherwise.

But all this has long since changed. In science Germany is no longer dependent upon her neighbors; in experimental investigation she is the peer of the foremost nations and keeps in the main well abreast of them, sometimes leading and sometimes following. This the country owes to the cooperation of many eager workers; but it

---

[1] [Here Hertz is referring to the *Naturphilosophie* of Schelling and Hegel, against which Helmholtz and Kirchhoff had rebelled, seeing it as a hindrance to the proper development of physics.]

naturally honors most those few whose names are most closely connected with the actual successes. Of these some have already left us forever; others still remain, and we hope to have them with us for many more years.

The greatest among all these, the acknowledged representative of this period of progress and well-earned fame, the scientific leader of Germany, is Hermann von Helmholtz, whose seventieth birthday we celebrate today, after he has for nearly half a century astonished the scientific world by the number, the depth, and the importance of his investigations. To the countless tributes of admiration and gratitude which will this day be laid at his feet we would with all modesty add our own. As Germans we are glad and proud to claim as our countryman one whose name we deem worthy to be placed among the noblest names of all times and all nations, confident that subsequent generations will confirm our judgment. As men we cherish the same feelings of admiration and gratitude. Other nations, too, will join us in paying honor to him today, as indeed they have in the past. For, although nations may appear narrow-minded in political affairs, men have not wholly lost the sense of a common interest in matters scientific: a Helmholtz is regarded as one of the noblest ornaments of humanity.

Let us try to recall the achievements for which we today do him honor. Here at once we feel how impossible it is to make others share fully in our admiration if they are not themselves in a position to appreciate his work. It is a mistake to suppose that the importance of an investigator's work can be gauged by stating what problem he has solved. A man must see a picture, and must see it with the eyes of an artist, before he can fully appreciate its value. Even so scientific investigations have a beauty of their own which can be enjoyed as well as understood; but in order to enjoy it a man must understand the investigation and steep himself in it.

Take one of Helmholtz's minor researches, e.g., the theoretical paper in which he discusses the formation of liquid jets. The problem is not one that appeals to the lay mind: its solution is only attained by the aid of assumptions which correspond but indifferently to the actual conditions; the influence of the investigation upon science and life can scarcely be called other than slight. And yet the manner in which the problem is solved is such that, in studying even a paper like this, one feels the same elevation and wonder as in beholding a pure work of art. Upon our comprehension of the difficulties to be surmounted depends the depth of this feeling. We see a man of surpassing strength leap across a yawning chasm apparently without effort, but in reality straining every nerve. Only after the jump do we clearly see how wide the chasm is. Instinctively we break out into applause. But we cannot expect the same spontaneous enthusiasm of spectators from whose standpoint the chasm is not visible, and who can only learn from our descriptions how difficult the feat really was.

To give a brief but fitting sketch of Helmholtz's work is difficult on account of its many-sidedness and its profundity. His scientific life interests us like an Odyssey through the whole region of exact investigation. He began as a doctor; he had to study the laws governing the human life to which he would minister, and this led him to the study of physiology, the scientific side of medicine. He found himself hampered by the gaps in our knowledge of inanimate nature; so he set about filling these and thus drifted more and more toward physics. For the sake of physics he became a mathematician, and in order to probe thoroughly the foundations of mathematical knowledge, and knowledge in general, he became a philosopher.[2] When we look through the technical literature of any of these sciences we meet his name; upon all of them he has left his mark. Without attending to chronological order we shall here only describe briefly three of those great achievements which constitute his title to fame.

I consider that the most beautiful and charming among these, although not the most profound, is the invention with which he has enriched practical medicine. I mean the ophthalmoscope.[3] Before him no one was able to investigate the living eye. Beyond the doubtful and unreliable feelings of the patient there were no means of diagnosing the disease or determining any defects in the refractive power of the eye. Before any cure was possible it was absolutely necessary for the surgeon to acquire an accurate knowledge of the disease; and this, in

---

[2][One of Helmholtz's obituary writers commented that, just as seven cities had claimed Homer as their own, so too seven sciences—medicine, physiology, chemistry, physics, mathematics, philosophy and aesthetics—each claimed Helmholtz as its glory.]

[3][Helmholtz invented the ophthalmoscope in 1851 when he was professor of physiology at the University of Koenigsberg. This discovery is described in Hermann von Helmholtz, "Beschreibung eines Augenspiegels zur Untersuchung der Netzhaut im lebenden Auge," *Annalen* 2, 229–260 (1851).]

the majority of cases, was only attainable after the invention of this simple instrument. Ophthalmic surgery rapidly rose to its present stature. Who can say how many thousands who have recovered their sight owe it to our investigator—to him personally, although they are unconscious of this and think that their thanks are simply due to the surgeon who has treated them!

The invention of the ophthalmoscope is like vaccination against smallpox, the antiseptic treatment of wounds, or the sterilization of children's food—one of those great gifts that enrich all without impoverishing any, one of those advances that are gratefully acknowledged everywhere by all men, and that keep alive in us the belief that there is such a thing as progress.

Equally powerful as a protection against intellectual blindness are the advances that physiology owes to Helmholtz, although their value may not be so easily or generally recognized. Here we may remind the reader in passing that he was the first to measure the speed with which sensations and volitions travel along the nerves; this would have sufficed to establish the fame of any other man, but it is not this that we now have in mind. His chief investigation in this science, the work of his mature years, is the development of the physiology of the senses, especially of sight and hearing. Within our consciousness we find an inner intellectual world of conceptions and ideas; outside our consciousness there lies the cold and alien world of actual things. Between these two stretches the narrow borderland of the senses. No communication between the two worlds is possible except across this narrow strip. No change in the external world can make itself felt by us unless it acts upon a sense organ and borrows form and color from this organ. In the external world, we can conceive no cause for our changing feelings until we have, however reluctantly, assigned to it sensible attributes. For a proper understanding of ourselves and of the world it is of the highest importance that this borderland should be thoroughly explored, so that we may not make the mistake of referring anything which belongs to it to one or the other of the worlds it separates.

When Helmholtz turned his attention to this borderland, it was not in a wholly uncultivated state; but he found the richest fields in it lying fallow, and on either side its limits were badly defined and hidden by a luxuriant growth of error. He left it carefully defined and well divided, and much of it had been transformed into a blooming garden.

His celebrated treatise *On the Sensations of Tone*[4] is known to a fairly wide circle of students. That which outside ourselves is a mere vibration of the air becomes within our minds a joyful harmony. What interests the physicist is the vibration of the air; what interests the musician and the psychologist is the harmony. The transition between the two is discovered in the sensation that connects the definite physical process with the definite mental process. What is there outside ourselves that corresponds to the quality of the tones of musical instruments, of human song, of vowels and consonants? What corresponds to consonance and dissonance? Upon what does the aesthetic opposition between the two depend? By what ideas of order within us were these codes of music, the musical scales, developed? Not all the questions that are prompted by a thirst for knowledge can be answered; but nearly all the questions that Helmholtz had to leave open thirty years ago remain unanswered to the present day.

In his *Physiological Optics*[5] he discusses similar questions relating to sight. How is it possible for vibrations of the ether to be transformed by means of our eyes into purely mental processes, which apparently can have nothing in common with the former, and whose relations nevertheless reflect with the greatest accuracy the relations of external things? In the formation of mental conceptions what part is played by the eye itself, by the form of the images that it produces, by the nature of its color sensations, accommodation, motion of the eyes, by the fact that we possess two eyes? Is the multiplicity of these relations sufficient to portray every conceivable variation in the external world, to justify all the diversity of the internal world?

We see how closely these investigations are connected with the possibility and the legitimacy of all natural knowledge. The heavens and the earth doubtless exist apart from ourselves, but for us they only exist insofar as we perceive them. Part of what we perceive, therefore, appertains to ourselves; part only has its origin in the properties of the heavens and of the earth. How are we to separate the two? Helmholtz's physiological investigations have cleared the ground for the answering of this question; they have supplied a firm fulcrum to which a lever can

---

[4] [Hermann von Helmholtz, *On the Sensations of Tone as a Physiological Basis for the Theory of Music* (Second English edition; Dover Publications, New York, 1954).]

[5] [Hermann von Helmholtz, *Handbook of Physiological Optics* (Dover Publications, New York, 1962).]

be applied. His own inclinations have led him to discuss these very questions in a series of philosophical papers, and no more competent judge could express an opinion on them. Will his philosophical views continue to be esteemed as a possession for all time? We should not forget that we have here passed beyond the bounds of the exact sciences; no appeal to nature is possible, and we have nothing but opinion against opinion and view against view.

As on the one hand Helmholtz was led by the study of the senses to the ultimate sources of knowledge, so on the other hand the same study led him to the glories of art. The rules that the painter and the musician instinctively observe were for the first time recognized as necessary consequences of the way we are put together, and were by this recognition of their necessity transformed into conscious laws of artistic creation.

As diverse and important as are these discoveries, they are all eclipsed by another with which the name of Helmholtz will ever be connected. This is a discovery that belongs to a more abstract field of knowledge, namely, to physics. Here the human observer with his sensations retires into the background; light and color fade away and sound becomes fainter; their place is taken by geometrical intuitions and general ideas, by time and space, matter and motion. Between these ideas relationships must be found, and these relationships must correspond to those between the things themselves.[6] The value of these relationships is measured by their generality. As relationships of the most general nature we may mention the conservation of matter, the inertia of matter, the mutual attraction of all matter. Of new relationships discovered in this century the most general is that which was first clearly recognized by Helmholtz. It is the law which he called the Principle of the Conservation of Force, but which is now better known to us as the Principle of the Conservation of Energy.[7] It had long before been suspected that in the unending succession of phenomena there was something else besides matter that persisted,

---

[6][This sentence is almost identical with one in Hertz's Introduction to his *Principles of Mechanics*. This is understandable, since Hertz was involved in writing his treatise on mechanics at the time (1891) he wrote the present article.]

[7][Hermann von Helmholtz, "The Conservation of Force: A Physical Memoir." There is a good English translation of this paper, which was first published in 1847, in Kahl (1971), pp. 3–55.]

which could neither be created nor destroyed, something immaterial and scarcely tangible. At one time it seemed to be the quantity of motion measured in this way or that, at another time the force, or again an expression compounded of both.

In place of these obscure guesses Helmholtz brought forward distinct ideas and fixed relations, which led immediately to a wealth of general and special connections. Magnificent were the vistas that the principle opened up to the past and the future of our planetary system; in every separate investigation, even the most restricted, its applications were innumerable. For forty years it has been so much expounded and extolled that no man of culture can be quite ignorant of it. It is noteworthy that about this time other heads began to think more clearly of these things; and it came about that as far as the phenomena of heat were concerned other men had anticipated Helmholtz by a few years without his knowing it.[8] It would be far from his wish to detract from the fame of these men; but it should not be forgotten that their researches were almost entirely restricted to the nature of heat, whereas the significance and value of the principle [of Conservation of Energy] lie precisely in the fact that it is not limited to this or that force of nature, but embraces all of them and can even serve as our guiding star in the discovery of unknown forces.[9]

It is not generally known that in his mature years Helmholtz has returned to the work of his youth and has still further developed it. The law of the conservation of energy, general though it is, nevertheless appears to be only one half of a still more comprehensive law. A stone projected into empty space would persist in a state of uniform motion, and thus its energy would remain constant; to this corresponds the conservation of energy in any system, however complicated that system may be. But the stone would also tend to retain its direction

---

[8][Hertz is referring here to Julius Robert Mayer (1814–1878), a German physician, who in 1842 suggested that heat and mechanical energy could be converted one into the other. A similar suggestion was made by James Prescott Joule (1818–1889) in Great Britain on the basis of a long series of careful experiments. Joule's work was completed in 1847, the same year that Helmholtz published his more detailed and more general Principle of Conservation of Energy.]

[9][For example, it was the application of the principles of conservation of energy and of momentum to the phenomenon of beta decay that in this century led to the prediction of the existence of the neutrino, and ultimately to its experimental detection.]

and to travel in a straight line; to this behavior there is a corresponding general behavior on the part of every moving system. In the case of purely mechanical systems it has long been known that every system, according to the conditions in which it is placed, arrives at its goal along the shortest path, in the shortest time, and with the least effort. This phenomenon has been regarded as the result of a designed wisdom; its general statement in the region of pure mechanics is known as the Principle of Least Action. To trace the phenomenon in its application to all forces, through the whole of nature, is the problem to which Helmholtz has devoted a part of the last decade. As yet the significance of these researches is not thoroughly understood. An investigator of this stamp treads a lonely path; years pass before even a single disciple is able to follow in his footsteps.[10]

It would be futile to try to enter into particulars on all of Helmholtz's researches. Our omissions might be divided among several scientific men and would amply suffice to make all of them famous. If one of them had carried out Helmholtz's electrical researches and nothing else, we should regard him as our chief authority on electricity. If another had done nothing but discover the laws of vortex motion in fluids, he could boast of having made one of the most beautiful discoveries in mechanics. If a third had only produced the speculations on the conceivable and the actual properties of space, no one would deny that he possessed a talent for profound mathematical thought. But we rejoice to find these discoveries united in one man instead of divided among several. The thought that one or other of them might be a mere lucky find is rendered impossible by this very union; we recognize them as proofs of an intellectual power far exceeding our own, and we are lost in admiration.

And yet how far the sum total of these obvious accomplishments is from exhausting the sum total of his complete personality! How can we estimate the intellectual value of the inspiration which he imparted, at first to his contemporaries, and afterwards to the pupils

---

[10][The same could be said about Hertz himself and the new ideas he expressed in his *Principles of Mechanics*. He never had a disciple to take up his ideas, fully comprehend their significance, and work to make them better appreciated by physicists. As a consequence, Hertz's contributions to mechanics are almost totally forgotten today.]

who flocked to him from far and near?[11] It is true that Helmholtz never had the reputation of being a brilliant university teacher, insofar as this depends upon communicating elementary facts to the beginners who usually fill the lecture halls. But it is quite another matter when we come to consider his influence upon well-prepared students and his preeminent fitness for guiding them in original research. Such guidance can only be given by one who is himself a master of the field, and the value of such guidance can be measured in terms of his own work. Here example is of more value than precept; a few occasional hints can point out the path better than formal and well-prepared lectures. The mere presence of the marvelous investigator helps the pupil to form a just estimate of his own efforts and those of his fellow students, and enables him to see things *sub specie aeterni*[12] instead of from his own narrow point of view.

Everyone who has had the good fortune to work even for a brief period under Helmholtz's guidance feels that in this sense he is above all things his pupil, and remembers with gratitude the consideration, the patience, and the good-will shown to him. Of the many pupils now scattered over the earth there is not one who will not today think of his master with love as well as admiration, and with the hope that he may yet see many years of useful work and of happy leisure.

---

[11][Helmholtz was one of the most successful mentors of young physicists in the history of physics. Among his students were Hertz, Max Planck, Heinrich Kayser, Eugen Goldstein, and Wilhelm Wien from Germany, and Henry Rowland, A.A. Michelson, and Michael Pupin from the United States. On this see Mulligan (1989b).]

[12][The more common Latin phrase is *sub specie aeternitatis*, meaning "in the light of eternity."]

PART III

Kiel University (1883–1885): Physics Instructor (*Privatdozent*)

> I think, however, that from the preceding we may infer without error that if the choice rests only between the usual system of electromagnetics and Maxwell's, the latter is certainly to be preferred....
>
> *Heinrich Hertz (1884)*

Heinrich Hertz in military uniform. This photograph was probably taken in 1877, when Hertz was twenty years of age and serving his year of military service with the First Railway Guards Regiment in Berlin. Courtesy of Deutsches Museum, Munich.

Introduction

Hertz had been appointed as *Privatdozent* for Theoretical Physics at Kiel University, in which position he was provided with no adequate space or equipment for experimental research. He therefore turned again to theoretical work and published three articles, two of them of minor importance, during his brief stay in Kiel. One piece of research, however, was of major significance for the development of electromagnetism. His 1884 article "On the Relations between Maxwell's Fundamental Electromagnetic Equations and the Fundamental Equations of the Opposing Electromagnetics" therefore is included in the present collection, despite the heavy demands it makes on the reader. In this paper Hertz starts from the older action-at-a-distance theories of electromagnetism of W. Weber and F. Neumann and proceeds to derive Maxwell's equations from first principles in a new way that avoids both the mechanical models Maxwell had originally used and Maxwell's concept of the displacement current. At that time both these approaches were out of favor with Hertz's colleagues in Germany.

This paper indicates two ways in which Hertz's ideas were developing at that time. He still appreciated Helmholtz's approach to electromagnetism but demonstrated a growing independence from the details of Helmholtz's treatment. He was also more and more convinced that electromagnetic waves travel at the speed of light and that an experimental proof of this fact was essential to validate Maxwell's theory. Therefore he devoted much thought to ways of carrying out such a measurement if the necessary equipment should become available, as it did in Karlsruhe a few years later.

The other article included in this section is a long review of Hertz's *Miscellaneous Papers* by the Irish physicist, George Francis FitzGerald, which appeared in *Nature* in 1896. *Miscellaneous Papers* was published in 1895, although all but three of its twenty-two articles were completed either in Berlin or at Kiel in the years 1878–1885. This, therefore, seems to be an apt place to introduce FitzGerald's review. It will serve as both a summary of the contents of those articles by Hertz omitted from this collection, and as a critical commentary on some of the most important articles from these seven years of Hertz's life.

George Francis FitzGerald (1851–1901) was the leader of the Maxwellians, a group of physicists from the British Isles who strongly supported Maxwell's ideas and spent the years from Maxwell's death in 1879 to the turn of the century refining, modifying and promoting his electromagnetic theory. FitzGerald's reviews of all three volumes of Hertz's collected works are included in this collection because of the close scholarly and personal relations that developed between Hertz and the Maxwellians in Great Britain and the high quality of FitzGerald's critical judgment and his eloquence in expressing it.

FitzGerald taught physics at Trinity College, Dublin, and was heavily burdened by teaching responsibilities throughout his academic career. As a consequence he seldom found time to follow his many brilliant ideas through to the definitive formulation they deserved. His published articles tend to be brief and incomplete, but they stimulated many of his students and colleagues to some very useful and important research. He was also an excellent critic who made his points either at a meeting or in a book review in an eloquent and trenchant fashion. These qualities mark the review of Hertz's *Miscellaneous Papers* included here.

PAPER NO. 3

On the Relations between Maxwell's Fundamental Electromagnetic Equations and the Fundamental Equations of the Opposing Electromagnetics

(Heinrich Hertz, "Über die Beziehungen zwischen den Maxwell'schen electrodynamischen Grundgleichungen und den Grundgleichungen der gegnerischen Electrodynamik," Wiedemann's *Annalen der Physik und Chemie* 23 [1884], pp. 84–103.)[1]

When Ampère heard of Oersted's discovery that the electric current sets a magnetic needle in motion, he suspected that electric currents would exhibit moving forces between themselves. Clearly his train of reasoning was somewhat as follows: The current exerts a magnetic force, for a magnetic pole moves when submitted to the action of the current; and the current is set in motion by a magnetic force, for by the principle of action and reaction a current-carrier will also move under the influence of a magnet. Unless we make the improbable assumption that different kinds of magnetic force exist, a current-carrier must also move under the action of the magnetic force that a second current exerts, and thus the interaction between currents follows.

The essential step in this reasoning is the assumption that only *one* kind of magnetic force exists; that therefore the magnetic forces exerted by currents are in all their effects equivalent to equal and equally directed forces produced by magnetic poles. But this assumption is well known to be sufficient to deduce not only the existence but also the precise magnitude of the electromagnetic

---

[1] [This paper of Hertz is the most mathematical and difficult of those in the present volume. It is included here because of its pivotal importance in Hertz's development as a physicist. It marks the beginning of Hertz's conversion to Maxwell's ideas, and forms the basis for all his future contributions, both theoretical and experimental, to electromagnetism. D'Agostino (1975) was the first one to point out the importance of this paper in the development of Hertz's ideas. For a recent, thorough discussion of the contents of Hertz's paper, see Buchwald (1990), pp. 286–303.]

actions of closed currents[2] from their magnetic actions. Whether Ampère actually started from this principle or not, he certainly stated it at the close of his investigations when he reduced the action of magnets directly to the action of supposed closed currents. At a later stage the principle was hardly mentioned, but was taken for granted as self-evident. After the discovery of the electric forces exerted by variable currents or moving magnets, a similar principle was added relative to these electric forces, and this, too, was not definitely expressed. It has perhaps nowhere been explicitly stated that the electric forces, which have their origin in inductive actions, are in every way equivalent to equal and equally directed electric forces of electrostatic origin; but this principle is the necessary presupposition and conclusion of the chief notions that we have formed of electromagnetic phenomena generally. According to Faraday's idea the electric field exists in space independently of and without reference to the method of its production; whatever therefore be the cause that has produced an electric field, the actions that the field produces are always the same. On the other hand, by those physicists who favor Weber's and similar views, electrostatic and electromagnetic actions are represented as special cases of one and the same action-at-a-distance emanating from electric particles. The statement that these forces are special cases of a more general force would be without meaning if we admitted that they could differ otherwise than in direction and magnitude, that is, according to their nature and method of acting. But, apart from all theory, the assumption we are speaking of is implicitly made in most electric calculations; it has never been directly rejected, and may thus be regarded as one of the fundamental ideas of all existing electromagnetics. Nevertheless, to my knowledge no one has yet drawn attention to certain consequences to which it leads, and which will be developed in what follows.

As premises we in the first place employ the two principles referred to, which we might designate as the principle of the unity of electric force and that of the unity of magnetic force. These may be regarded as generally accepted, even if not as self-evident. In the second place, we use the principle of the conservation of energy; that of action and reaction as applied to systems of closed currents; that of

---

[2][By a "closed current" Hertz means a current flowing around a complete electrical circuit. A circuit containing a break, such as a capacitor or spark gap, he calls an "open current."]

the superposition of electric and magnetic actions; and lastly, the well-known laws of the magnetic and electromotive actions of closed currents and of magnets. The investigation throughout refers to closed currents, even where this is not specifically stated.

1. Suppose a ring-magnet,[3] whose cross-section we shall for simplicity take as small compared with its other dimensions, to lose its magnetism. Then it will exert a force on all electricity in its neighborhood, which causes this electricity to circulate around the body of the magnet. The magnitude of this force is proportional to the [time] rate of loss of magnetization, and may be constant during a short but finite time, if during this time the magnetization diminishes at a constant rate. The distribution of force in space is precisely the same as [the distribution of magnetic force] that would be produced by a current flowing in the body of the magnet. Like the latter, the electric force considered has a potential which is many-valued and, apart from its multiplicity, is the same as that due to an electric double layer of uniform moment bounded by the axis of the magnet. The potential of the ring-magnet at an electric pole can, apart from its multiplicity, be represented by the potential of the double layer at the pole; or, taking the multiplicity into account, it can be represented by the solid angle subtended by the magnet at the pole, multiplied by a suitable constant.

Now this potential determines the action of the magnet on the [electric] pole as well as that of the pole on the magnet. If we have not a single pole but a whole system of electric charges, the potential of the diminishing magnetization on this system can be found by a simple summation. In particular, when the electric forces that act on the ring-magnet are due, not to electric charges, but to a second ring-magnet of diminishing moment, their distribution is the same as if they were due to an electric double layer. Hence, according to our assumption of the unity of electric force, interaction occurs between our two ring-magnets of diminishing moment; and the potential determining this interaction is the mutual potential of two electric double layers which are bounded by the bodies of the magnets. As in electromagnetics the mutual potential of two magnetic double layers is reduced to an integral to be taken along their boundaries, so here we can bring into the same form the potential of the electric layers, that is, of the two magnets of diminishing moment. We thus find that this potential is the

---

[3] [A ring-magnet is a magnetized toroid.]

product of the factor[4] A^2 of the rates of diminution of the moments of the magnets per unit length measured in electromagnetic units, and of the integral

$$\iint \left(\cos\frac{\varepsilon}{r}\right) dl\, dl',$$

where dl and dl' denote elements of the axes of the magnets, and ε is the inclination of these elements to each other.

The potential thus determined is of the same form as the mutual potential of electric currents, and therefore represents the same actions. Two ring-magnets, which are placed close together and side by side, will attract each other at the moment when they both lose their magnetism, if they are magnetized in the same direction; they will repel each other if oppositely magnetized. In the usual[5] electromagnetics this action is missing. To describe it more simply we shall introduce a new name. We call the change of magnetic polarization a magnetic current, and take as unit that magnetic current intensity which corresponds to unit change per unit time of the polarization per unit volume measured in absolute magnetic units. So far as we can conclude from the phenomena of unipolar induction as now known, magnetic poles distributed continuously along a closed curve and moving along it exert the same electromagnetic action at outside points as a ring-magnet coinciding with that curve and of suitably changing moment. If this relation can be looked upon as true in general, the name "magnetic current" includes all the different cases of magnetism in motion; and we may speak of constant magnetic currents just as we speak of constant electric currents. But here that name is only to be regarded as a simple contraction for "changing polarization." Our result may now be stated in this form: Magnetic currents act on each other according to the same laws as electric currents; in absolute magnitude the action between magnetic currents of S magnetic units is equal to that between electric currents of S

---

[4]A is, as usual, the reciprocal of the velocity of light. We get this factor by a quantitative investigation of the case which above is considered only qualitatively. In this regard see the paragraph marked 2. below.

[5]By "usual" I mean, here and in what follows, that electromagnetics which regards the forces deduced from Neumann's laws of the potential as exactly applicable, even when we consider the attraction of variable currents. Every such system of electromagnetics is necessarily opposed to Maxwell's.

Fundamental Equations of Electromagnetism 131

electrical units. This theorem may not be capable of experimental verification. It may be possible to show that electrically charged bodies are set in motion by a ring-magnet whose moment is diminishing; perhaps even that the magnet itself is turned by electrostatic force so that its plane sets itself normal to that force; but even with very powerful electrostatic forces these actions will lie at the limits of observation, and hence it is hopeless to expect to see a ring-magnet set itself under the action of the weak forces produced by a second ring-magnet when the moments of both are diminishing.

But our premises permit of our drawing further inferences. It is known that a knowledge of the mutual electromagnetic potential of two currents, together with the principle of the conservation of energy, enables us to predict the existence and absolute value of the inductive action. Similar conclusions may be drawn for magnetic circuits (rings of soft iron). A determinate expenditure of work is necessary to maintain in such a circuit a magnetic current, which we may suppose to be alternating. If the amount of this work were the same, whether the magnet were at rest free from any electrical influence, or did work in moving through the electric field, nothing could be simpler than the infinite production of work from this motion. Hence such an independence is impossible. The work done must depend on the nature and velocity of the circuit's motion and on the changes in the electric field; and thus the magnetic (magnetomotive) force which produces this uniform current must also depend upon these circumstances. This may be expressed by saying that a magnetic force, produced by the motion and the changes in the field, is superimposed upon the magnetic forces due to other causes; this added force we may describe as induced. Its magnitude is given by the condition that for any displacement whatever of the circuit the external work done in this displacement must be compensated by an equal additional amount of work which in consequence of the displacement must be done in the circuit. This reasoning is in form the same as that used to deduce the inductive actions in electric circuits; and since also the forces between magnetic circuits are of the same form as those between electric circuits, the final result must in form be the same in both cases. In the laws of electric induction we need only interchange the words "electric" and "magnetic" throughout in order to obtain the inductive actions in magnetic circuits. Thus we find that a plane magnetic circuit, e.g., a plane ring of soft iron, whose plane is perpendicular to the lines of force in an electric field, is traversed by a magnetizing force at the instant when the intensity of the field is reduced to zero; and that the same ring is subject to an alternating polarization when we turn it about an axis which is perpendicular to the direction of the electric force. It does not appear impossible that such actions may become capable of experimental detection. Again, a ring-magnet whose polarization is continually changing its direction must by induction call forth alternating polarizations in all neighboring iron rings; but this action is certainly too small to reach an observable value.

2. It may at first sight seem as if the actions here deduced from generally accepted premises could be incorporated without disturbance into the usual system of electromagnetics; but this is not the case. In fact, suppose that in place of the ring-magnets we have so far considered we have endless electric solenoids, in which the current-intensity is variable; then the induced electric forces produced by these solenoids are certainly quite analogous to those exerted by the variable magnets. From these latter forces we deduced magneto-dynamic attractions, and we must therefore infer corresponding electrodynamic attractions between the variable solenoids. But as long as the currents in them are constant no action takes place. Hence in general the electromagnetic attraction between currents must depend on their variations and not merely on their momentary intensities. This statement is in opposition to an assumption uniformly accepted in the usual electromagnetics.[6] The correction which must be made in the laws of the magnetic actions of constant currents to make them applicable to variable currents may be calculated from our premises. But this correction requires, on account of the principle of the conservation of energy, a correction in the induced electrical forces as well. This again requires a second correction in the magnetic forces, and so on; so that we obtain an infinite series of successive approximations. We shall now calculate these separate terms. We assume that they are simply to be added to yield the total result, and that, if only the infinite sums converge to definite limits, then these limiting values are those corresponding to the actual case. In the

---

[6]Cf. v. Helmholtz, "On the Theory of Electrodynamics, Part 3: The Electromagnetic Forces on Moving Conductors" [in Helmholtz (1882–1995), vol. 1, pp. 702–762; on p. 729.]

Fundamental Equations of Electromagnetism

calculation[7] we use the following special notation: $\bar{u}$ is to denote a function U for which, throughout infinitely extended space,

$$\nabla^2 U = -4\pi u;$$

Hence, in general,

$$U = \bar{u} = \int \frac{u}{r} d\tau,$$

the integral being taken over all space.

As regards the electric currents, let u, v, w be the components. As we only consider closed circuits we have

$$\frac{\partial u}{\partial x} + \frac{\partial v}{\partial y} + \frac{\partial w}{\partial z} = 0.$$

Further let

$$U_1 = \bar{u}, \quad V_1 = \bar{v}, \quad W_1 = \bar{w}.$$

Then the components L_1, M_1, and N_1 of the magnetic force exerted by the currents are, according to the usual electromagnetics, given by the equations

$$L_1 = A\left(\frac{\partial V_1}{\partial z} - \frac{\partial W_1}{\partial y}\right)$$
$$M_1 = A\left(\frac{\partial W_1}{\partial x} - \frac{\partial U_1}{\partial z}\right), \qquad \frac{\partial U_1}{\partial x} + \frac{\partial V_1}{\partial y} + \frac{\partial W_1}{\partial z} = 0 \quad (1)$$
$$N_1 = A\left(\frac{\partial U_1}{\partial y} - \frac{\partial V_1}{\partial x}\right)$$

From the existence of these forces, and from the principle of the conservation of energy, it may be, and has been, concluded that

---

[7][Hertz's calculation is given clearly, in modern vector notation, by D'Agostino (1975), pp. 284–296. On this calculation see also Whittaker (1951), vol. 1, pp. 319–321; Zatzkis (1965); Havas (1966); and Buchwald (1990). Whittaker's conclusion about the validity of this calculation is the following: "That Hertz's deduction is ingenious and interesting will readily be admitted. That it is conclusive may scarcely be claimed, for the argument of Helmholtz regarding the induction of currents is not altogether satisfactory, and Hertz in following his master, is on no surer ground." Whittaker (1951), vol. 1, p. 321.]

changes of u, v, and w produce electrical forces whose components X_1, Y_1, and Z_1 are

$$X_1 = -A^2 \frac{dU_1}{dt}, \quad Y_1 = -A^2 \frac{dV_1}{dt}, \quad Z_1 = -A^2 \frac{dW_1}{dt} \qquad (2)$$

These expressions hold good inside the conductors conveying the currents u, v, and w as well as for the space outside. The forces (2) have been deduced from the forces (1) on the assumption that the latter were due to electric currents. But on account of our premises we may affirm that even if the forces (1) are caused by any system whatever of variable currents and variable magnets, then their variation must equally give rise to the forces (2). Let A denote the system that produces the forces (1). We superimpose a system B, consisting only of electric currents, which still neutralizes the forces of system A everywhere. Such a system is possible; we need only choose as current-components u, v, and w, where

$$4\pi u = \nabla^2 U_1, \quad 4\pi v = \nabla^2 V_1, \quad 4\pi w = \nabla^2 W_1.$$

If we now move electric currents about under the action of both systems A and B, there is no work done in this motion. Hence the electromotive force necessary to maintain the currents must be independent of the motion, so that the induced electromotive force is zero. But the system B by itself exerts inductive actions; hence the system A must exert inductive actions equal and opposite to those of B, and therefore equal to those of a purely electrical system which exerts the same magnetic forces as A. What is true of inductive actions due to motion must also be true of those due to variations of intensity; both are most simply determined in terms of each other by the principle of the conservation of energy. Hence from the existence of magnetic forces of the form (1) we may directly infer the existence of electric forces of the form (2), whatever may be the origin of these magnetic forces.

Let us now consider magnetic currents. Let λ, μ, ν be the components of magnetization throughout space, and let

$$\frac{\partial \lambda}{\partial x} + \frac{\partial \mu}{\partial y} + \frac{\partial \nu}{\partial z} = 0, \quad \text{and} \quad \Lambda = \bar{\lambda}, \quad M = \bar{\mu}, \quad N = \bar{\nu}.$$

Fundamental Equations of Electromagnetism

These quantities are to be measured in absolute magnetic units. It follows from the forces (1) by the principle of the conservation of energy, and is indeed generally accepted in electromagnetics, that the electric force produced by variation of λ, μ, ν has for components[8]

$$X_1 = A\frac{d}{dt}\left(\frac{\partial N}{\partial y} - \frac{\partial M}{\partial z}\right), \quad Y_1 = A\frac{d}{dt}\left(\frac{\partial \Lambda}{\partial z} - \frac{\partial N}{\partial x}\right),$$

$$Z_1 = A\frac{d}{dt}\left(\frac{\partial M}{\partial x} - \frac{\partial \Lambda}{\partial y}\right).$$

We now put, in accordance with our notation,

$$p = \frac{d\lambda}{dt}, \quad q = \frac{d\mu}{dt}, \quad r = \frac{d\nu}{dt},$$

and call p, q, r the components of the magnetic current.

Further we put $P_1 = \bar{p}, Q_1 = \bar{q}, R_1 = \bar{r}$, and call P_1, Q_1, R_1 the components of the vector-potential of this current. Then the electric forces produced by the magnetic current are

$$X_1 = A\left(\frac{\partial R_1}{\partial y} - \frac{\partial Q_1}{\partial z}\right)$$
$$Y_1 = A\left(\frac{\partial P_1}{\partial z} - \frac{\partial R_1}{\partial x}\right), \qquad \frac{\partial P_1}{\partial x} + \frac{\partial Q_1}{\partial y} + \frac{\partial R_1}{\partial z} = 0 \qquad (3)$$
$$Z_1 = A\left(\frac{\partial Q_1}{\partial x} - \frac{\partial P_1}{\partial y}\right)$$

The reasoning which allows us to infer from the forces (1) that the mutual potential of two electric current-systems u_1, v_1, w_1, and u_2, v_2, w_2, has the form

$$A^2 \iint \frac{1}{r}(u_1 u_2 + v_1 v_2 + w_1 w_2) d\tau_1 d\tau_2,$$

leads to the conclusion, using forces (3), that the magnetic current-systems p_1, q_1, r_1, and p_2, q_2, r_2, have the mutual potential

---

[8] Cf. v. Helmholtz, "On the Theory of Electrodynamics, Part 1: On the Equations of Motion of Electricity in Conducting Bodies at Rest" [Helmholtz (1882–1895), vol. 1, pp. 545–628; on p. 619.]

$$A^2 \iint \frac{1}{r}(p_1 p_2 + q_1 q_2 + r_1 r_2) d\tau_1 d\tau_2.$$

The same considerations which led us from that potential of electric currents to the inductive forces (2) allow us, from the potential of magnetic currents, to infer the existence of induced magnetic forces of the form

$$L_1 = -A^2 \frac{dP_1}{dt}, \quad M_1 = -A^2 \frac{dQ_1}{dt}, \quad N_1 = -A^2 \frac{dR_1}{dt}, \tag{4}$$

Here also we may affirm that these forces act inside the magnetic bodies as well as in the space outside; and we easily convince ourselves that we cannot well confine the connection between the forces (3) and (4) to the case where the forces (3) are due to magnetic currents alone. We must conclude that when a system of currents or magnets gives rise to electrical forces of the form (3), then a variation of this system will give rise to magnetic forces of the form (4).

So far we have merely repeated in precise form the results of the preceding paragraph. We now go further and conclude that a system of variable currents exerts electric forces of the form (2). These may be represented in the form (3). Hence unless they are constant they will give rise to magnetic forces of form (4). And these must be added as a correction to the known magnetic forces of form (1). To arrive at the expression of the forces (2) in the form (3) we put

$$-A^2 \frac{dU_1}{dt} = A\left(\frac{\partial R}{\partial y} - \frac{\partial Q}{\partial z}\right),$$

$$-A^2 \frac{dV_1}{dt} = A\left(\frac{\partial P}{\partial z} - \frac{\partial R}{\partial x}\right),$$

$$-A^2 \frac{dW_1}{dt} = A\left(\frac{\partial Q}{\partial x} - \frac{\partial P}{\partial y}\right).$$

Assuming for the present that

$$\frac{\partial P}{\partial x} + \frac{\partial Q}{\partial y} + \frac{\partial R}{\partial z} = 0 \tag{a}$$

we get, by differentiating the second equation with respect to z, the third with respect to y, and subtracting the results,

Fundamental Equations of Electromagnetism

$$-A\frac{d}{dt}\left(\frac{\partial V_1}{\partial z} - \frac{\partial W_1}{\partial y}\right) = \nabla^2 P,$$

and thence

$$P = \frac{1}{4\pi} A \frac{d}{dt}\left(\frac{\partial}{\partial z}\overline{V}_1 - \frac{\partial}{\partial y}\overline{W}_1\right).$$

We get similar expressions for Q and R. It is easy to see that these satisfy the equation (a), and the assumption of the truth of this equation is justified.

From the values of P, Q, R follow the magnetic forces produced by their variation. The x-component is

$$-A^2 \frac{dP}{dt} = -\frac{1}{4\pi} A^3 \frac{d^2}{dt^2}\left(\frac{\partial}{\partial z}\overline{V}_1 - \frac{\partial}{\partial y}\overline{W}_1\right).$$

This term we must add to the component L_1 of the previously assumed magnetic force. Let us call the component thus corrected L_2 and form the similarly corrected components M_2, N_2: then these forces may be represented by the system of equations

$$L_2 = A\left(\frac{\partial V_2}{\partial z} - \frac{\partial W_2}{\partial y}\right),$$
$$M_2 = A\left(\frac{\partial W_2}{\partial x} - \frac{\partial U_2}{\partial z}\right), \qquad \frac{\partial U_2}{\partial x} + \frac{\partial V_2}{\partial y} + \frac{\partial W_2}{\partial z} = 0 \quad (5)$$
$$N_2 = A\left(\frac{\partial U_2}{\partial y} - \frac{\partial V_2}{\partial x}\right),$$

where we have put

$$U_2 = U_1 - \frac{1}{4\pi} A^2 \frac{d^2}{dt^2}\overline{U}_1, \quad V_2 = V_1 - \frac{1}{4\pi} A^2 \frac{d^2}{dt^2}\overline{V}_1,$$

$$W_2 = W_1 - \frac{1}{4\pi} A^2 \frac{d^2}{dt^2}\overline{W}_1,$$

From what precedes we may at once conclude that the electromotive forces produced by a variation of the current-system no longer have exactly the form (2), but have these corrected values

$$X_2 = -A^2 \frac{dU_2}{dt}, \quad Y_2 = -A^2 \frac{dV_2}{dt}, \quad Z_2 = -A^2 \frac{dW_2}{dt} \tag{6}$$

Exactly similar reasoning compels us to correct the actions of magnetic systems presented by the equations (3) and (4). The results may be represented by the following sample equations

$$X_2 = A\left(\frac{\partial R_2}{\partial y} - \frac{\partial Q_2}{\partial z}\right), \text{ etc.} \tag{7}$$

$$L_2 = -A^2 \frac{dP_2}{dt}, \text{ etc.} \tag{8}$$

where

$$P_2 = P_i - \frac{1}{4\pi} A^2 \frac{d^2}{dt^2} \overline{P}_1, \text{ etc.}$$

If we wish to represent the forces by which the corrected equations (5) and (7) differ from the usually accepted ones (1) and (3), as distinct from these latter in nature, we need only form a system of electric or magnetic currents in which the forces (1) or (3), as the case may be, are zero. Any endless electric or magnetic solenoid will serve as an example.

We see at once that we cannot regard the result obtained as final. Indeed we deduced forces (5) from forces (2); but now the forces (2) have been found inexact and have been replaced by the forces (6). Hence we must repeat our reasoning with these latter forces. The result is easily seen; we obtain it if we everywhere replace the index 2 by 3 and put

$$U_3 = U_1 - \frac{A^2}{4\pi}\frac{d^2}{dt^2}\overline{U}_2 = U_1 - \frac{A^2}{4\pi}\frac{d^2}{dt^2}\overline{U}_1 + \frac{A^4}{16\pi^2}\frac{d^4}{dt^4}\overline{\overline{U}}_1,$$

with similar expressions for the other components of the vector-potential.[9] The terms in A^5, which here appear in the expressions for the magnetic forces of electric currents, and the electric forces of magnetic currents, may be perceived apart from the terms of lower

---

[9] [Note that, in accordance with Hertz's statement in footnote 4, A is here the reciprocal of the velocity of light in vacuo.]

Fundamental Equations of Electromagnetism

order. We need only take an ordinary electric or magnetic solenoid, which may be called a solenoid of the first order, and roll it up into a solenoid which may be called a solenoid of the second order, in order to get a system in which the forces here calculated are the largest of those occurring. From the consideration of such solenoids we may demonstrate the existence of the separate terms, independently of the fact of our admitting or not admitting that they are simply added together to give the final result.

Our reasoning prevents us from stopping at any stage and constantly adds, as before, more and more terms, thus leading to an infinite series. To represent the final result we denote by L, M, N, X, Y, Z the completely corrected forces and obtain

$$L = A\left(\frac{\partial V}{\partial z} - \frac{\partial W}{\partial y}\right)$$
$$M = A\left(\frac{\partial W}{\partial x} - \frac{\partial U}{\partial z}\right) \quad (9)$$
$$L = A\left(\frac{\partial U}{\partial y} - \frac{\partial V}{\partial x}\right)$$

$$X = -A^2 \frac{dU}{dt}$$
$$Y = -A^2 \frac{dV}{dt} \quad (10)$$
$$Z = -A^2 \frac{dW}{dt}$$

where we now have for U, V, W

$$U = \bar{u} - \frac{A^2}{4\pi} \frac{d^2}{dt^2} \bar{\bar{u}} + \frac{A^4}{16\pi^2} \frac{d^4}{dt^4} \bar{\bar{\bar{u}}} - \ldots$$
$$V = \bar{v} - \frac{A^2}{4\pi} \frac{d^2}{dt^2} \bar{\bar{v}} + \frac{A^4}{16\pi^2} \frac{d^4}{dt^4} \bar{\bar{\bar{v}}} - \ldots$$
$$W = \bar{w} - \frac{A^2}{4\pi} \frac{d^2}{dt^2} \bar{\bar{w}} + \frac{A^4}{16\pi^2} \frac{d^4}{dt^4} \bar{\bar{\bar{w}}} - \ldots$$
$$\frac{\partial U}{\partial x} + \frac{\partial V}{\partial y} + \frac{\partial W}{\partial z} = 0.$$

Corresponding equations hold for the magnetic currents. If the series are convergent, there is no reason to doubt that they give us the true values. But in general they will converge. For let us consider that element of the integral U which is due to the current u in a certain element of space. We resolve this current into a series of simple harmonic functions of the time and suppose $u_0 \sin nt$ to be the term involving sin nt. Then the element of U due to this term will be given by the equation

$$dU = d\tau \frac{u_0 \sin nt}{r}\left(1 - \frac{1}{1.2}\frac{A^2}{4\pi}n^2 r^2 + \frac{1}{1.2.3.4}\frac{A^4}{16\pi^2}n^4 r^4 + \ldots\right).$$

This series converges to a limit easily found. If n and r are not very large, then every term after the first few will be infinitesmal compared with the preceding one. Hence also the integral of the elements of U will have a determinate value. Since the same is true of V, W, P, Q, and R, we may expect to find, in the equations (9) and (10) and the corresponding ones for magnetic currents, a system of forces in complete agreement with all our requirements.

3. It is obvious that this system may be represented, or in technical terms described, more simply than by the equations (9) and (10). By these equations we have

$$\nabla^2 U = -4\pi u + A^2 \frac{d^2}{dt^2}\overline{u} - \ldots$$

and

$$A^2 \frac{d^2 U}{dt^2} = A^2 \frac{d^2}{dt^2}\overline{u} - \ldots,$$

Hence

$$\nabla^2 U - A^2 \frac{d^2 U}{dt^2} = -4\pi u.$$

The other components of the vector-potentials, both of electric and magnetic currents, satisfy analogous differential equations. Since u, v, w, p, q, r vanish in empty space, the distribution of these potentials is there given by the equations

Fundamental Equations of Electromagnetism

$$\nabla^2 U - A^2 \frac{d^2 U}{dt^2} = 0, \qquad \nabla^2 P - A^2 \frac{d^2 P}{dt^2} = 0$$

$$\nabla^2 V - A^2 \frac{d^2 V}{dt^2} = 0, \qquad \nabla^2 Q - A^2 \frac{d^2 Q}{dt^2} = 0$$

$$\nabla^2 W - A^2 \frac{d^2 W}{dt^2} = 0, \qquad \nabla^2 R - A^2 \frac{d^2 R}{dt^2} = 0$$

$$\frac{\partial U}{\partial x} + \frac{\partial V}{\partial y} + \frac{\partial W}{\partial z} = 0, \qquad \frac{\partial P}{\partial x} + \frac{\partial Q}{\partial y} + \frac{\partial R}{\partial z} = 0,$$

(11)

The vector-potentials now show themselves to be quantities which are propagated with finite velocity—the velocity of light—and indeed according to the same laws as the vibrations of light and of radiant heat. Riemann in 1858 and Lorenz in 1867,[10] with a view to associating optical and electrical phenomena with one another, postulated the same or quite similar laws for the propagation of the potentials. These investigators recognized that these laws involve the addition of new terms to the forces which actually occur in electromagnetics; and they justify this by pointing out that these new terms are too small to be experimentally observable. But we see that the addition of these terms is far from needing any apology. Indeed their absence would necessarily involve a contradiction of principles which are quite generally accepted.

The vector-potentials of electric and magnetic currents have hitherto occurred as quite separate, and from them the electric and magnetic forces were deduced in an unsymmetric manner. This contrast between the two kinds of forces disappears as soon as we attempt to determine the propagation of these forces themselves, i.e.,

---

[10][In a brief paper, presented to the Göttingen Academy in 1858, but not published until 1867, Bernhard Riemann (1826–1866) discussed the process by which electric forces are propagated through space, and suggested that electric effects travel at the speed of light. Ludwig Lorenz (1829–1891) of Copenhagen in 1867 followed the procedure Riemann had suggested and modified the accepted formulas of electrodynamics to include terms that, although too minute to be observable in the laboratory, would explain the propagation of electromagnetic effects through space with a finite velocity. On this see Bernard Riemann, "Ein Beitrag zur Elektrodynamik," *Annalen* 131 (1867), 237–243; and Ludwig Lorenz, "Über die Identität der Schwingungen des Lichts mit den elektrischen Strömen," *Annalen* 131 (1867), 243–263.]

as soon as we eliminate the vector-potentials from these equations. This may be performed by differentiating equations (9) with respect to t and removing the differential coefficients of U, V, W with respect to t by equations (10). It may also be accomplished by differentiating equations (10) with respect to t, remembering that, e.g.,

$$A^2 \frac{d^2 U}{dt^2} = \nabla^2 U = \frac{\partial}{\partial y}\left(\frac{\partial U}{\partial y} - \frac{\partial V}{\partial x}\right) - \frac{\partial}{\partial z}\left(\frac{\partial W}{\partial x} - \frac{\partial U}{\partial z}\right),$$

and removing the functions of U, V, W in the brackets by means of equations (9). In this way we get six equations connecting together the values of L, M, N, X, Y, Z in empty space, namely, the following:

$$A\frac{dL}{dt} = \frac{\partial Z}{\partial y} - \frac{\partial Y}{\partial z}, \qquad A\frac{dX}{dt} = \frac{\partial M}{\partial z} - \frac{\partial N}{\partial y}$$

$$A\frac{dM}{dt} = \frac{\partial X}{\partial z} - \frac{\partial Z}{\partial x}, \qquad A\frac{dY}{dt} = \frac{\partial N}{\partial x} - \frac{\partial L}{\partial z} \qquad (12)$$

$$A\frac{dN}{dt} = \frac{\partial Y}{\partial x} - \frac{\partial X}{\partial y}, \qquad A\frac{dZ}{dt} = \frac{\partial L}{\partial y} - \frac{\partial M}{\partial x}$$

These same equations connect together the forces produced by magnetic currents, for they are obtained by eliminating P, Q, R as well as U, V, W. Hence they connect together the electric and magnetic forces in empty space quite generally, whatever the origin of these forces. The electric and magnetic forces are now interchangeable. If we eliminate first one set and then the other, we obtain the following system, which, however, does not completely represent the system (12):

$$\nabla^2 L - A^2 \frac{d^2 L}{dt^2} = 0, \qquad \nabla^2 X - A^2 \frac{d^2 X}{dt^2} = 0$$

$$\nabla^2 M - A^2 \frac{d^2 M}{dt^2} = 0, \qquad \nabla^2 Y - A^2 \frac{d^2 Y}{dt^2} = 0$$

$$\nabla^2 N - A^2 \frac{d^2 N}{dt^2} = 0, \qquad \nabla^2 Z - A^2 \frac{d^2 Z}{dt^2} = 0 \qquad (13)$$

$$\frac{\partial L}{\partial x} + \frac{\partial M}{\partial y} + \frac{\partial N}{\partial z} = 0, \qquad \frac{\partial X}{\partial x} + \frac{\partial Y}{\partial y} + \frac{\partial Z}{\partial z} = 0$$

Now the system of forces [fields] given by the equations (12) and (13) is just that given by Maxwell. Maxwell found it by considering the ether to be a dielectric in which a changing polarization produces the same effect as an electric current. We have reached it by means of other premises, generally accepted even by opponents of the Faraday-Maxwell view.[11] The equations (12) and (13) appear to us to be a necessary complement of the equations (1), (2), and (3), which are usually regarded as exact. From our point of view, the Faraday-Maxwell view does not furnish the basis of the system of equations (12) and (13), although it affords the simplest interpretation of them. In Maxwell's theory the equations (12) and (13) apply not merely to empty space but also to any other dielectric.[12] Starting from our premises we can also show that these laws hold in every homogeneous medium. We must assume the fact as experimentally demonstrated that the magnetic forces which surround a current-system placed in a homogeneous medium are distributed according to the equations (1) in the same way as in empty space. Hence we need only imagine the conductors and masses of iron which we have considered to be completely immersed in the given medium. In this medium we must define the units of electricity and magnetism in the same terms as in empty space. We must then determine the constant A, which gives the absolute value of the magnetic force produced by unit current in the new electrostatic measure. All further forces follow from the assumed experimental facts and the general premises; and since all the propositions are the same as those for empty space, the final result is the same. It is true that the value of the constant A will not be the same as in empty space, and that it will have different values in different media. Its reciprocal is always the velocity of propagation of electric and magnetic changes. It is an internal constant, and the only internal electromagnetic constant of the medium. The two constants of which it is generally built up, namely, the specific inductive capacity and the magnetic permeability, should in contrast to it be termed external constants. Not only the measurement, but even the definition of these

---

[11][Hertz here means most German physicists who still accepted the theories of Wilhelm Weber and Franz Neumann and paid little attention to Maxwell's electromagnetic theory.]

[12][A crucial feature of Hertz's theory is the equivalence of empty space to any other dielectric, since it differs from other dielectrics only in the values of its electric permittivity and magnetic permeability.]

latter constants, requires the specification of at least two media (one of which may be empty space).[13]

In what precedes I have attempted to demonstrate the truth of Maxwell's equations by starting from premises that are generally admitted in the opposing system of electromagnetics, and by using propositions which are familiar in it. Consequently I have made use of the conceptions of the latter system; but, excepting in this connection, the deduction given is in no sense to be regarded as a rigid proof that Maxwell's system is the only possible one. It does not seem possible to deduce such a proof from our premises. The exact may be deduced from the inexact as the most fitting from a given point of view, but never as the necessary.[14] I think, however, that from the preceding we may infer without error that if the choice rests only between the usual system of electromagnetics and Maxwell's, the latter is certainly to be preferred;[15] and that for the following reasons:

1. The system of the electromagnetic action of closed currents founded on direct action-at-a-distance is in its present state certainly incomplete. Either it must introduce different kinds of electric force, which it has never done, or it must admit the existence of actions

---

[13][In present-day terminology Hertz is saying that the dielectric constant of any material is the ratio of its electric permittivity to the permittivity of free space; and its relative magnetic permeability is the ratio of its magnetic permeability to the magnetic permeability of free space.]

[14]The mode in which we have deduced conclusions from the principle of the conservation of energy clearly marks at each stage the point at which our deductions are only the most fitting, but not the necessary ones. This mode is the most fitting from the standpoint of the usual system of electromagnetics, for it corresponds exactly to the accepted proposition in which Helmholtz in 1847 and Sir W. Thomson [Lord Kelvin] in 1848 deduced induction from electromagnetic action. But perhaps it may not be the only possible method; for just as in that proposition, so we have in ours made tacit assumptions besides the principle of conservation of energy. That proposition also is not valid if we admit the possibility that the motion of metals in the magnetic field may of itself generate heat; that the resistance of conductors may depend on that motion; and other such possibilities.

[15][This is Hertz's most important conclusion in this paper. Jungnickel and McCormmach have pointed out that Hertz's paper did not have any major impact on the development of electromagnetic theory in the years immediately following 1884. See Jungnickel and McCormmach (1986), vol. 2, pp. 46–47. But it certainly had a great influence on Hertz's approach to Maxwell's theory and to the experiments he performed to validate its predictions.]

Fundamental Equations of Electromagnetism

which hitherto it has not taken into account. Maxwell's system does not in the same way contain within itself the proof of its incompleteness.

2. When we attempt to complete the usual system of electromagnetics, we always arrive at laws that are very complicated and very difficult to handle. And either we refuse to admit the accumulated conclusions given in the second paragraph [of this article],[16] in which case we end up with an unfruitful declaration of incompetence; or, as from the standpoint of the system seems more reasonable, we accept them as being valid, and so arrive at forces which in fact are the same as those demanded by Maxwell's system. But then the latter offers by far the simplest exposition of the results.

3. The objections which may perhaps be raised to the further conclusions of the second paragraph do not apply to the reasoning especially exhibited in the first paragraph [of this article], which proved the attraction between magnetic currents. This latter depends directly on the premises: it stands or falls with them alone. But it is sufficient to show the superiority of Maxwell's system; for it is predicted in the latter, whereas it is unknown in the opposing system.[17]

---

[16][Here Hertz is primarily referring to the principle of the unity of electric force, which is the key element in his derivation of Maxwell's equations, as Jed Buchwald has pointed out. On this see Buchwald (1990), p. 287.]

[17]For further discussion of this paper, and of the reception it received from physicists, see pp. 21-23 in the Introductory Biography.

PAPER NO. 4

Review of Hertz's *Miscellaneous Papers* by George Francis FitzGerald

(G.F. Fitzgerald, "Hertz's *Miscellaneous Papers*," *Nature* 55 [Nov. 5, 1896], pp. 6–9.[1] Reprinted with permission from *Nature* 55. Copyright ©1896 Macmillan Magazines Limited.)

Anything written by Hertz is of interest; and these papers are of interest, not only on this account, but also on account of their suggestiveness. It is always a question as to the desirability of republishing and translating papers published some years ago. Most valuable papers of ten years' standing have produced their effect. Their vitality has been transmitted to and reproduced in subsequent work, but what the scientific world requires is advance rather than revision. The work of pioneers is, however, largely an exception to this rule. They are generally in advance of their times, and much of their work is of value long after it was done. Such a one was Hertz. Most of his papers are suggestive of questions which still require answers, and they all breathe a spirit that, as he says himself of Helmholtz's work, evokes "the same elevation and wonder as in beholding a pure work of art." His papers are not mere enumerations of observations, nor mathematical gymnastics. Each has a definite purpose and an artistic unity. A life-giving idea pervades it. It is no mere dry bones, but an organic whole that lives for a purpose, and does some work for science.

Prof. Lenard has earned much gratitude for his Introduction.[2] It gives a charming picture of Hertz, of his simplicity, his devotion to science, his loving regard for his parents. There is just enough added to the very well-selected letters to give the reader a continuous view of

---

[1][This review is reprinted in FitzGerald (1902), pp. 433–442. All page references given by FitzGerald in this review, and those of the editor, are to the English edition of *Miscellaneous Papers* published by Macmillan in London in 1896.]

[2][Paper No. 1 in the present collection. Lenard's Introduction and Planck's tribute to Hertz after his death (Paper No. 18) also discuss many of the same papers considered by FitzGerald in this review. FitzGerald is, however, more critical and penetrating than either Lenard or Planck in his reaction to Hertz's papers.]

Hertz's work, and enable him to follow its development, and hence feel an interest in it and sympathy with the worker, thus fulfilling the best ideal of the biographer.

One of Hertz's first investigations was on the kinetic energy of an electric current.[3] The question is still of great interest. It is known that the magnetic induction that accompanies an electric current behaves exactly as if it were a mass moving with inertia. This is the inertia of magnetic induction. Hertz was, however, looking for a different inertia. He looked at the subject from the flow of electricity point of view. He thought that there might be some phenomenon corresponding to an inertia of the electric charges, which upon this theory are supposed to be flowing in opposite directions through a conductor. He supposed that these might have some inertia *in addition* to the magnetic inertia which accompanied their motion. To test this he tried two different forms of experiment, and obtained results which showed that if there were inertia of this kind, it must be small compared with that of the magnetic kind. The first method consisted essentially in a careful comparison of the extra current in a conductor with its calculated value; the second consisted in observing whether anything like the action by which the trade winds are deflected from a due northerly and southerly flow by the rotation of the earth, could be observed in a rotating conductor when traversed by an electric current. That there is some directed inertia in the conductor when traversed by an electric current is very probable, and in some cases we can be sure it exists. Hertz himself remarks that the inertia of the motion of the ions in electrolysis is considerably greater than what he was looking for in a metallic conductor. He could not make sufficiently delicate experiments with his apparatus to detect it, however, when using the small densities of current that were available in liquids; but the question is of great interest, and deserves further investigation. There can be no doubt that in gaseous discharges, cathode rays, as well as in electrolysis of liquids, there is a directed motion of matter accompanying the electric current which would be of the nature of the inertia Hertz was looking for, but failed to find. There seems much reason for thinking that in metallic conductors some similar actions are also taking place.

---

[3][This is the paper for which Hertz received a prize just a few months after commencing his studies in Berlin. It is in *Misc.Pprs.*, pp. 1–34.]

Besides all this, there is the question as to how far the theory that all electricity is molecular and consists of electrons, involves the supposition of an inertia of this kind. Is the inertia of an electron completely specified by the magnetic force [field] accompanying it? Does it occupy no space itself, and is its external field its all? We are hardly in a position to answer such questions. We might, however, be able to answer the former question as to the inertia of the directed matter movements accompanying the current, and as to another interesting question of a similar kind, namely, as to how far we can legitimately assume the current inside a conductor to be absolutely homogeneous. The self-induction of a single wire of a square mm in section is not exactly the same as that of, say, a hundred wires each of the thousandth of a square mm in section, and distributed over the square mm. Subdividing the current would increase its self-induction. Outside the wire the distribution of magnetic force would be practically the same as before, but inside we would have it concentrated into a hundred small wires, instead of being uniformly distributed, and the effect of this would be to slightly increase the self-induction, and the more so the smaller the section of each wire into which the square mm were subdivided. Hence we conclude that if the current in a real wire be from molecule to molecule, and so be concentrated on certain lines, its inertia should be somewhat greater than that calculated from the hypothesis that it is uniformly distributed over the section of the conductor. The difference between these two views is most clear when we consider the case of a Leyden [jar] discharging by its insulating medium becoming a conductor. If the Leyden [jar] be completely closed, and the medium become a conductor in such a way that the strain in each cubic cm is there destroyed by conductivity, there will be no magnetic force anywhere accompanying this discharge of the Leyden, and consequently no magnetic inertia, if the conduction be perfectly homogeneous. Now it seems almost impossible that any directed change can take place without some accompanying inertia, and we may consequently conclude that either (a) an electric current has inertia such as Hertz was looking for, or (b) electric conduction currents are essentially heterogeneous, or (c) electric conduction is essentially accompanied by material inertia, or (d) two or all three of these are true. That (c) certainly exists in this case is incontestable in view of the known directed strains that Kerr[4] and Duter have proved to exist in matter subject to electric stress. What is the complete answer, is the important question. It is still unsolved. It lies at the foundation of every theory of electric conduction. Hertz attacked it. It is still waiting solution.

The papers on the contact of elastic spheres and on hardness[5] are most valuable contributions to the subject. They place the question of hardness on a scientific basis, and lay the foundation for a quantitative study of this most variable property of matter. There is no quality in which different materials differ more than in hardness. Electric conductivity is perhaps as various as hardness, compressibility, and viscosity, but hardly any other quality of matter is at all comparable with these in variety of range. Of these hardness is one of the most important and least known and, since Hertz's work on it, can be scientifically studied. Innumerable subsidiary questions arise in connection with it. Why are some bodies so easily polished? Is the polishing of marble connected with the ease with which crystals of calcspar can be twinned by pushing over one corner? What is the essential difference between polishing and grinding? What is the effect of impurities on hardness? Is it comparable to their effect on electric conductivity? What is the cause of this effect?

In considering the cracking of a material like glass, Hertz seems to think that its cracking will depend only on the tension; that it will crack where the tension exceeds a certain limit. He does not seem to consider whether it might not crack by shearing with hardly any tension. It is doubtful whether a material in which there were sufficient general compression to prevent any tensions at all, would crack. Rocks seem capable of being bent about and distorted to almost any extent without cracking, and this might very well be expected if they were at a sufficient depth under other rocks to prevent their parts being under tension. It is an interesting question whether a piece of glass could be bent without breaking if it were strained at the bottom of a sufficiently

---

[4] [The reference here is to the work of the Scottish physicist, John Kerr (1824-1907). In 1875 he discovered the Kerr Electro-optic Effect, the appearance of birefringence in certain isotropic substances when placed in a strong electric field. This led to the development of the Kerr cell, which is still used today in physics laboratories to chop a light beam at extremely high frequencies.]

[5] [*Misc.Pprs.*, pp. 146-183.]

deep ocean. On the other hand, there seems very little doubt that the parts of a body might slide past one another under the action of a shear, and would certainly crack unless there were a sufficiently great compressive stress to prevent the crack; and that consequently a body might crack, even though the tensions were not by themselves sufficiently great to cause separation, and might crack where the shear was greatest, and not where the tensions were greatest.

Then follow some papers on hygrometry and evaporation.[6] A very interesting point is raised in this latter connection. Can a liquid evaporate at an unlimited rate if the vapor produced is removed as rapidly as it is evolved? From two points of view Hertz shows that there is a limit, and by his experiments went far to show that there was no other cause limiting the rate of evaporation. The first point of view was that a limit is imposed by the difficulty of supplying heat sufficiently rapidly to keep the surface temperature constant. He does not seem in his experiments to have attempted to supply this by radiation, but was content to allow the liquid to supply itself by conduction and convection from below. The second point of view was that the molecules could not leave the surface faster than they would be moving in the vapor that was formed. Hertz's investigation of this case only assumes an average velocity for the molecules; he does not consider the distribution of velocity among the molecules, nor whether they escape equally easily in all directions. The experimental investigation of the conditions of evaporation is extremely difficult; and until some more satisfactory method of studying these conditions be invented, this rough approximation seems to be sufficient to explain the observations. It might be interesting to see whether there was any difference between the superficial friction of a gas and a liquid which did not evaporate, and of a vapor in contact with its own liquid. In one case there would be no exchange of molecules between the two bodies that were sliding past one another, while in the second case there would be an exchange. A study of the conduction of heat between a gas and a liquid might also help to elucidate the nature of the exchange which takes place between a liquid and its vapor.

In the paper on the vapor pressure of mercury,[7] there are some very rough approximations which are hardly sufficiently accurate for

---

[6][*Misc.Pprs.*, pp. 184–199.]

[7][*Misc. Pprs.*, pp. 200–206.]

general application. One is as to the extent to which a saturated vapor obeys the laws of a perfect gas. Hertz assumes that this is more nearly true the lower the temperature. This is not generally so. For each kind of material there is a particular temperature at which its saturated vapor most nearly obeys these laws, and below as well as above this temperature it departs from these laws. Again, there is a process, described at the bottom of p. 203 and top of p. 204, which cannot possibly be carried out. He says: "Let a quantity of liquid at temperature T be brought to any other temperature. At this temperature it is converted into vapor without external work." This is absolutely impossible, and the equation he deduces from all this is not true, though it is sometimes a rough approximation to the truth.

There is a very interesting paper on the floating of bodies by thin sheets of rigid material like ice.[8] Hertz shows that if the sheet be large enough it would be possible to cause a thin sheet of iron, which by itself would sink, to float by placing weights at its center. The weights might so depress the center and make the sheet so boat-shaped as to float both themselves and it.

In 1883 Hertz published a deduction from first principles of Maxwell's equations for the electromagnetic field in the symmetrical form,[9] afterwards used by himself in his investigations on oscillatory discharge waves. He applies the very same arguments by which Helmholtz, Lord Kelvin, and others had argued, from the work done by one electric current on another, that there must be a corresponding reaction of the second on the first current, and hence deduced electromagnetic induction. Hertz applies this argument to the case of a ring magnet changing in strength and producing magnetic force on another ring magnet in the neighborhood, and doing work there, and shows thereby that there should be a magnetic force due to a changing electric field exactly corresponding to the electric force due to a changing magnetic field. This, of course, is what Maxwell describes as the magnetic effect of the changing electric displacement, and its effects are expressed by the very same equations as Maxwell deduces. The argument is no more and no less conclusive than in the corresponding application of the principle of the conservation of

---

[8] [*Misc.Pprs.*, pp. 266–272.]

[9] [*Misc.Pprs.*, pp. 273–290; Paper No. 3 in this book. The publication date of this paper is actually 1884, not 1883 as given by FitzGerald.]

energy to deduce ordinary electromagnetic induction. Hertz is careful to point this out, for he was early imbued by Helmholtz with the fact that the principle of the conservation of energy is by itself utterly inadequate as a complete explanation of physical phenomena. He specially mentions himself Helmholtz's interest in this problem of the simplest basis for dynamics, and Hertz's last great work was to place general dynamics on a sound basis.[10] The simplest of all cases is the easiest in which to see how the principle of the conservation of energy fails to give a complete solution. A body moving without any action from other bodies describes a right [straight] line at a constant velocity. The principle of the conservation of energy requires the constant velocity. But, why the right [straight] line? Conservation of energy cannot solve even the simplest of all examples. It would be well if some modern chemists would mark, learn, and inwardly digest this.[11]

The part of Hertz's work which is of greatest interest just now is that in connection with cathode rays.[12] He began with some very interesting observations on the aura accompanying spark discharges. It appeared to be projected from the positive electrode, and occasionally formed a vortex ring of incandescent gas, which lasted for an appreciable time between the electrodes of a [Leyden] jar discharging in air. Goldstein[13] had noticed similar effects; and some recent experiments on the discharge of large Leyden batteries, in which some of the phenomena of globular lightning seem to have been reproduced, make it appear possible that this latter is a spherical vortex of incandescent air.

Hertz's study of cathode rays, in 1883,[14] set finally at rest two questions. In the first place he showed that the discharge in a gas may

---

[10][FitzGerald is referring here to Hertz's last work: *The Principles of Mechanics Presented in a New Form.*]

[11][FitzGerald's reference here is presumably to Friedrich Ostwald and his school of "Energetics," which was fiercely debated in Germany at the time this was written. As an example, there was a "stiff fight" involving Ostwald, Boltzmann, Nernst, Felix Klein, and Sommerfeld at the 1895 meeting of the German Association of Natural Scientists and Physicians in Lübeck. On this see Jungnickel and McCormmach (1986), vol. 2, pp. 219–222.]

[12][*Misc.Pprs.*, pp. 216–223]

[13][Eugen Goldstein (1850–1930), a fellow student and colleague of Hertz in Berlin.]

[14][*Misc.Pprs.*, pp. 224–254.]

be as continuous as any other form of current. In no case are we absolutely certain that the current is absolutely continuous. On the large scale it certainly is; but all we know of electrolysis seems to show that on a sufficiently small scale the current is carried in detachments, and is consequently essentially discontinuous. This, however, was not the question at issue, and so far as a continuity of the same kind as that in any liquid electrolyte is concerned, Hertz showed that the discharge through a gas might be equally continuous. The second question he decided was as to the direction of flow of the average current in an exhausted space. He showed that the average flow at any point was nearly the same as if the whole space were a conductor: that there was no connection between the cathode rays and the flow of the current. From experiments on cathode rays projected down a tube, and quite away from both electrodes he deduced that they produce no magnetic action outside the tube, although they are deflected by the magnet. His conclusion, that the cathode rays are not streams of electrified particles, was largely founded on this, and on another experiment on the action of electrostatic force on the particles. This experiment on the magnetic action of cathode rays is quite inconclusive, and it is very remarkable that Hertz should have attributed much importance to it.[15] Whatever current was carried down his tube by the cathode ray[s] must have come back the tube by the surrounding gas, and these two opposite currents should have produced no magnetic force outside the tube; and this is exactly what Hertz observed. In a similar way, what he observed in the case of a flat box was the average direction of the current, and he showed that this average direction was approximately the same as in a conducting sheet. This proved that if there were any concentration of the current along the direction of the cathode rays, this concentration was neutralized by a corresponding return current, so that the average current was as described. At the same time there does not seem much doubt but that the cathode rays only carry a very small part of the current.

---

[15][Helmholtz also attributed great importance to Hertz's results. The high regard German physicists had for both Hertz and Helmholtz retarded the correct understanding of the nature of cathode rays in Germany for years. On this see the letter of Helmholtz to Hertz under date of July 29, 1883 in Paper No. 1. Hertz's reply to Helmholtz may be found in Koenigsberger (1902–1903), vol. 2, pp. 306–309.]

The third part of the paper is concerned with the electrostatic effects due to cathode rays. The experiments do not seem to fully justify the conclusion drawn, that cathode rays cannot be charged molecules. Sufficient account does not seem to have been taken of the shielding action of the conducting gas surrounding the cathode ray[s], nor of the way in which the potential is distributed between two electrodes in a gas. Hertz describes an experiment with two plates inside the tube kept at a considerable difference of potential. He says: "The phosphorescent image of the Ruhmkorff coil discharge appeared somewhat distorted through deflection in the neighborhood of the negative strip; but the part of the shadow in the middle between the two strips was not visibly displaced." Now this is exactly what one might expect, because the fall of potential between two such strips is very small indeed, except close to the negative strip, and there the electric force *did* deflect the rays. Hence the conclusion is just the reverse of the one Hertz gives. From the experiment it appears that cathode rays do behave like electrified particles.[16] It is very remarkable that in all these investigations Hertz does not once even mention, as a thing to be explained, the repulsive actions which Crookes observed, and which have been almost universally attributed to the impact of gas particles.[17]

---

[16][This was the opinion of most British physicists at the time, but was resisted by the Germans, who thought that cathode rays must be waves or some other kind of phenomenon in the ether.]

[17][William Crookes (1832–1919) believed that cathode rays were molecules which became negatively charged by striking the cathode of a discharge-tube, and were then violently repelled from the negatively-charged cathode. These experiments were performed by Crookes in the late 1870s in England. His attempts to improve the radiometer, which he had invented, led Crookes to improve the vacuum in a Geissler tube by a factor of 75,000. With such a high-vacuum tube (often called a Crookes tube), he was able to show that cathode rays travelled in straight lines from the cathode, cast sharp shadows, and could turn a small wheel when they struck the blades of the wheel. He also showed that cathode rays could be deflected by a magnetic field. He therefore argued strongly that cathode rays were charged particles and not electromagnetic radiation or other phenomena in the ether. Two decades later J.J. Thomson proved convincingly that Crookes was right, and Hertz wrong, by measuring the charge and mass of electrons, the particles now known to make up the cathode-ray beam.]

The other important paper, on the transmission of cathode rays through thin metallic films,[18] is particularly interesting as the starting-point for Lenard's work, which has resulted in the discovery of the x-rays. A good deal of what Hertz observed would be accounted for by the production of x-rays where the cathode rays meet the diaphragms, and by the reproduction of cathode rays mixed with x-rays on the other side of the diaphragm, which would thus act as a sort of local electrode.[19] That something exists in a vacuum on the far side of such a thin film, which does not ordinarily exist in x-rays in air, seems conclusively proved by there being something there which can be deflected by a magnet. There seems no doubt that cathode rays themselves are quite invisible, and that it is only where they are interfered with by gaseous molecules or by phosphorescent solids that they are sources of light. This is very much what one would expect. An electrified atom would not in general be a source of light unless its free movement were interfered with by impact.

The concluding article, on his master Helmholtz's seventieth birthday,[20] is a noble and generous tribute to that great teacher's abilities and character. How truly he portrays the important characteristics of a University Professor! "It is true that Helmholtz never had the reputation of being a brilliant university teacher, as far as this depends upon communicating elementary facts to the beginners who usually fill the lecture-rooms. But it is quite another matter when we come to consider his influence on trained students, and his pre-eminent fitness for guiding them in original research." The most important duty of a University is to increase the knowledge of mankind, and to train up a new generation who may be able to continue the good work. It is thus mankind has advanced since the dawn of civilization in Egypt. He who produced the most enthusiastic

---

[18][Paper No. 13.]

[19][It has been said, with some justice, that Hertz and Lenard were kept from the discovery of x-rays because they did not understand the nature of cathode rays.]

[20][Paper No. 2. It is worth noting that the one important paper on which FitzGerald does not comment in his review of *Miscellaneous Papers* is the address Hertz gave in 1889 "On the Relation between Light and Electricity" (Paper No. 10). The reason for FitzGerald's neglect was probably that the paper was a popular one and too well-known to the readers of *Nature* to need further elaboration or criticism.]

disciples has most advanced the well-being and the well-living of the race.

PART IV

Karlsruhe (1885–1889): Professor of Physics, *Technische Hochschule*

> Optics is no longer restricted to minute ether-waves a small fraction of a millimeter in length; its dominion is extended to waves which are measured in decimeters, meters, and kilometers.
>
> *Heinrich Hertz (1889)*

Apparatus that Heinrich Hertz used to study the polarization and refraction of electromagnetic waves. The wire frame, almost 2 m high and strung with parallel wires, acted as an analyzer for the polarized radiation produced by Hertz's spark transmitter. The 30° prism, when filled with hard pitch, weighed more than half a ton. Courtesy of Deutsches Museum, Munich.

Introduction

The four years Hertz spent as Professor of Physics at the Karlsruhe *Technische Hochschule* were the climax of his scientific career. He came to Karlsruhe with the desire to test experimentally the validity of Maxwell's electromagnetic theory—something he had been unable to do in Kiel for lack of adequate laboratory space and essential equipment. His first year in Karlsruhe was devoted to adjusting to his new duties as a professor, to a study of Maxwell's papers, and to a careful consideration of how best to test Maxwell's theory experimentally. In the years 1886–1888 Hertz's pent-up intellectual energy finally burst forth in the form of nine research articles on electric waves, which revolutionized physics and established Hertz's undying reputation. It was this research that persuaded physicists of the superiority of Maxwell's theory over the older theories of W. Weber, F. Neumann and even of H. von Helmholtz, Hertz's mentor and inspiration. It also encouraged others to repeat his experiments, modify and improve them, and eventually bring them to a point where Marconi could capitalize on their potential for world-wide wireless communications.

In this section the first paper chosen for inclusion is Hertz's introduction to the 1893 German edition of his *Untersuchungen über die Ausbreitung der elektrischen Kraft*, a collection of all his experimental papers on electromagnetic radiation and a number of his theoretical papers on the same subject. Since this introduction was written five years after he had completed his research at Karlsruhe, it provides a fine summary of all his experimental work on electromagnetic radiation and includes some later developments in the form of supplementary notes to the original papers.

Four of the most significant of Hertz's nine papers follow: those that relate how he was able to produce electric waves in air with wavelengths of a few meters; how he employed a spark-micrometer as a detector for these waves; how he was able to measure their wavelength by setting up standing waves in space; and how he was able to show that waves of 66-cm wavelength exhibited all the fundamental properties of light waves: reflection, refraction, diffraction, interference and polarization. Thus he was finally able to conclude that electromagnetic waves with wavelengths a million times longer than light waves obey the same laws that had long since been known for light and that Maxwell's electromagnetic theory was equally valid for both.

One of the four Karlsruhe papers chosen for inclusion here is somewhat peripheral to Hertz's research plan; this is the first paper ever written on the photoelectric effect (Paper No. 7). Hertz hit upon the phenomenon by accident but immediately realized its importance and dedicated six months of research time to a thorough investigation of the interaction of light with electricity.

This section also includes three other articles and addresses relating to Hertz's work on electric waves. Number 10 is his own address, "On the Relations between Light and Electricity," delivered in Heidelberg in 1889 to the annual conference of German scientists and medical doctors. This is a popular, but profound, account of his research in Karlsruhe, presented in a wonderfully clear and interesting fashion. It is an exemplar of what a scientific talk to a mixed group of experts and novices should be. A brief review by George Francis FitzGerald (No. 11) of the original German edition of *Electric Waves* follows, and finally Helmholtz's 1889 speech nominating Hertz as a member of the Berlin Academy of Sciences (No. 12), a speech that stresses Hertz's work on electric waves as the primary reason that he merited inclusion in the Academy.

These eight articles provide a valuable account of Hertz's scientific accomplishments during his years in Karlsruhe and of the impact his research had on physics at the end of the nineteenth century.[1]

---

[1] Additional information on Hertz's research in Karlsruhe may be found in Part IV of the Introductory Biography.

PAPER NO. 5

Introduction to *Electric Waves, Being Researches on the Propagation of Electric Action with Finite Velocity through Space* by Heinrich Hertz[1]

A. Experimental

I have often been asked how I was first led to carry out the experiments which are described in the following pages. The general inducement was this. In the year 1879 the Berlin Academy of Science had offered a prize for research on the following problem: to establish experimentally any relation between electromagnetic forces and the dielectric polarization of insulators—that is to say, either an electromagnetic force exerted by polarizations in non-conductors, or the polarization of a non-conductor as an effect of electromagnetic induction. As I was at that time engaged upon electromagnetic researches at the Physical Institute in Berlin, Herr von Helmholtz drew my attention to this problem, and promised that I should have the assistance of the Institute in case I decided to take up the work.[2] I reflected on the problem, and considered what results might be expected under favorable conditions by using the oscillations of Leyden jars or of open induction coils. The conclusion at which I arrived was certainly not what I had wished for; it appeared that any decided effect could scarcely be hoped for, but only an action lying just within the limits of observation. I therefore gave up the idea of working at the problem; nor am I aware that it has been attacked by anybody else. But in spite of having abandoned the solution at that time, I still felt ambitious to discover it by some other method; and my interest in everything connected with electric oscillations had become keener. It was scarcely possible that I should overlook any new form of such

---

[1][*El. Waves*, pp. 1–28. The German title of this paper is actually *Einleitende Übersicht*, i.e., "Introductory Survey."]

[2][There is a good discussion of Helmholtz's influence on Hertz's work on electromagnetic waves, and of the present paper, in D'Agostino (1971). See also the section "Helmholtz and Hertz's Researches on Electromagnetism" at the end of Part IV of the Introductory Biography.]

oscillations, in case a happy chance should bring such within my notice.

Such a chance occurred to me in the spring of 1886, and brought with it the special inducement to take up the following researches. In the collection of physical instruments at the *Technische Hochschule* in Karlsruhe (where these researches were carried out),[3] I had found and used for lecture purposes a pair of so-called Riess or Knochenhauer spirals. I had been surprised to find that it was not necessary to discharge large batteries through one of these spirals in order to obtain sparks in the other; that small Leyden jars amply sufficed for this purpose, and that even the discharge of a small induction coil would do, provided it had to spring across a spark-gap. In altering the conditions I came upon the phenomenon of side-sparks which formed the starting-point of the following research. At first I thought the electrical disturbances would be too turbulent and irregular to be of any further use; but when I had discovered the existence of a neutral point in the middle of a side-conductor, and therefore of a clear and orderly phenomenon, I felt convinced that the problem of the Berlin Academy was now capable of solution. My ambition at the time did not go further than this. My conviction was naturally strengthened by finding that the oscillations with which I had to deal were regular. The first of the papers here republished ("On Very Rapid Electric Oscillations")[4] gives, mostly in the actual time order, the course of the investigation as far as it was carried out up to the end of the year 1886 and the beginning of 1887.

While this paper was in press I learned that its contents were not as new as I had believed them to be. The Geographical Congress of April, 1887, brought Herr W. von Bezold to Karlsruhe and into my laboratory.[5] I spoke to him about my experiments; he replied that years ago he had observed similar phenomena, and he drew my attention to his "Researches on the Electric Discharge," in vol. 140 of Poggendorff's *Annalen*. This paper had entirely escaped me, inasmuch as its external appearance seemed to indicate that it related to matters

---

[3] [A good discussion of Hertz's experiments on electromagnetic waves in Karlsruhe may be found in Friedburg (1988).]

[4] [Paper No. 6. In the title of this paper, "rapid" really means "high-frequency."]

[5] [Wilhelm von Bezold (1837–1907), a well-known meteorologist at the time Hertz wrote, had formerly been a teacher of Hertz in Munich.]

quite other than electric oscillations, namely to Lichtenberg figures;[6] indeed, it does not appear to have attracted such attention as the importance of its contents merited. In an appendix to my paper I acknowledged Herr von Bezold's prior claim to a whole series of observations. In place of this appendix, I here, with Herr von Bezold's kind consent, include as the second of these papers that part of his communication which is of the most immediate interest in the present connection. It may now well be asked with surprise how it was possible that results so important and so definitely stated should have exercised no greater influence upon the progress of science? Perhaps the fact that Herr von Bezold described his communication as a preliminary one may have something to do with this.

I may here be permitted to record the good work done by two English colleagues who at the same time as myself were striving towards the same end. In the same year in which I carried out the above research, Professor Oliver Lodge,[7] in Liverpool, investigated the theory of the lightning-rod, and in connection with this carried out a series of experiments on the discharge of small condensers that led him on to the observation of oscillations and waves on wires. In as much as he entirely accepted Maxwell's views, and eagerly strove to verify them, there can scarcely be any doubt that if I had not anticipated him he would also have succeeded in observing waves in air, and thus also in proving the propagation in time of electric force. Professor FitzGerald,[8] in Dublin, had some years before endeavored to

---

[6]["Lichtenberg figures" were much used in the nineteenth century to investigate the distribution of static charges on solid surfaces by means of certain colored powders that were preferentially attracted to the positive and negative charges on the surface. The patterns produced were called "Lichtenberg figures," and the process used to produce them employs the basic principle of modern xerographic copying. Georg Lichtenberg (1742–1799) was Professor of Physics at Göttingen in the years 1775 to 1799. He was also well known as a satirical writer, ridiculing the romantic and metaphysical excesses of the time.]

[7][Oliver Lodge (1851–1940), one of the British "Maxwellians," who later made important contributions to the field of wireless-telegraphy (radio).]

[8][George Francis FitzGerald (1857–1901), another "Maxwellian," was born in the same year as Hertz, and died seven years after Hertz, at the age of only 44. He is best known for his suggestion of the "Lorentz-FitzGerald contraction" as an explanation of the negative results of the Michelson-Morley experiment. On FitzGerald, Lodge and the other Maxwellians, see Hunt (1991).]

predict, with the aid of theory, the possibility of such waves, and to discover the conditions for producing them. My own experiments were not influenced by the researches of these physicists, for I only knew of them subsequently. Nor, indeed, do I believe that it would have been possible to arrive at a knowledge of these phenomena by the aid of theory alone. For their appearance upon the scene of our experiments depends not only upon their theoretical possibility, but also upon a special and surprising property of the electric spark which could not have been foreseen by any theory.

By means of the experiments already mentioned I had succeeded in obtaining a method of exciting more rapid electric disturbances than were hitherto at the disposal of physicists. But before I could proceed to apply this method to the examination of the behavior of insulators, I had to finish with another investigation. Soon after starting the experiments I had been struck by a noteworthy reciprocal action between simultaneous electrical sparks. I had no intention of allowing this phenomenon to distract my attention from the main object that I had in view; but it occurred in such a definite and perplexing way that I could not altogether neglect it. For some time, indeed, I was in doubt whether I had not before me an altogether new form of electrical action-at-a-distance. The supposition that the action was due to light seemed to be excluded by the fact that glass plates cut it off; and naturally it was some time before I came to experiments with plates of rock-crystal. As soon as I knew for certain that I was only dealing with an effect of ultraviolet light, I put aside this investigation so as to direct my attention once more to the main question. Inasmuch as a certain acquaintance with the phenomenon is required in investigating the oscillations, I have reprinted the communication relating to it ("On an Effect of Ultraviolet Light upon the Electric Discharge") as the fourth of these papers.[9] A number of investigators, more especially Herren Righi, Hallwachs, and Elster and Geitel, have helped to make our knowledge of the phenomenon more accurate; nevertheless, its mechanism has not yet been completely disclosed to our understanding.

---

[9][This paper of Hertz, which records the first observation of the photoelectric effect, appears here as Paper No. 7. Hertz downplays too much its significance and the effort involved in this research. He devoted a solid six months of intense effort to it before he felt comfortable in handing it over to other investigators so that he could get back to his work on electric waves.]

Hertz's Introduction to Electric Waves

The summer of 1887 was spent in fruitless endeavors to establish the electromagnetic influence of insulators by the aid of the new class of oscillations. The simplest method consisted in determining the effect of dielectrics upon the position of the neutral point of a side-circuit. But in that case I should have had to include the electrostatic forces in the bargain, whereas the problem consisted precisely in investigating the electromagnetic induction alone. The plan which I adopted was the following: The primary conductor[10] had the form shown in Fig. 1; between the plates A and A' at its ends was introduced a block BB of sulfur or paraffin, and this was then quickly removed. I placed the secondary conductor C in the same position with respect to the primary as before (the only position which I had taken into consideration), and expected that when the block was in place very strong sparks would appear in the secondary, and that when the block was removed there would only be feeble sparks. This latter expectation was based upon the supposition that the electrostatic forces could in no case induce a spark in the almost closed circuit C, for since these forces have a potential, it follows that their integral

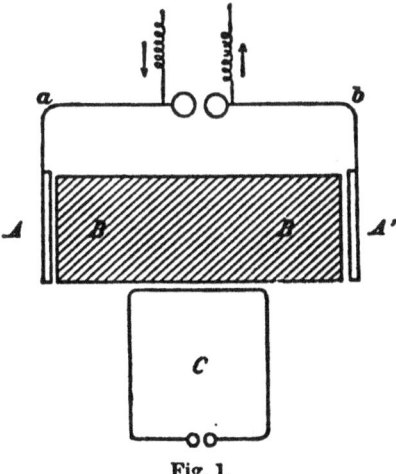

Fig. 1.

---

[10] The reader is assumed to be already acquainted with the papers referred to.

over a nearly closed circuit is vanishingly small. Thus in the absence of the insulator we should only have to consider the inductive effect of the more distance wire ab. The experiment was frustrated by the invariable occurrence of strong sparking in the secondary conductor, so that the moderate strengthening or weakening effect which the insulator must exert did not make itself felt. It only gradually became clear to me that the law which I had assumed as the basis of my experiment did not apply here; that on account of the rapidity [i.e., high frequency] of the disturbance, even forces that possess a potential are able to induce sparks in the nearly closed conductor; and, in general, that the greatest care has to be observed in applying here the general ideas and laws that form the basis for the usual electrical theories. These laws all relate to statical or stationary states; whereas here I had truly before me a variable state. I perceived that I had in a sense attacked the problem too directly. There was yet an infinite number of other positions of the secondary with respect to the primary conductor, and among these there might well be some more favorable for my purpose. These various positions had first to be examined. Thus I came to discover the phenomena which are described in the fifth paper ("On the Action of a Rectilinear Electric Oscillation upon a Neighboring Circuit"),[11] and which surprised me by their variety and regularity. The finding out and unravelling of these extremely orderly phenomena gave me peculiar pleasure. The paper certainly does not include all the discoverable details; whoever may extend the experiments to various other forms of conductor will find that the task is not an ungrateful one. The observations at greater distances are also probably very inaccurate, for they are affected by the disturbing influence of reflections which were not at that time suspected. What especially surprised me was the continual increase of the distance up to which I could perceive the action; up to that time the common view was that electric forces decreased according to the Newtonian law [i.e., as the inverse square of the distance], and therefore rapidly tended to zero as the distance increased.

Now during the course of this investigation I had made sure of other positions of the secondary conductor in which it was possible, by bringing an insulator near, to cause the appearance and disappearance of sparks, instead of simply altering their size. The problem which I

---

[11][*El.Waves*, pp. 80–94.]

was investigating was now solved directly in the manner described in the sixth paper ("On Electromagnetic Effects produced by Electrical Disturbances in Insulators").[12] On 10 Nov. 1887 I was able to report the successful issue of the work to the Berlin Academy.

The particular problem of the Academy which had been my guide thus far was evidently proposed at the time by Herr von Helmholtz in the following connection: If we start from the electromagnetic laws which in 1879 enjoyed universal recognition, and make certain further assumptions, we arrive at the equations of Maxwell's theory, which at that time [in Germany] were by no means universally recognized. These assumptions are: first, that changes of dielectric polarization in non-conductors produce the same electromagnetic forces as do the currents which are equivalent to them; secondly, that electromagnetic forces as well as electrostatic are able to produce dielectric polarizations; thirdly, that in all these respects air and empty space behave like all other dielectrics. In the latter part of his paper, "On the Equations of Motion of Electricity for Conducting Bodies at Rest,"[13] von Helmholtz has deduced Maxwell's equations from the older views and from hypotheses which are equivalent to those just stated. The problem of proving all three hypotheses, and thereby establishing the correctness of the whole of Maxwell's theory, appeared to be an unreasonable demand; the Academy, therefore, contented itself with requiring a confirmation of one of the first two.

The first assumption was now shown to be correct. I thought for some time of attacking the second. To test it appeared by no means impossible; and for this purpose I cast closed rings of paraffin. But while I was at work it struck me that the center of interest in the new theory did not lie in the consequences of the first two hypotheses. If it were shown that these were correct for any given insulator, it would follow that waves of the kind expected by Maxwell could be propagated in this insulator with a finite velocity which might perhaps differ widely from that of light. This should not surprise us very much, no more, perhaps, than the well known fact that the electric excitations are

---

[12][*El. Waves*, pp. 95–106.]
[13][Helmholtz (1882–1895), vol. 1, pp. 545–628.]

propagated in wires with a greater but still finite velocity.[14] I felt that the third hypothesis contained the gist and the special significance of Faraday's, and therefore of Maxwell's, view and that it would thus be a more worthy goal for me to aim at. I saw no way of testing separately the first and second hypotheses for air;[15] but both hypotheses would be proved simultaneously if one could succeed in demonstrating both a finite rate of propagation and the existence of waves in air. Certainly some of the first experiments in this direction failed; these are described in the paper referred to, and they were carried out at short distances. But in the meantime I had succeeded in detecting the inductive action at distances up to 12 meters. Within this distance the phase of the motion must have been reversed more than once; and now it only remained to detect and prove this reversal. Thus the scheme was conceived which was carried out as described in the research "On the Finite Velocity of Propagation of Electromagnetic Actions."[16] The first step that had to be taken was easy. In straight stretched wires surprisingly distinct stationary waves were produced with nodes and antinodes, and by means of these it was possible to determine the wavelength and the change of phase along the wire. Nor was there any greater difficulty in producing interference between the effect which had travelled along the wire and that which had travelled through the air, and thus in comparing their phases. Now if both effects were propagated, as I expected, with one and the same finite velocity, they must at all distances interfere with the same phase. A simple qualitative experiment which, with the experience I had now gained, could be finished within an hour, should decide this question and lead at once to the goal. But when I had carefully set up the apparatus and carried out the experiment, I found that the phase of the interference was obviously different at different distances, and that the alternation was such as would correspond to an infinite rate of propagation in air. Disheartened, I gave up experimenting.

Some weeks passed before I began again. I reflected that it would be quite as important to find out that electric force was

---

[14][This sentence was omitted in the 1893 English translation of *Electric Waves*. This omission was first pointed out by D'Agostino (1975), pp. 310–311.]

[15]The expressions air (*Luftraum*) and empty space (*leerer Raum*) are here used as synonyms, inasmuch as the influence of the air itself in these experiments is negligible.

[16][*El.Waves*, pp. 107–123.]

propagated with an infinite velocity and that Maxwell's theory was wrong, as it would be, on the other hand, to prove that his theory was correct, provided only that the result arrived at should be definite and certain. I therefore confirmed with the greatest care, and without heeding what the outcome might be, the phenomena observed; the conclusions arrived at are given in the paper. When I then proceeded to consider more closely these results, I saw that the sequence of the interferences could not be harmonized with the assumption of an infinite rate of propagation; that it was necessary to assume that the velocity [in air] was finite, but greater than that in the wire. As shown in the paper, I endeavored to bring into harmony the various possibilities; and although the difference in the velocities appeared to me to be somewhat improbable, I could see no reason for mistrusting the experiments. And it was not by any means impossible that the motion in the wire might be retarded by some unknown causes, as, for example, by the essential inertia of the free electricity.

I have entered into these details here in order that the reader may be convinced that my desire has not been simply to establish a preconceived idea in the most convenient way by a suitable interpretation of the experiments. On the contrary, I have carried out with the greatest possible care these experiments (by no means easy ones), although they were in opposition to my preconceived views. And yet, although I may have been lucky elsewhere, in this research I have been decidedly unlucky. For instead of reaching the right goal with little effort, as a properly devised plan might have enabled me to do, I seem to have taken great pains and still to have fallen into error.

In the first place, the research has been marred by an error of calculation. The time of oscillation is overestimated in the ratio of 1.41 to 1. M. Poincaré first drew attention to this error.[17] As a matter of fact, this error affects the form of the research more than the substance of it. My reliance on the correctness of the calculation was mainly due to its supposed accordance with the experiments of Siemens and Fizeau and with my own.[18] If I had used the correct value for the capacity, and so

---

[17] H. Poincaré, *Comptes Rendus* 111 (1890). p. 322.

[18] See the remark at the end of the second part of the paper (p. 114 in *Electric Waves*).

[The remark Hertz is referring to seems to be the following: "Our result comes in well between the above experimental values (of Fizeau and Gounelle, and of Siemens). Since it was obtained with the aid of a doubtful theory, we are

found out the discrepancy between calculation and experiment, I would have placed less reliance on the calculation; the investigation would have been somewhat altered in form, but the content would have remained unaltered.

In the second place (and this is the more important point), one of the principal conclusions of the investigation can scarcely be regarded as correct—namely, that the velocities in air and in the wire are different. Such further knowledge as has been gained respecting waves in wires, instead of confirming this result, tends to make it more and more improbable. It now seems fairly certain that, if the experiment had been carried out quite correctly, and without any disturbing causes, it would have given almost exactly the result which I expected at the start. There is no doubt that the phase of the interference must have changed sign once (and this I had not expected beforehand); but the interference should have exhibited no second change of sign; and yet the experiments without exception pointed to this. It is not easy to point to any disturbing cause which could imitate in such a deceptive way the effect of a difference in velocity; but there is no reason why we should not admit the possibility of such a deception. While performing the experiments, I never in the least suspected that they might be affected by the neighboring walls. I remember that the wire along which the waves travelled was carried past an iron stove, and only 1.5 meters from it. A disturbance caused in this way, and always acting at the same point, might have given rise to the second change of phase of the interference. However this may be, I should like to express a hope that these experiments may be repeated by some other observer under the most favorable conditions possible, i.e., in a room as large as possible. If the plan of the experiment is correct, as I think it is, then it must, when properly carried out, give the result which it should have given at first; it would then prove, without need for measurement, the finite velocity of propagation of the waves

---

not justified in publishing it as a new measurement of this same velocity; but, on the other hand, we may conclude, from the accordance between the experimental results, that our calculated value of the period of oscillation is of the right order of magnitude."

The experimental values Hertz refers to are found in the following papers:

Fizeau and Gounelle, *Annalen*, 80 (1850), p. 158; W. Siemens, *Annalen* 157 (1876), p. 309.]

Hertz's Introduction to Electric Waves 173

in air, and at the same time the equality between this and the velocity of the waves in the wire.

I might also mention here some further considerations which at that time strengthened my conviction that the waves in the wire suffered a retardation. If the waves in the wire run along at the same speed as waves in air, then the lines of electric force must be perpendicular to the wire. Thus a straight wire traversed by waves cannot exert any inductive action upon a neighboring parallel wire. But I found that there was such an action, even though it was only a weak one. I concluded that the lines of force were not parallel to the wire, and that the velocity of the waves was not the same as that of light. Further, if the lines of force are perpendicular to the wire, it can be shown by a simple calculation that the energy propagated by a wave in a single wire becomes logarithmically infinite. I therefore concluded that such a wave was *a priori* impossible. Lastly, it seemed to me that it could have no effect upon the rate of propagation in a straight conductor, whether that conductor was a smooth wire or a wire with side projections, or a crooked wire, or a spiral wire with small convolutions, provided always that these deviations from the straight line were small compared with the wavelength, and that their resistance did not come into consideration. But now I found that all these alterations produced a very noticeable effect upon the velocity. Hence I concluded that here again there was some obscure cause at work that caused a retardation and would also make itself felt in simple smooth wires. At the present moment, these and other reasons do not appear to me to be of decisive weight; but at that time they so far satisfied me that I asserted without any reserve that there was a difference between the velocities, and I regarded this decision as one of the most interesting of my experimental results. Soon I was to discover what appeared to be a confirmation of my opinion; and at that time it was very welcome.

While investigating the action of my primary oscillation at great distances, I came across something like a formation of shadows behind conducting masses, which did not strike me as very surprising. Somewhat later on I thought that I noticed a peculiar reinforcement of the action in front of such shadow-forming masses and of the walls of the room. At first it occurred to me that this reinforcement might arise from a kind of reflection of the electric force from the conducting masses; but although I was familiar with the conceptions of Maxwell's theory, this idea appeared to me to be almost inadmissible—so utterly was it at variance with the conceptions then current as to the nature of the electric force. But when I had established with certainty the existence of actual waves, I returned to the mode of explanation which I had at first abandoned, and so arrived at the phenomena which are described in the paper "On Electromagnetic Waves in Air and their Reflection."[19]

No objection can be urged against the qualitative part of this research; the experiments have been frequently repeated and confirmed. But the part of the research that relates to the measurements is doubtful, inasmuch as it also leads to the very unlikely result that the velocity in air is considerably greater than that of waves on wires. Assuming that this result is incorrect, how are we to explain the error that has crept in? Certainly it is not due to simple inaccuracy of observation. The error of observation may perhaps be about a decimeter, but certainly not a meter. I can only here attribute the mistake in a general sense to the special conditions of resonance in the room used. The vibrations natural to it may possibly have been aroused, and I may have observed the nodes of such a vibration when I thought that I was observing the nodes of the waves of the primary conductor. There was certainly a substantial difference between the distances between the nodes in air that I measured, and the wavelengths in the wire. I especially directed my attention to the question whether or not such a difference existed. As far as any exact agreement with the first series of experiments is concerned, I freely allow that in the interpretation of the experimental results I may have allowed myself to be influenced by a desire to establish an agreement between the two sets of measurements. I put back the first node a certain distance behind the wall, and an exact control of the amount of this setback cannot be deduced from the experiments. If I had wished to combine the experiments otherwise, I might indeed have been able to calculate a ratio of the velocities which would come out nearer to unity; but I certainly could not infer from them that the velocities were the same.

Now, if the experiments which I made at that time all agree in pointing to a difference between the velocities, it will naturally be asked what reasons now induce me to allow that there may have been unknown sources of error in the experiments, rather than to abide by

---

[19][Paper No. 8.]

the statement made as to the difference of velocities. Is it the objection which has been raised in several quarters as to the lack of agreement between the experimental results and the theory? Certainly not. The theory was known to me at the time; and furthermore, it must be subordinated to the experiments. Is it the experiment performed in this connection by Herr Lecher?[20] This, too, I must deny, although I fully recognize the value of the work that Herr Lecher has done in this direction. In working out his results Herr Lecher assumes that the calculation is correct, and therefore in a certain sense that the theory itself is correct.[21] Is it then the results of MM. Sarasin and de la Rive,[22] who carefully repeated the experiments and arrived at conclusions which were completely in accord with the theory? In a certain sense, yes; in another sense, no. The Genevan physicists worked in a much smaller room than my own; the greatest distance which they had available was only 10 meters, and over this distance the waves could not develop freely enough. Their mirror was only 2.8 meters high. Care in carrying out the observations cannot compensate for the unfavorable nature of the room. In my experiments, on the other hand, the waves had perfectly free play up to 15 meters. My mirror was 4 meters high. If the decision rested simply and solely with the experiments, I could not attribute greater weight to those of MM. Sarasin and de la Rive than to my own.[23] So far, then, I again say no. But certainly the Genevan experiments show that my experiments are subject to local variations; they show that the phenomena are different if the reflecting walls and the room are different, and also that under certain conditions the wavelengths have the values required by theory. But if the experiments furnish information that is ambiguous and contradictory, they obviously contain sources of error that are not understood; and hence they cannot be brought forward as arguments against a theory that is supported by so many reasons based on

---

[20]E. Lecher, "Eine Studie über elektrische Resonanzerscheinungen," *Annalen* 41 (1837), p. 850.

[21]The same remark holds good for the work recently published by M. Blondlot, *Comptes Rendus* 113 (1891), p. 628 (Cp. Note 15 at end of book). [It appears that the correct reference here should be to note 16, not 15.]

[22]E. Sarasin and L. de la Rive, *Comptes Rendus* 112 (1891), p. 658.

[23]Mr. Trouton, in a room of which the dimensions are not exactly given, found like myself that the wavelength of my primary conductor in air was about 10 meters.—*Nature* 39 (1888–89), p.391.

probability. Thus the Genevan experiments deprive my own of their force, and so far they restore the balance of probability to the theoretical side.

Still I must acknowledge that the reasons which decided me were of a more indirect kind. When I first thought that I had found a retardation of waves on the wires, I hoped soon to discover the cause of this retardation, and to find some gradual change in its value. This hope has not been realized. I found no such change, and as my experience increased, instead of coming across an explanation, I met with increasing discrepancies; these at last appeared to me to be insoluble, and I had to give up all hope of proving the correctness of my first observation. My own discovery, that for short waves the difference between the velocities very nearly disappears, tended in the same direction. Before one of my scientific colleagues had attacked this question, I had stated my opinion in the following words:[24] "Thus I found that for long waves the wavelength is greater in air than on wires, whereas for short waves both appear to be practically equal. This result is so surprising that we cannot regard it as certain. The decision must be reserved until further experiments are made."[25] The only experiments of the kind referred to that have hitherto been performed are those by MM. Sarasin and de la Rive; and inasmuch as these were carried out in small rooms, they may more properly be regarded as a confirmation of the second part of my statement than as a refutation of the first part. Decisive experiments for long waves seem to me to be still wanting.[26] I have little doubt that they will decide in favor of equal velocities in all cases.

---

[24]*Archives de Genève* (3), 21 (1889), p. 302.

[25][In retrospect it appears that Hertz's results can be explained by the greater effect at longer wavelengths of reflections from the room's walls and objects, and of diffraction.]

[26]Since the above was written, the wish expressed has been amply satisfied by the experiments which MM. Sarasin and de la Rive have carried out in the great hall of the Rhône waterworks at Geneva (see *Archives de Genève* 29 (1893), pp. 358 and 441). These experiments have proved the equality of the velocity in air and in wires, and have thus established the full agreement between theory and experiment. I consider these experiments to be conclusive, and submit to them now with as much readiness as I then felt hesitation in submitting to experiments which were not superior to my own. I gladly avail myself of the opportunity of thanking MM. Sarasin and de la Rive for the great kindness and goodwill that they have invariably exhibited in this whole

The reader may, perhaps, ask why I have not endeavored to settle the doubtful point myself by repeating the experiments. I have indeed repeated the experiments, but have only found, as might be expected, that a simple repetition under the same conditions cannot remove the doubt, but rather increases it. A definite decision can only be arrived at by experiments carried out under more favorable conditions. More favorable conditions here mean larger rooms, and such were not at my disposal. I again emphasize the statement that care in making the observations cannot make up for want of space. If the long waves cannot develop, they clearly cannot be observed.

The experiments hitherto described on the reflection of waves were finished in March 1888. In the same month I attempted, by means of reflection at a curved surface, to prevent the spreading of the radiation for large distances. I constructed a concave parabolic mirror of 2 meters aperture and 4 meters in height for my large oscillator. Contrary to my expectations I found that the effect was considerably weakened. The large mirror acted like a protecting screen surrounding the oscillator. I concluded that the wavelength of the oscillation was too large in comparison with the focal length of the mirror. A moderate reduction in the size of the primary mirror did not improve the result. I therefore tried to work with a conductor which was geometrically similar to the larger one, but smaller in the proportion 10:1. Perhaps I did not persevere sufficiently in this attempt; at any rate I failed entirely at that time to produce and observe such short oscillations, and I abandoned these experiments in order to turn my attention to other questions.

In the first place, it was important to devise a clearer theoretical treatment of the experiments. In the researches to which I have hitherto referred, the experiments were interpreted from the standpoint which I absorbed through studying von Helmholtz's papers.[27] In these papers Herr v. Helmholtz distinguishes between two forms of electric force—the electromagnetic and the electrostatic—to which, until the contrary is proved by experience, two different velocities are attributed. An interpretation of the experiments from this point of view would in no way be incorrect, but might perhaps be unnecessarily

---

controversy—a controversy which has now been decided entirely in their favor. [This footnote first appeared in the 1893 English edition of *El.Waves*.]

[27][Helmholtz (1882–1898), vol. 1, pp. 545–628.]

complicated. In a special limiting case Helmholtz's theory becomes considerably simplified, and its equations in this case become the same as those of Maxwell's theory; only one form of the force remains, and this is propagated with the velocity of light. I had to see whether the experiments would agree with these much simpler assumptions of Maxwell's theory. The attempt was successful. The results of the calculation are given in the paper on "The Forces of Electric Oscillations, treated according to Maxwell's Theory."[28] Some of this research that relates to interference between waves in air and on wires could clearly be adapted without difficulty to any other form of such interference that might arise from more complete experiments.

Side by side with the theoretical discussions I continued the experimental work, directing the latter again more to waves on wires. In doing so, my primary object was to find out the cause of the supposed retardation of these waves. Secondly, I wished to test the correctness of the view according to which the location and field of action of the waves is not the interior of the conductor, but rather the surrounding space. I now made the waves travel in the space between two wires, between two plates, and in cylindrical pipes, instead of along a single wire; no longer in different metals, but in various interposed insulators. This research on "The Propagation of Electric Waves by Means of Wires"[29] was, for the most part, carried out in the summer of 1888, although it was only completed and published at a later date.

For in the autumn a singular phenomenon stole my attention away from the experiments with wires. For the investigation of waves in the narrow space between two wires I was using resonators of small external dimensions, and was engaged in tuning these. I found that I obtained distinct nodes at the end of the wires even when I used resonators which were much too small. Even when I decreased the size of the circles to a few centimeters diameter, I still obtained nodes; these were situated at a small distance from the end of the wires, and I could observe half wavelengths as small as 12 cm. Thus good fortune put me on the track of the short waves, which had been previously

---

[28][*El.Waves*, pp. 137–159. This is the paper that anticipated so much of the development of wireless telegraphy, as Arnold Sommerfeld has pointed out. See Sommerfeld (1954), p. 6.]

[29][*El.Waves*, pp. 160–171.]

unknown.[30] I at once followed up this track, and soon succeeded in finding a form of the primary conductor which could be used with the small resonators.

I paid no special attention to the phenomenon which led me back to the observation of short waves; and, as no suitable occasion arose for doing so, I have not mentioned it in my papers. Clearly it was a special case of the same phenomenon that was later discovered by MM. Sarasin and de la Rive,[31] called by the name of "Multiple Resonance," and explained by saying that the primary conductor did not possess any definite period of oscillation, but that it performed simultaneously all possible oscillations lying within wide limits. If I paid little attention myself to this phenomenon, it was partly because I was soon led on to other researches. It arose no less from the fact that I had from the start conceived an interpretation of the phenomenon that lent much less interest to it than the interpretation given by MM. Sarasin and de la Rive. I regarded the phenomenon as a consequence of the rapid damping of the primary oscillation—a necessary consequence and one that could be foreseen. M. Sarasin was good enough to communicate at once to me the results of his research, and I told him my doubts as to his explanation of the phenomenon, and gave him my own explanation of it; but although he received my explanation with the readiest goodwill, we did not succeed in coming to a common understanding as to the interpretation of the experiment. With M.H. Poincaré such an understanding was secured at once; he had formed a conception of the phenomenon that was practically identical with my own, and had communicated it to me in a letter. This conception he has worked out mathematically and published in his book *Electricité et Optique*.[32] Herr V. Bjerknes has worked out the mathematical developments simultaneously and independently.[33] That the explanation given by MM. Poincaré and Bjerknes is not only a possible one, but is the only possible one, appears to me to be proved by an investigation by Herr Bjerknes,[34] which has appeared and which

---

[30][These waves would today be called microwaves or radar waves.]

[31]E. Sarasin and L. de la Rive, *Archives de Genève* (3), 23 (1890), p. 113.

[32]H. Poincaré, *Electricité et Optique*, vol. 2, *Les Théories de Helmholtz et les Expériences de Hertz* (Paris: Georges Carré, 1891), p. 249.

[33]V. Bjerknes, *Annalen* 44 (1891), p. 92. [On Bjerknes see section V. of the Introductory Biography.]

[34]V. Bjerknes, *Annalen* 45 (1891), p 513.

makes it certain that the oscillation of the primary conductor is, at any rate to a first approximation, a uniformly damped sine-wave of determinate period. Hence the careful investigations of MM. Sarasin and de la Rive are of great value in completing our knowledge of this part of the work, but they in no way contradict any statement made by me. The authors themselves regard their experiments in this light. Nevertheless, these experiments gave occasion to an adverse criticism of my work from a distinguished French physicist who had not, however, repeated the experiments. I hope it will now be allowed that there was no cause for such a criticism.[35]

I may be permitted to take this opportunity to refer to the doubts that have recently been expressed by Herrn Hagenbach and Zehnder as to what my experiments really prove.[36] Perhaps I ought not yet consider their work as completed. The authors reserve to themselves the right to return to the explanation and the formation of nodes and antinodes in my experiments. But it is just precisely upon these phenomena that my experiments and the whole interpretation of them rest.

After I had succeeded (as already described) in observing very short waves, I chose waves about 30 cm long[37] and, first of all, repeated all the earlier experiments with them. I now found, contrary to my expectation, that these short waves travelled along wires with very nearly the same velocity as in air. As it was easy to obtain for such short waves room for them to propagate freely, no doubt could arise in this case as to the correctness of the results. After I had become quite used to managing these short waves, I returned to the experiment with the concave reflector. The old large reflector was no longer at my disposal, so I had a smaller one made, about 2 meters high and a little more than 1 meter in aperture. It worked so remarkably well that, directly after the first trial, I ordered not only a second concave reflector, but also a plane reflecting surface and a large prism. The experiments which are described in the paper "On Electric Radiation"[38] now followed each other in rapid succession, and without difficulty; they had been

---

[35]M.A. Cornu, *Comptes Rendus*, 110 (1890), p. 72.

[36]E. Hagenbach and L. Zehnder, *Annalen* 43 (1891), p. 610.

[37][This is the half-wavelength; the actual wavelength used in these experiments was 66 cm.]

[38][Paper No. 9.]

considered and prepared for long beforehand, with the exception of the polarization experiments, which only occurred to me during the progress of the work. These experiments with concave mirrors soon attracted attention; they have frequently been repeated and confirmed. The approval with which they have been received has far exceeded my expectations.[39] A considerable part of this approval was due to reasons of a philosophical nature. The old question as to the possibility and nature of forces acting at a distance was again raised. Such forces have long been highly regarded in science, but have always been accepted with reluctance by ordinary common sense; in the domain of electricity these forces now appeared to be dethroned from their position by simple and striking experiments.

Although in the last-mentioned experiments my research had, in a certain sense, come to its natural end, I still felt that there was one thing wanting. The experiments related only to the propagation of the electric force.[40] It was desirable to show that the magnetic force was also propagated with a finite velocity. According to theory it was not necessary for this purpose to produce special magnetic waves; the electric waves should at the same time be waves of magnetic force; the only important thing was really to detect in these waves the magnetic force in the presence of the electric force. I hoped that it would be possible to do this by observing the mechanical forces which the waves exerted upon ring-shaped conductors. So experiments were planned which (for other reasons) were only carried out later on, and then incompletely; these are described in the last piece of experimental research "On the Mechanical Action of Electric Waves in Wires."[41]

Casting now a glance backwards we see that by the experiments above sketched the propagation in time of a supposed action-at-a-distance is for the first time proved. This fact forms the philosophical result of the experiments; and, indeed, in a certain sense the most

---

[39] These experiments gave occasion to the lecture "On the Relations between Light and Electricity," which I delivered to the *Naturforscherversammlung* at Heidelberg in 1889, and in which I gave a general account of my experiments in a popular form (published by E. Strauss, Bonn). [This lecture is included in the present volume as Paper No. 10.]

[40] [On what Hertz means by "electric force," here and throughout the rest of his papers, see footnote 52 below.]

[41] [*El.Waves*, pp. 186–194. This research was completed and published when Hertz was Professor of Physics in Bonn.]

important result. The proof includes a recognition of the fact that the electric forces can disentangle themselves from material bodies, and can continue to subsist as changes in the properties of space. The details of the experiments further prove that the particular manner in which the electric force is propagated exhibits the closest analogy with the propagation of light;[42] indeed, that it corresponds almost completely to it. The hypothesis that light is an electrical phenomenon is thus made highly probable. To give a strict proof of this hypothesis would logically require experiments upon light itself.

What we here indicate as having been accomplished by the experiments is accomplished independently of the correctness of particular theories. Nevertheless, there is an obvious connection between the experiments and the theory in connection with which they were really undertaken. Since the year 1861 science has been in possession of a theory which Maxwell constructed upon Faraday's ideas, and which we therefore call the Faraday-Maxwell theory. This theory affirms the possibility of the class of phenomena here discovered just as positively as the remaining electrical theories are compelled to deny it. From the outset Maxwell's theory excelled all others in elegance and in the abundance of the relations between the various phenomena that it explained. The probability of this theory, and therefore the number of its adherents, has increased from year to year. But as long as Maxwell's theory depended solely on the probability of its results, and not on the certainty of its hypotheses, it could not completely displace the theories that were opposed to it. The fundamental hypotheses of Maxwell's theory contradicted the usual views, and did not rest upon the evidence of decisive experiments. In this connection we can best characterize the object and the result of our experiments by saying: The object of these experiments was to test the fundamental hypotheses of the Faraday-Maxwell theory, and the result of the experiments is to confirm the fundamental hypotheses of the theory.[43]

---

[42]The analogy does not consist only in the agreement between the more or less accurately measured velocities. The approximately equal velocity is only one factor among many others. [These other factors include reflection, refraction, interference and polarization, as Hertz shows in Paper No. 9.]

[43][For this reason Hertz's research serves as one of the best instances of the interplay between theory and experiment that characterizes the progress of physics.]

B. Theoretical

And now, to be more precise, what is it that we call the Faraday-Maxwell theory? Maxwell has left us as the result of his mature thought a great treatise on Electricity and Magnetism;[44] it might therefore be said that Maxwell's theory is the one which is propounded in that work. But such an answer will scarcely be regarded as satisfactory by all scientific men who have considered the question closely. Many a man has thrown himself with zeal into the study of Maxwell's work and, even when he has not stumbled upon unusual mathematical difficulties, has nevertheless been compelled to abandon the hope of forming for himself an altogether consistent conception of Maxwell's ideas. I have fared no better myself. Notwithstanding the greatest admiration for Maxwell's mathematical conceptions, I have not always felt quite certain of having grasped the physical significance of his statements. Hence it was not possible for me to be guided in my experiments directly by Maxwell's book. I have rather been guided by Helmholtz's work, as indeed may plainly be seen from the manner in which the experiments are set forth. But unfortunately, in the special limiting case of Helmholtz's theory which leads to Maxwell's equations, and to which the experiments pointed, the physical basis of Helmholtz's theory disappears, as indeed it always does, as soon as action-at-a-distance is disregarded. I therefore endeavored to form for myself in a consistent manner the necessary physical conceptions, starting from Maxwell's equations, but otherwise simplifying Maxwell's theory as far as possible by eliminating or simply leaving out of consideration those portions [that I did not understand and][45] that could be dispensed with, inasmuch as they could have no possible effect on any observed phenomena. This explains how the two theoretical papers (forming the conclusion of this collection)[46] came to be written. Thus the representation of the theory in Maxwell's own work, its representation as a limiting case of Helmholtz's theory, and its representation in these two papers of mine—however different in form—have substantially the same inner significance. This common

---

[44][Maxwell (1873).]

[45][The words in brackets are in the original German text, but have been omitted in the English translation by D.E. Jones.]

[46][*El. Waves*, pp. 195–268.]

significance of the different modes of representation (and others can certainly be found) appears to me to be the undying part of Maxwell's work. This, and not Maxwell's peculiar conceptions or methods, would I designate as "Maxwell's Theory." To the question, "What is Maxwell's theory?" I know of no shorter or more definite answer than the following: "Maxwell's theory is Maxwell's system of equations."[47] Every theory that leads to the same system of equations, and therefore includes the same possible phenomena, I would consider as being a form or special case of Maxwell's theory; every theory that leads to different equations, and therefore to different possible phenomena, is a different theory. Hence in this sense, and in this sense only, may the two theoretical papers in the present volume be regarded as representations of Maxwell's theory. In no sense can they claim to be a precise rendering of Maxwell's ideas. On the contrary, it is doubtful whether Maxwell, were he alive, would acknowledge them as representing his own views in all respects.

The very fact that different modes of representation have essentially the same content, renders the proper understanding of any one of them all the more difficult. Ideas and conceptions that are akin and yet different may be symbolized in the same way in different modes of representation. Hence for a proper comprehension of any one of them, the first essential is that we should endeavor to understand each representation by itself without introducing into it the ideas that belong to another [representation]. Perhaps it may be of service to many of my colleagues if I here briefly explain the fundamental conceptions of the three representations of Maxwell's theory to which I have already referred. I shall thus have an opportunity to state wherein lies, in my opinion, the special difficulty of Maxwell's own representation. I cannot agree with the oft-stated opinion that this difficulty is of a mathematical nature.

When we see bodies acting upon one another at a distance, we can form for ourselves various conceptions of the nature of this interaction. We may regard the effect as being that of a direct action-at-a-distance, jumping across space, or we may regard it as the consequence of an action that is propagated from point to point in a hypothetical medium. Meanwhile, in applying these conceptions to

---

[47][This sentence is probably the most frequently quoted statement from Hertz's writings.]

Hertz's Introduction to Electric Waves

electricity, we can make a series of finer distinctions. As we pass from the pure conception of direct attraction to the pure conception of indirect [*vermittelten*] attraction, we can distinguish four standpoints.

From the first standpoint we regard the attraction of two bodies as a kind of spiritual affinity between them. The force which each of the two exerts is bound up with the presence of the other body. In order that force should be present at all, there must be at least two bodies present. In some way a magnet only exerts a force when another magnet [or an electric current] is brought into its neighborhood. This conception is the pure conception of action-at-a-distance, the conception of Coulomb's law. In the theory of electricity it has almost been abandoned, but it is still used in the theory of gravitation. Mathematical astronomy speaks of the attraction between the sun and a planet, but with any attraction in empty space it has no concern.

From the second standpoint we still regard the attraction of the bodies as a kind of spiritual influence of each upon the other. But, although we admit that we can only notice this action when we have at least two bodies, we further assume that each of the acting bodies continually strives to excite at all surrounding points attractions of definite magnitude and direction, even if no other similar bodies happen to be in the neighborhood. With these strivings, varying always from point to point, we fill (according to this conception) the surrounding space. At the same time we do not assume that there is any change at the place where the action is exerted; the acting body is still both the seat and the source of the force. This is approximately the standpoint of the potential theory. It obviously is also the standpoint of certain chapters in Maxwell's work, although it is not the standpoint of Maxwell's theory. In order to compare these conceptions more easily with one another, we represent from this standpoint (as in Fig. 2) two oppositely-charged condenser plates. The diagrammatic representation will be easily understood; upon the plates are seen the positive and negative electricities (as if they were material); the force between the plates is indicated by arrows. From this standpoint it is irrelevant whether the space between the plates is full or empty. If we admit the existence of the light-ether, but suppose that it is removed from a part B of the space, the force will still remain unaltered in this space.

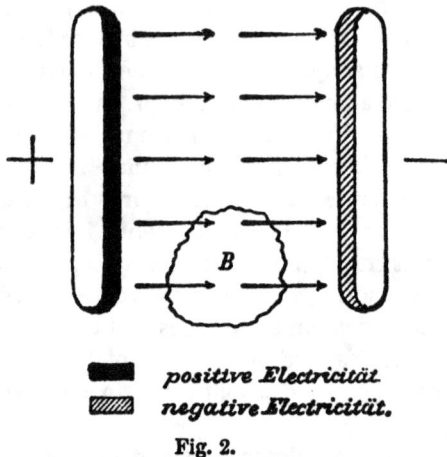

■ *positive Electricität.*
▨ *negative Electricität.*
Fig. 2.

The third standpoint retains the conceptions of the second, but adds to them a further complication. It assumes that the action of the two separate bodies is not determined solely by forces acting directly at a distance. It rather assumes that the forces induce changes in the space (supposed to be nowhere empty), and that these again give rise to distance-forces (*Fernkräften*). The attractions between the separate bodies depend, then, partly upon their direct action and partly upon the influence of the changes in the medium. The change in the medium itself is regarded as an electric or magnetic polarization of its smallest parts under the influence of the acting force. This view has been developed by Poisson with respect to static phenomena in magnetism, and has been transferred by Mosotti to electrical phenomena. In its most general development, and in its extension over the whole domain of electromagnetism, it is represented by Helmholtz's theory.[48]

Fig. 3 illustrates this standpoint for the case in which the medium plays only a small part in the resultant action. Upon the plates are seen the free electricities, and in the dielectric [between the plates] the electrical fluids which are separated, but which cannot be

---

[48]At the end of the paper "On the Equations of Motion of Electricity for Conducting Bodies at Rest." [Helmholtz (1882–1898), vol. 1, pp. 545–628.]

Hertz's Introduction to Electric Waves

completely detached from each other. Let us suppose that the space between the plates contains only the light-ether, and let a space, such as B, be hollowed out of this; the forces will then remain in this space, but the polarization will disappear.

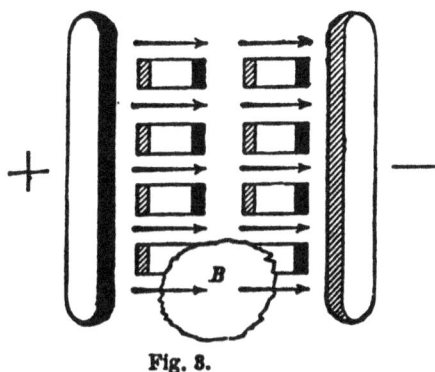

Fig. 8.

One limiting case of this mode of conception is of special importance. As closer examination reveals, we can split up the resultant action (which alone can be observed) of material bodies upon one another into an influence due to direct action-at-a-distance, and an influence due to the intervening medium. We can increase that part of the total energy which has its seat in the electrified bodies at the expense of that part which is situated in the medium, and conversely. Now in the limiting case we locate the whole of the energy in the medium. Since no energy corresponds to the electricities that exist upon the conductors, the distance-forces must become infinitely small. But for this it is a necessary condition that no free electricity should be present. The electricity must therefore behave like an incompressible fluid. Hence we have only closed currents; and so there arises the possibility of extending the theory to all kinds of electrical disturbances in spite of our ignorance of the laws of open currents.

The mathematical treatment of this limiting case leads us to Maxwell's equations. We therefore call this treatment a form of Maxwell's theory. The limiting case is also designated in this way by v. Helmholtz. But in no sense must this be taken as meaning that the physical ideas on which it is based are Maxwell's ideas.

Fig. 4 indicates the state of the space between two electrified plates in accordance with the conceptions of this theory. The distance-forces have become merely nominal. The electricity on the conductors is still present, and is a necessary part of the conception, but its action-at-a-distance is completely neutralized by the opposite electricity of the medium which is displaced towards it. The pressure which this medium exerts, on account of the attraction of its internal electrifications, tends to draw the plates together. In the empty space B there are only vanishingly small distance-forces.

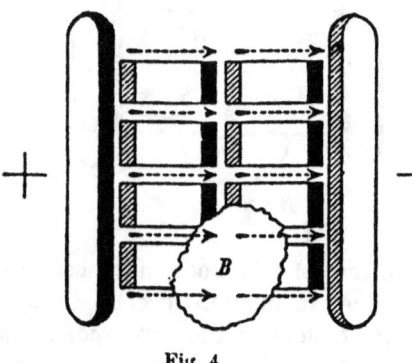

Fig. 4.

The fourth standpoint belongs to the pure conception of action through a medium. From this standpoint we acknowledge that the changes in space assumed from the third standpoint are actually present and that it is by means of them that material bodies act upon one another. But we do not admit that these polarizations are the result of distance-forces; indeed, we altogether deny the existence of these distance-forces; and we discard the electricities from which these forces are supposed to proceed. Rather we now regard the polarizations as the only things that are really present; they are the cause of the movements of ponderable bodies, and of all the phenomena that enable us to perceive changes in these bodies. The explanation of the nature of the polarizations, of their relations and effects, we either defer or look for in mechanical hypotheses; but we refuse to recognize in the previously-employed electricities and distance-forces a satisfactory explanation of these relations and effects.

The expressions "electricity," "magnetism," etc., have no further value for us beyond their use as abbreviations.

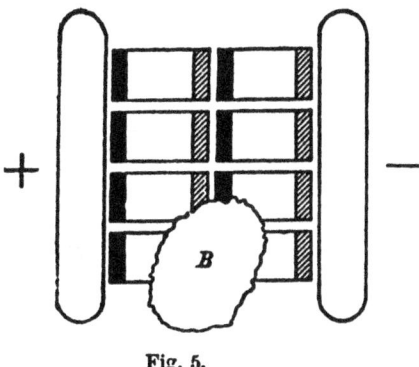

Fig. 5.

Considered from a mathematical point of view, this fourth mode of treatment may be regarded as coinciding completely with the limiting case of the third. But from the physical point of view the two differ fundamentally. It is impossible to deny the existence of distance-forces, and at the same time to regard them as the cause of the polarizations. Whatever we may designate as "electricity" from this standpoint does not behave like an incompressible fluid.[49] If we consider Fig. 5, which brings symbolically before us the view presented from this standpoint, we are struck by another distinction. The polarization of the space is represented by the same symbolic method as was employed in discussing the third standpoint. But whereas in Figs. 3 and 4 this mode of representation explained the nature of the polarization through the nature of electricity (assumed to be known), we have here to regard the mode of representation as defining the nature of an electric charge through the state of polarization of the space (regarded as known). Each particle of the dielectric here appears charged in opposite senses with electricity, just as it did from our third standpoint. If we again remove the ether from the space B, there remains nothing whatsoever in this space which could remind us of the electrical disturbance in the neighborhood.

---

[49][At the time Hertz was writing this, the electron had not yet been discovered, and the nature of electric currents was still unclear.]

Now this fourth standpoint, in my opinion, is Maxwell's standpoint. The general explanations in his work leave no room for doubting that he wished to discard distance-forces entirely. He expressly says that if the force and the "displacement" in a dielectric is directed towards the right hand, we must conceive each particle of the dielectric as being charged with negative electricity on the right-hand side, and with positive electricity on the left-hand side. But it cannot be denied that other statements made by Maxwell appear at first sight to contradict the conceptions of this standpoint. Maxwell assumes that electricity also exists in conductors; and that this electricity always moves in such a way as to form closed currents with the displacements in the dielectric. That electricity moves like an incompressible fluid is a favorite statement of Maxwell. But these statements do not fit in with the conceptions of the fourth standpoint; they lead one to suspect that Maxwell rather viewed things from the third point of view. My own opinion is that this was never really the case; that the contradictions are only apparent and arise from a misunderstanding of his words. The following, if I am not mistaken, is the state of affairs: Maxwell originally developed his theory with the aid of very definite and special conceptions as to the nature of electrical phenomena. He assumed that the pores of the ether and of all bodies were filled with an attenuated fluid, which, however, could not exert forces at a distance. In conductors this fluid moved freely, and its motion formed what we call an electric current. In insulators this fluid was confined to its place by elastic forces, and its "displacement" was regarded as being identical with electric polarization. The fluid itself, as being the cause of all electrical phenomena, Maxwell called "electricity." Now, when Maxwell composed his great treatise [in 1873], the accumulated hypotheses of this earlier mode of conception no longer suited him, or else he discovered contradictions in them and so abandoned them. But he did not eliminate them completely; quite a number of expressions remained that were derived from his earlier ideas. And so, unfortunately, the word "electricity," in Maxwell's work, obviously has a double meaning. In the first place, he uses it (as we also do) to denote a quantity that can be either positive or negative, and that forms the starting-point of distance-forces (or what appear to be such). In the second place, it denotes that hypothetical fluid from which no distance-forces (not even apparent ones) can proceed, and the amount of which in any given space must, under all circumstances, be a positive quantity. If we read Maxwell's explanations and always interpret the meaning of the word "electricity" in a suitable way, nearly all the contradictions, which at first are so surprising, can be made to disappear. Nevertheless, I must admit that I have not succeeded in doing this completely, or to my entire satisfaction; otherwise, instead of hesitating, I would speak more definitely.[50]

Whether this is so or not, an attempt has been made, in the two theoretical papers here printed,[51] to exhibit Maxwell's theory, i.e., Maxwell's system of equations, from this fourth standpoint. I have endeavored to avoid from the beginning the introduction of any conceptions foreign to this standpoint and which might afterwards have to be removed.[52] I have further endeavored in the exposition to limit as far as possible the number of those conceptions that are arbitrarily introduced by us, and only to admit such elements as cannot be removed or altered without at the same time altering possible experimental results. It is true, that in consequence of these endeavors, the theory acquires a very abstract and colorless appearance. It is not particularly pleasing to hear general statements made about "directed changes of state," where we used to have placed before our eyes pictures of electrified atoms. It is not particularly satisfactory to see equations set forth as direct results of observation and experiment, where we used to get long mathematical deductions as apparent proofs of them. Nevertheless, I believe that we cannot, without deceiving ourselves, extract much more from known facts than is asserted in the papers referred to. If we wish to lend more color to the theory, there is nothing to prevent us from supplementing all this and aiding our powers of imagination by concrete representations of the various conceptions as to the nature of electric polarization, the

---

[50]M. Poincaré, in his treatise *Électricité et Optique*, vol. 1, *Les Théories de Maxwell)*, expresses a similar opinion. Herr L. Boltzmann, in his *Vorlesungen über Maxwell's Theorie*, appears like myself to aim at a consistent development of Maxwell's system rather than an exact rendering of Maxwell's thoughts. But no definite opinion can be given, inasmuch as the work is not yet completed.

[51]*[El.Waves*, pp. 195–268.]

[52]The expression "electric force" in these papers is only another name for a state of polarization of space. It would perhaps have been better, in order to prevent misconceptions, if I had replaced it by another expression, such, for example, as "electric field-intensity," or "*elektrische Intensität*," which Herr E. Cohn proposes in his paper that refers to the same subject ("Zur Systematik der Elektricitätslehre," *Annalen*, 40 (1890), p. 625.

electric current, etc. But scientific accuracy requires of us that we should in no wise confuse the simple and homely figure, as it is presented to us by nature, with the gay garment that we use to clothe it. Of our own free will we can make no change whatever in the form of the one, but the cut and color of the other we can choose as we please.[53]

Such further remarks as I may wish to make on points of detail will be found at the end of the book as supplementary notes.[54]

---

[53][Olivier Darrigol has made a very apt comment on this metaphor of Hertz: "For Hertz the motion of a pervasive, energy-carrying ether belongs to the homely figure, and electric flow to the gay garment. A few years later, Lorentz and Einstein would revert the situation: electric flow would become a realistic electronic motion, and ether a mere name." See Darrigol (1993), p. 257.]

[54][In the present collection all Hertz's notes are included as footnotes to the relevant papers. This should be kept in mind in reading the papers that follow in this volume. We know from Hertz's *Erinnerungen* (p. 321) that his Introduction to *Electric Waves* was written in October 1891. The notes he added to his already published papers were probably written at about the same time. The first German edition of *El. Waves* was published in the middle of 1892.]

PAPER NO. 6

On Very Rapid Electric Oscillations

(Heinrich Hertz, "Über sehr schnelle elektrische Schwingungen," Wiedemann's *Annalen der Physik und Chemie* 31 [1887], pp. 421–448; 543–544.)[1]

The electric oscillations of open induction coils have a period of vibration that is measured in ten-thousandths of a second. The vibrations in the oscillatory discharges of Leyden jars, such as were observed by Feddersen,[2] follow each other about a hundred times as rapidly. Theory admits the possibility of oscillations even more rapid than these in open wire circuits of good conductivity, provided that the ends are not loaded with large capacities; but at the same time theory does not enable us to decide whether such oscillations can be actually excited on such a scale as to admit of their being observed. Certain phenomena led me to expect that oscillations of the latter kind do really occur under certain conditions, and that they are of such strength as to allow of their effects being observed. Further experiments confirmed my expectation, and I propose to give here an account of the experiments made and the phenomena observed.

---

[1] [*El.Waves*, pp. 29–53.]
[2] For the literature see Colley, *Annalen* 26 (1885), p. 432.

[In 1891, when Hertz edited his papers for publication in book form, he added the following note.]

It was v. Helmholtz, in his paper "Über die Erhaltung der Kraft," who first stated (in 1847) that the discharge of a Leyden jar is oscillatory. He arrived at this conclusion from its varying and opposite magnetic effects, and from the fact that when one endeavors to decompose water by electric discharges, both gases are developed at both electrodes. Sir William Thomson [Lord Kelvin] arrived independently at the same result from theoretical considerations. The mathematical treatment of the problem given by him in the year 1853 (*Philosophical Magazine* [4], vol. 5, p. 393) still holds good today.

[The famous paper by Helmholtz here referred to by Hertz is translated in Kahl (1979), pp. 3–55. The suggestion that the discharge of a Leyden jar is oscillatory is on page 31. Hertz then gives a lengthy list of early papers on electrical oscillations, which is omitted here for reasons of space.]

The oscillations which are here dealt with are about a hundred times as rapid as those observed by Feddersen.[3] Their period of oscillation—estimated, it is true, only by the aid of theory—is of the order of a hundred-millionth of a second.[4] Hence, according to their period, these oscillations range themselves in a position intermediate between the acoustic oscillations of ponderable bodies and the light-oscillations of the ether. In this, and in the possibility that a closer observation of them may be of service in the theory of electrodynamics, lies their interest to us.

Preliminary Experiments

If, in addition to the ordinary spark-gap of an induction coil, there be introduced in its discharging circuit a Riess spark-micrometer,[5] the poles of which are joined by a long metallic shunt, the discharge follows the path across the air-gap of the micrometer in preference to the path along the metallic conductor, as long as the length of the air-gap does not exceed a certain limit. This is already known, and the construction of lightning-protectors for telegraph lines is based on this experimental fact. It might be expected that, if the metallic shunt were only made short and of low resistance, the sparks in the micrometer would then disappear. As a matter of fact, the length of the sparks obtained does not diminish with the length of the shunt, and the sparks themselves can scarcely be made to disappear entirely under any circumstances. Even when the two knobs of the micrometer are connected by a few centimeters of thick copper wire, sparks can still be observed, although they are exceedingly short.

---

[3] [Feddersen's publications on the discharge of Leyden jars have been reprinted in *Ostwald's Klassiker der exacten Wissenschaften*, vol. 166 (Leipzig, Geest und Portig, 1908).]

[4] [This is the half-period; the wavelength is about 6 m.]

[5] [The Riess spark-micrometer is a small measuring micrometer modified for use as the gap in an open electric circuit. Sparks jump across the gap, the length of which can be adjusted with the movable screw of the micrometer. The length of the spark can then be read off the micrometer scale. Peter Riess (1804–1883) was the physics professor at Berlin who developed this instrument.]

This experiment shows directly that at the instant when the discharge occurs, the potential along the circuit must vary in value by hundreds of volts even in a few centimeters; indirectly it proves with what extraordinary rapidity the discharge takes place. For the difference of potential between the knobs of the micrometer can only be regarded as an effect of self-induction in the metallic shunt. The time in which the potential of one of the knobs is appreciably changed is of the same order as the time in which such a change is transmitted to the other knob through a short length of a good conductor. The potential difference between the micrometer-knobs might indeed be supposed to be determined by the resistance of the shunt, the current density during the discharge being possibly large. But a closer examination of the quantitative relations shows that this supposition is inadmissible; and the following experiment shows independently that this conjecture cannot be put forward.

We again connect the knobs of the micrometer by a good metallic conductor, say by a copper wire 2 mm in diameter and 0.5 m long, bent into rectangular form; we do not, however, introduce this into the discharging-circuit of the induction-coil, but we simply place one pole of it in communication with any point of the discharging circuit by means of a connecting wire. (Fig. 6 shows the arrangement of the apparatus; A represents diagrammatically the induction-coil, B the discharger, and M the micrometer.) Thereupon we again observe, while the induction-coil is working, a stream of sparks in the

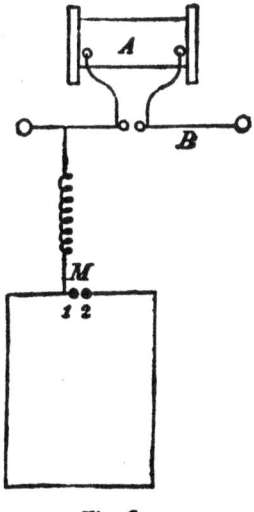

Fig. 6.

micrometer which may, under suitable conditions, attain a length of several millimeters. Now this experiment shows, in the first place, that at the instant when the discharge takes place, violent electrical disturbances occur not only in the actual discharging circuit, but also in all conductors connected with it. But, in the second place, it shows more clearly than the preceding experiment that these disturbances run on so rapidly that even the time taken by electrical waves in rushing through short metallic conductors becomes of appreciable importance. For the experiment can only be interpreted in the sense that the change of potential proceeding from the induction-coil reaches the knob 1 in an appreciably shorter time than it takes to reach knob 2. This phenomenon may well cause surprise when we consider that, as far as we know, electric waves in copper wires are propagated with a velocity that is approximately the same as that of light. So it appeared to me to be worthwhile to endeavor to determine what conditions were most favorable for the production of brilliant sparks in the micrometer. For the sake of brevity we shall, in order to distinguish them from the discharge proper, speak of these sparks as the side-sparks (*Nebenfunken*), and of the micrometer discharging-circuit as the side-circuit (*Nebenkreis*).

First of all it became evident that powerful discharges are necessary if side-sparks of several millimeters in length are desired. I therefore used in all the following experiments a large Ruhmkorff coil, 52 cm long and 20 cm in diameter, which was provided with a mercury interrupter and was excited by six large Bunsen cells. Smaller induction-coils gave the same qualitative results, but the side-sparks were shorter, and it was therefore more difficult to observe differences between them. The same held true when discharges from Leyden jars or from batteries were used instead of the induction-coil. It further appeared that even when the same apparatus was used, a good deal depended upon the nature of the exciting spark in the discharger (B). If this takes place between two points, or between a point and a plate, it only gives rise to very weak side-sparks; discharges in rarefied gases or through Geissler tubes were found to be equally ineffective. The only kind of spark that proved satisfactory was that between two knobs (spheres), and this must neither be too long nor too short. If it is shorter than 0.5 cm, the side-sparks are weak; if it is longer than 1.5 cm, they disappear entirely.

On Very Rapid Electric Oscillations

In the following experiments I used, as being the most suitable, sparks three-quarters of a centimeter long between two brass knobs of 3 cm diameter. Even these sparks were not always equally efficient; the most insignificant details, often without any apparent connection, resulted in useless sparks appearing instead of active ones. After some practice one can judge from the appearance and noise of the sparks whether they are such as are able to excite side-sparks. The active sparks are brilliant white, slightly jagged, and are accompanied by a sharp crackling. That the spark in the discharger is an essential condition for the production of side-sparks is easily shown by drawing the discharger knobs so far apart that the distance between them exceeds the sparking distance of the induction-coil; every trace of the side-sparks then disappears, although the differences of potential now present are greater than before.

The length of the micrometer circuit naturally has great influence upon the length of the sparks produced in it. For the greater this distance, the greater is the retardation that the electric wave suffers between the time of its arrival at the one knob and at the other. If the side-circuit is made very small, the side-sparks become extremely short; but it is scarcely possible to prepare a circuit in which sparks will not show themselves under favorable circumstances. Thus, if you file the ends of a stout copper wire, 4 to 6 cm long, to sharp points, bend it into an almost closed circuit, insulate it and now touch the discharger with this small wire circuit, a stream of very small sparks between the pointed ends generally accompanies the discharges of the induction-coil. The thickness and material (and therefore the resistance) of the side-circuit have very little effect on the length of the side-sparks. We were therefore justified in declining to attribute to the resistance the differences of potential that arise. And, according to our conception of the phenomenon, the fact that the resistance is of scarcely any importance can cause us no surprise; for, to a first approximation, the rate of propagation of an electric wave along a wire depends solely upon its capacitance and inductance, and not upon its resistance. The length of the wire that connects the side-circuit to the principal circuit also has little effect, provided that it does not exceed a few meters. We must assume that the electric disturbance that proceeds from the principal circuit travels along it without suffering any real change of intensity.

On the other hand, the position of the point at which contact with the side-circuit is made has a very noteworthy effect upon the length of the sparks in it. We should expect this to be so if our interpretation of the phenomenon is correct. For if the point of contact is so placed that the paths from it to the two knobs of the micrometer are of equal length, then every variation that passes through the connecting wire will arrive at the two knobs in the same phase, so that no difference of potential between them can arise. Experiment confirms this supposition. Thus, if we shift the point of contact on the side-circuit, which we have hitherto supposed near one of the micrometer knobs, farther and farther away from this knob, the spark length diminishes, and in a certain position the sparks disappear completely or very nearly so; they become stronger again in proportion as the contact approaches the second micrometer knob, and in this position attain the same length as they had in the first position. The point at which the spark length is a minimum may be called the null-point [*Indifferenzpunkt*]. It can generally be determined to within a few centimeters. It always divides the length of the wire between the two micrometer knobs into very nearly equal parts. If the conductor is symmetrical on the right and left of the line joining the micrometer and the null-point, the sparks always disappear completely. The phenomenon can be observed even with quite short side-circuits.

Fig. 7 shows a convenient arrangement for the experiment; *abcd* is a rectangle of bare copper wire 2 mm in diameter, insulated upon sealing-wax supports;[6] in my experiments it was 80 cm wide and 125 cm long. When the connecting wire is attached to either of the knobs 1 or 2, or either of the points *a* and *b*, sparks 3 or 4 mm long pass between 1 and 2; no sparks can be obtained when the connection is at the point *e* in the figure; shifting the contact a few centimeters to right or left causes the sparks to reappear. It should be remarked that we consider sparks as being perceptible when they are only a few hundredths of a millimeter in length.

---

[6]At first I insulated carefully with sealing-wax, etc., But I always found that, for all such experiments as are here considered, the insulation afforded by dry wood is amply sufficient. In the subsequent experiments no other means of insulation was used. [This and all subsequent footnotes by Hertz were added in 1891 when he was preparing the German edition of his *Electric Waves* for publication.]

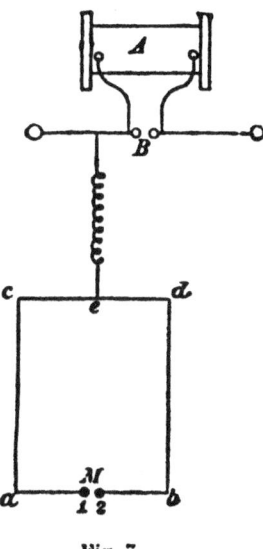

Fig 7.

This experiment shows that the above is not a complete representation of the way in which things go on. For if, after the contact has been adjusted so as to make the sparks disappear, we attach to one of the micrometer knobs another conductor projecting beyond it, active sparking again occurs. This conductor, being beyond the knob, cannot affect the simultaneous arrival of the waves travelling from e to 1 and 2. But it is easy to see what the explanation of this experiment is. The waves do not come to an end after rushing once towards *a* and *b*; they are reflected and traverse the side-circuit several, perhaps many, times and so give rise to stationary oscillations in it. If the paths *eca*1 and *edb*2 are equal, the reflected waves will again arrive at 1 and 2 simultaneously. If, however, the wave reflected from one of the knobs is missing, as in the last experiment, then, although the first disturbance proceeding from *e* will not give rise to sparks, the reflected waves will. We must therefore imagine the abrupt variation that arrives at *e* as creating in the side-circuit the oscillations which are natural to it, much as the blow of a hammer produces in an elastic rod its natural vibrations. If this idea is correct, then the condition for disappearance of sparks in M must be primarily the equality of the vibration periods of the two portions $e1$ and $e2$. These periods of vibration are determined by the product of the inductance of those parts of the conductor by the capacitance of their ends;[7] they are practically independent of the resistance of the branches. The following experiments may be applied to test these considerations and are found to agree with them.

If the connection is placed at the null-point and one of the micrometer knobs is touched with an insulated conductor, sparking begins again because the capacitance of the branch is increased. An insulated sphere of 2 to 4 cm diameter is quite sufficient. The larger the capacitance that is thus added, the more energetic becomes the sparking. Touching at the null-point has no influence since it affects both branches equally. The effect of adding capacitance to one branch is annulled by adding an equal capacitance to the other. It can also be compensated by shifting the connecting wire in the direction of the loaded branch, i.e., by diminishing the inductance of the latter. The addition of a capacitance produces the same effect as increasing the inductance. If one of the branches be cut and a few centimeters or decimeters of coiled copper wire introduced into it, sparking begins again. The change thus produced can be compensated for by inserting an equal length of copper wire in the other branch, or by shifting the copper wire toward the branch that was altered, or by adding a suitable capacitance to the other branch. Nevertheless, it must be remarked that when the two branches are not of like kind, a complete disappearance of the sparks cannot generally be secured, but only a minimum of the sparking distance.

The results are but little affected by the resistance of the branch. If the thick copper wire in one of the branches was replaced by a much thinner copper wire or by a wire of German silver, the equilibrium was not disturbed, although the resistance of the one branch was a hundred times that of the other. Very large fluid resistances certainly made it impossible to secure a disappearance of the sparks, and short air-spaces introduced into one of the branches had a like effect.

---

[7] [In 1853 Lord Kelvin had derived the following formula for the natural period of a circuit containing inductance L and capacitance C:

$$T = 2\pi\sqrt{LC}$$

For the reference see footnote 2 above.]

The self-induction of iron wires for slowly alternating currents is about eight to ten times as great as that of copper wires of equal length and thickness. I therefore expected that short iron wires would produce equilibrium with longer copper wires. This expectation was not confirmed; the branches remained in equilibrium when a copper wire was replaced by an iron wire of equal length. If the theory of the observations here given is correct, this can only mean that the magnetism of iron is quite unable to follow oscillations as rapid as those with which we are concerned here, and that it, therefore, is without effect. A further experiment, which will be described below, appears to point in the same direction.

Induction Effects of Unclosed Currents

The sparks that occur in the preceding experiments owe their origin, according to our supposition, to self-induction. But if we consider that the induction effects in question are derived from exceedingly weak currents in short, straight conductors, there appears to be good reason to doubt whether this really does account satisfactorily for the sparks. In order to settle this doubt I tried to see whether the observed electrical disturbances did not manifest effects of corresponding magnitude in neighboring conductors. I therefore bent some copper wire into the form of rectangular circuits, about 10 to 20 cm on a side and containing only very short spark-gaps. These were insulated and brought near to the conductors in which the disturbances took place, and in such a position that one side of the rectangle was parallel to the conductor. When the rectangle was brought sufficiently close, a stream of sparks always accompanied the discharges of the induction-coil. These sparks were most brilliant in the neighborhood of the discharger, but they could also be observed along the wire leading to the side-circuit as well as in the branches of the latter. The absence of any direct discharge between the inducing and induced circuits was carefully verified, and was also prevented by the introduction of a solid insulator [between the two circuits]. Thus it is scarcely possible that our conception of the phenomenon is erroneous. That the induction between two simple straight lengths of wire, traversed by only small quantities of electricity, can yet become strong enough to produce sparks, shows again the extraordinary shortness of the time in which these small quantities of electricity must pass backwards and forwards along the conductors.

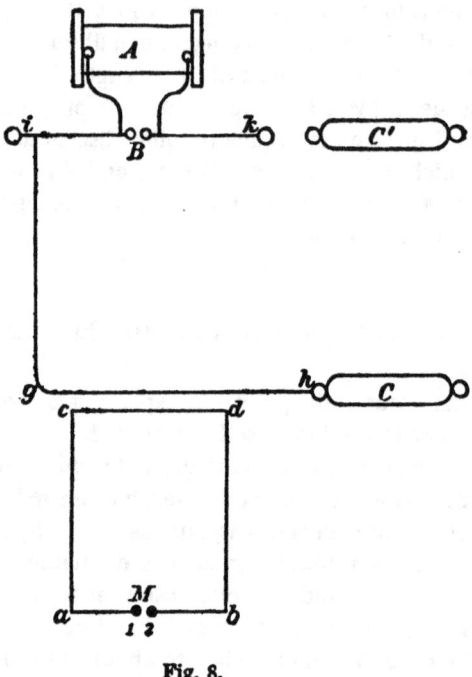

Fig. 8.

In order to study the phenomena more closely, the rectangle which was at first employed as the side-circuit was again brought into use, but this time as the circuit in which induction occurred. Along the short side of this (as indicated in Fig. 8) and at a distance of 3 cm from it was stretched a second copper wire *gh*, which was placed in connection with some part of the discharger. As long as the end *h* of the wire *gh* was free, only weak sparks appeared in the micrometer gap M, and these were due to the discharge current of the wire *gh*. But if an insulated conductor C—one taken from an electrical machine—was then attached to *h*, so that larger quantities of electricity had to pass through the wire, sparks up to two millimeters long appeared at M. This was not caused by an electrostatic effect of the conductor, for if it was attached to *g* instead of to *h*, it was without effect; and the action was not due to the charging current of the conductor, but to the sudden discharge brought about by the sparks. For when the knobs of the discharger were drawn so far apart that sparks could no longer spring across it, then the sparks disappeared completely from the circuit containing the gap M. Not every kind of spark produced a sufficiently energetic discharge; here, again, only such sparks as were before found to occasion powerful side-sparks were found to be effective in exciting the inductive action. The sparks excited in the secondary circuit passed not only between the knobs of the micrometer but also from these to other insulated conductors held near. The length of the sparks was notably diminished by attaching to the knobs conductors of somewhat large capacity or touching one of them with the hand; clearly the quantities of electricity set in motion were too small to charge conductors of rather large capacity to the full potential. On the other hand, the sparking was not much affected by connecting the two micrometer knobs by a short wet thread. No physiological effects of the induced current could be detected; the secondary circuit could be touched or completed through the body without experiencing any shock.

Certain related phenomena led me to suspect that the reason why the electric disturbance in the wire gh produced such a powerful inductive action lay in the fact that it did not consist of a simple charging current, but was rather of an oscillatory nature. I therefore endeavored to strengthen the induction by modifying the conditions so as to make them more favorable to the production of powerful oscillations. The following arrangement of the experiment suited my purpose particularly well. I attached the conductor C as before to the wire gh and then separated the micrometer knobs so far from each other that only single sparks passed over [the gap]. I then attached to the free pole of the discharger k (Fig. 8) a second conductor C' of about the same size as the first. The sparking then again became very active, and on drawing the micrometer knobs still farther apart decidedly longer sparks than at first could be obtained. This cannot be due to any direct action of the portion of the circuit ik, for this would diminish the effect of the portion gh; it must, therefore, be due to the action of the conductor C' upon the discharge current of C. Such an action would be incomprehensible if we assumed that the discharge of the conductor C was aperiodic. It becomes, however, intelligible if we assume that the inducing current in gh consists of an electric oscillation which, in the one case, takes place in the circuit C –wire *gh* –discharger, and in the other in the system Ċ –wire *gh* –wire *ik*–C'. It is clear in the first place that the natural oscillations of the latter system would be more powerful, and in the second place that the position of the spark in it is more suitable for exciting the vibration.

Further confirmation of these views may be deferred for the present. But here we may bring forward in support of them the fact that they enable us to give a more correct explanation of the part that the Ruhmkorff coil plays in the experiment. For if oscillatory disturbances in the circuit C–C' are necessary for the production of powerful induction effects, it is not sufficient that the spark in this circuit should be established in an exceedingly short time, but it must also reduce the resistance of the circuit below a certain value, and in order that this may be the case the current density from the very start must not fall below a certain limit. For this reason the inductive effect is exceedingly feeble when the conductors C and C' are charged by means of an electrical machine[8] (instead of a Ruhmkorff coil) and then allowed to discharge themselves; and it is also very feeble when a small coil is used, or when too large a spark-gap is introduced; in all these cases the motion is aperiodic. On the other hand, a powerful discharge from a Ruhmkorff coil gives rise to oscillations, and therefore to powerful disturbances all around, by performing the following functions: In the first place, it charges the ends C and C' of the system to a high potential; secondly, it gives rise to the disruptive discharge; and thirdly, after starting the discharge, it keeps the resistance of the air-gap so low that oscillations can take place. It is known that if the capacity of the ends of the system is large—if, for example, they consist of the plates of a battery of Leyden jars—the discharge current from these capacitors is able of itself to reduce the resistance of the spark-gap considerably; but when the capacities are small, this function must be performed by some extraneous discharge, and for this reason the

---

[8]I expect that the action of the induction coil partly depends on the fact that directly before the discharge it allows the potential to rise very rapidly. Several related phenomena lead me to believe that when this rapid rise takes place, the difference of potential is forced beyond the point at which sparking occurs when the difference of potential increases slowly; and that for this reason the discharge takes place more suddenly and energetically than when a static charge is released.

discharge of the induction coil is, under the conditions of our experiment, absolutely necessary for exciting oscillations.

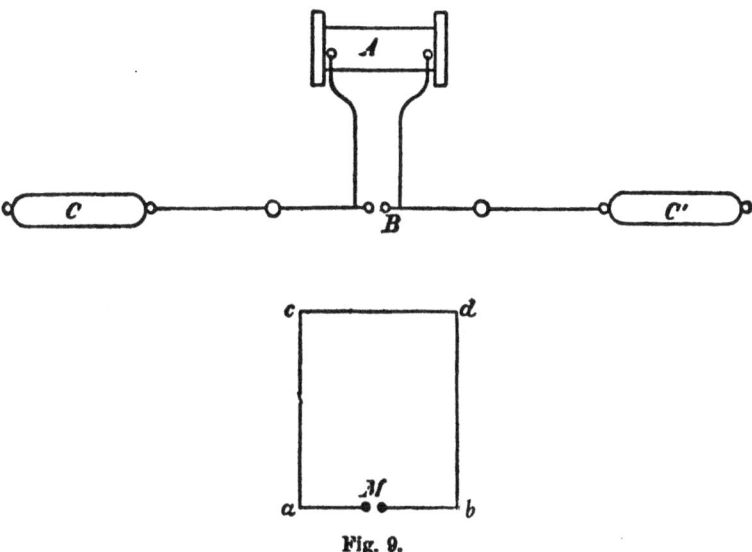

Fig. 9.

As the induced sparks in the last experiment were several millimeters long, I had no doubt that it would be possible to obtain sparks even when the wires used were much farther apart; I therefore tried to arrange a modification of the experiment that appeared interesting. I gave the inducing circuit the form of a straight line (Fig. 9). Its ends were formed by the conductors C and C'. These were 3 meters apart, and were connected by a copper wire 2 mm thick, at the center of which was the discharger of the induction coil. The circuit in which induction was produced was the same as in the preceding experiment, 120 cm long and 80 cm wide. If the shortest distance between the two systems was now made equal to 50 cm, induced sparks 2 mm in length could still be obtained; at greater distances the spark length decreased rapidly, but even when the shortest distance was 1.5 meters, a continuous stream of sparks was perceptible. If the observer moved between the inducing circuit and the one in which induction was produced, no disruption of the observed phenomenon occurred. A few control experiments again established the fact that the phenomena observed were really caused by the current in the rectilinear portion. If one or both halves of this were removed, the sparks in the micrometer ceased, even when the coil was still in action. They also ceased when the knobs of the discharger were drawn so far apart as to prevent any sparking in it. Inasmuch as the differences of electrostatic potential at the ends of the conductors C and C' are in this case greater than before, this shows that these differences of potential are not the cause of the sparks in the micrometer.

Hitherto the induced circuit was closed; it was, however, to be supposed that the induction would take place equally well in an open circuit. A second insulated copper wire was therefore stretched parallel to the straight wire in the preceding arrangement, and at a distance of 60 cm from it. This second wire was shorter than the first; two insulated spheres 10 cm in diameter were attached to its ends and the spark-micrometer was introduced at its center. When the coil was now started, the stream of sparks from it was accompanied by a similar stream in the secondary conductor. But this experiment should be interpreted with caution, for the sparks observed are not solely due to electromagnetic induction. The alternating motion in the system C–C' is indeed superposed upon the Ruhmkorff discharge itself. But during its whole course the latter determines an electrification of the conductor C and an opposite electrification of the conductor C'. These electrifications had no effect on the closed circuit in the preceding experiment, but in the present discontinuous conductor they induce by purely electrostatic action opposite electrifications in the two parts of the conductor, and thus produce sparks in the micrometer. In fact, if we draw the knobs of the discharger so far apart that the sparks in it disappear, the sparks in the micrometer, although weakened, still remain. These sparks represent the effect of electrostatic induction, and conceal the only effect we are really interested in demonstrating.

There is, however, an easy way of getting rid of these disturbing sparks. They die away when we interpose a bad conductor between the knobs of the micrometer, which is most simply done by means of a wet thread. The thread's conductivity is obviously good enough to allow the current to follow the relatively slow alternations of the discharge from the coil; but in the case of the exceedingly rapid oscillations of the rectilinear circuit it is, as we have already seen, not good enough to bring about an equalization of the electrifications. If, after placing the thread in position, we again start the sparking in the primary circuit, vigorous sparking begins again in the secondary circuit, and is now solely due to the rapid oscillations in the primary circuit. I have tested to what distance this action extended. Up to a distance of 1.2 meters between the parallel wires the sparks were easily perceptible; the greatest perpendicular distance at which regular sparking could be observed was 3 meters. Since the electrostatic effect diminishes more rapidly with increasing distance than the electromagnetic induction, it was not necessary to complicate the experiment by using the wet thread at greater distances, for, even without this, only those discharges that excited oscillations in the primary wire were attended by sparks in the secondary circuit.

I believe that the mutual action of rectilinear open circuits, which plays such an important part in the theory, is as a matter of fact illustrated here for the first time.

Resonance Phenomena

We may now regard it as experimentally proven that rapidly varying currents, capable of producing powerful induction effects, are present in conductors connected to the discharge circuit. The existence of regular oscillations, however, was only assumed for the purpose of explaining a comparatively small number of phenomena, which might perhaps be accounted for in other ways. But it seemed to me that the existence of such oscillations might be proved by showing, if possible, syntonic[9] [or resonance] relations between the mutually reacting circuits. According to the principle of resonance, a regularly alternating current must (other things being similar) act with much stronger inductive effect upon a circuit having the same period of oscillation than upon one of only slightly different period.[10] If, therefore, we allow two circuits, which may be assumed to have approximately the same period of vibration, to react on each other, and if we vary continuously the capacitance or inductance of one of them, the resonance should show that for certain values of these

---

[9] [D.E. Jones, the British translator of *Electric Waves*, here translates *resonanzartige* as "symphonic," which seems a poor translation. A reasonable assumption is that he meant "syntonic," a word frequently used at the end of the nineteenth century to describe resonance effects. See, for example, Lodge (1900), pp. 6–8; and Aitken (1985), p. 24.]

[10] Cf. Oberbeck, *Annalen* 26 (1885), p. 245.

quantities the induction is perceptibly stronger than for neighboring values on either side.

The following experiments were devised in accordance with this principle, and after a few trials they fully achieved my goal. The experimental arrangement was very nearly the same as that of Fig. 9, except that the circuits were made somewhat different in size. The primary conductor was a perfectly straight copper wire 2.6 meters long and 5 mm thick. This was divided in the middle so as to include the spark-gap. The two small knobs between which the discharge took place were mounted directly on the wire and connected with the poles of the induction coil. To the ends of the wire were attached two spheres, 30 cm in diameter, made of strong sheet zinc. These could be shifted along the wire. As they formed (electrically) the ends of the conductor, the circuit could easily be shortened or lengthened. The secondary circuit was proportioned so that it was expected to have a somewhat smaller period of oscillation than the primary; it was in the form of a square 75 cm on a side, and was made of copper wire 2 mm in diameter. The shortest distance between the two systems was made equal to 30 cm, and at first the primary current was allowed to remain of full length. Under these circumstances the length of the biggest spark in the induced circuit was 0.9 mm. When two insulated metal spheres of 8 cm diameter were placed in contact with the two poles of the circuit, the spark length increased and could be made as large as 2.5 mm by suitably diminishing the distance between the two spheres. On the other hand, if two conductors of very large surface area were placed in contact with the two poles, the spark length was reduced to a small fraction of a millimeter. Exactly similar results followed when the poles of the secondary circuit were connected with the plates of a Kohlrausch condenser. When the plates were far apart the spark length was increased by increasing the capacitance, but when they were brought closer together the length of the spark again fell to a very small value. The easiest way to adjust the capacitance of the secondary circuit was to hang over its two ends two parallel bits of wire and to alter the lengths of these and their distance apart. By careful adjustment the sparking distance was increased to 3 mm, after which it diminished, not only when the wires were lengthened but also when they were shortened. That an increase of the capacitance should diminish the spark length appeared only natural; but that it should have the effect of increasing it can scarcely be explained except by the principle of resonance.

If our interpretation of the above experiment is correct, the secondary circuit had a somewhat shorter period than the primary before its capacity was increased. Resonance should therefore have occurred when the rapidity of the primary oscillations was increased. And, in fact, when I reduced the length of the primary circuit in the manner above indicated, the sparking distance increased, again reached a maximum of 3 mm when the centers of the terminal spheres were 1.5 meters apart, and again diminished when the spheres were brought closer together. It might be supposed that the spark length would now increase still further if the capacitance of the secondary circuit were again increased as before. But this is not the case; on attaching the same wires, which previously had the effect of increasing the spark length, this latter falls to about 1 mm. This is in accordance with our conception of the phenomenon; that which at first brought about an equality between the periods of oscillation now upsets an equality that had been achieved in another way. The experiment was most convincing when carried out as follows: The spark-micrometer was adjusted for a fixed sparking distance of 2 mm. If the secondary circuit was in its original condition, and the primary circuit 1.5 meters long, sparks passed regularly. If a small capacitance is added to the secondary circuit in the way already described, the sparks are completely extinguished; if the primary circuit is now lengthened to 2.6 meters, they reappear. They are extinguished a second time if the capacitance added to the secondary circuit is doubled; and by continuously increasing the capacitance of the already-lengthened primary circuit they can be made to appear and disappear again and again. This experiment shows us quite plainly that effective action is determined, not by the condition of either of the circuits, but by the resonance [*Harmonie*] between the two.

The length of the induced sparks increased considerably beyond the values given above when the two circuits were brought closer together. When the two circuits were at a distance of 7 cm from one another and were adjusted to exact resonance, it was possible to obtain induced sparks 7 mm long; in this case the electromotive forces induced in the secondary were almost as great as those in the primary.

In the above experiments resonance was secured by altering the inductance and the capacitance of the primary circuit, as well as the capacitance of the secondary circuit. The following experiments show that an alteration of the inductance of the secondary circuit can also be used for this purpose. A series of rectangles *abcd* (Fig. 9) were prepared in which the sides *ab* and *cd* were kept of the same length, but the sides *ac* and *bd* were made of wires varying in length from 10 cm to 250 cm. A marked maximum of the sparking distance was apparent when the length of the rectangle was 1.8 meters. In order to get an idea of the quantitative relations, I measured the longest sparks that appeared with various lengths of the secondary circuit. Fig. 10a shows the results.[11] Abscissae represent the total length of the induced circuit and ordinates the maximum spark length. The points indicate the observed values. Measurements of sparking distances are always very uncertain, but this uncertainty cannot be such as to vitiate the general nature of the result. In another set of experiments not only the lengths of the sides *ab* and *cd*, but also their distance apart (30 cm) and their position were kept constant; but the sides *ac* and *bd* were formed of wires of gradually increasing length coiled into loose spirals. Fig. 10b shows the results obtained.[12] The maximum here corresponds to a somewhat greater length of wire than before. Probably this is because the lengthening of the wire in this case increases only the inductance, whereas in the former case it increased the capacitance as well.

Some further experiments were made in order to determine whether any different result would be obtained by altering the resistance of the secondary circuit. With this intention the wire *cd* of the rectangle was replaced by various thin copper and German silver wires, so that the resistance of the secondary circuit was made about a hundred times as large. This change had very little effect on the sparking distance, and none at all on the resonance or, in other words, on the period of the oscillation.

---

[11] These curves should be compared with the corresponding resonance curves that Herr V. Bjerknes has obtained by more accurate experimental measurements (*Annalen* 44 {1891}, p. 74.)

[12] {The graphs presented by Hertz in Fig. 10 appear to be the first of their kind ever published for electromagnetic waves. In modern terms they show the tuning of a radio receiver to a radio transmitter. Of course, resonance phenomena in sound and acoustics had been well explored by Helmholtz, Kundt and others long before this date {1887}.]

On Very Rapid Electric Oscillations

The effect of the presence of iron was also examined. The wire *cd* was in some experiments surrounded by an iron tube, in others

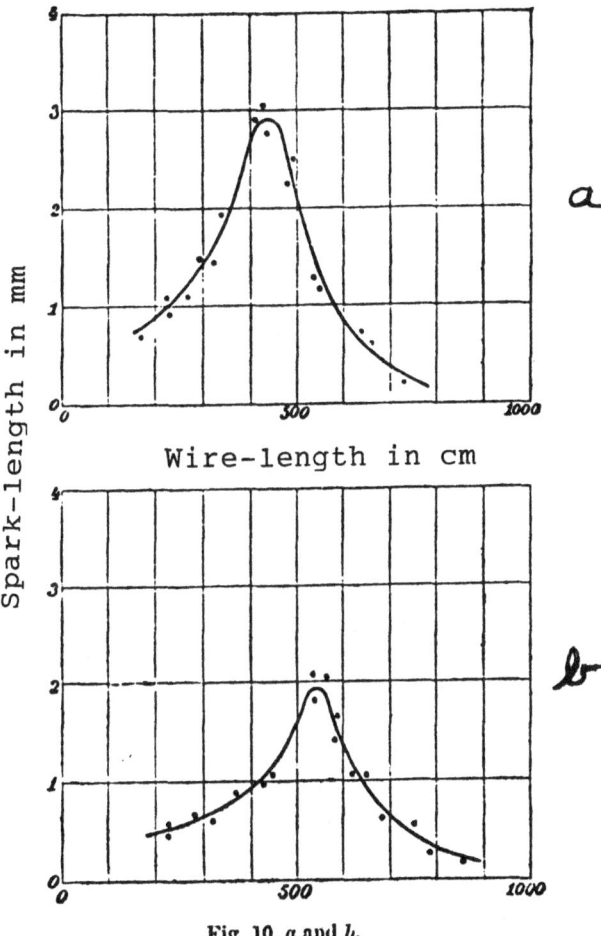

Fig. 10, *a* and *b*.

replaced by an iron wire. Neither of these changes produced an observable effect in any direction. This again confirms the supposition that the magnetism of iron cannot follow such exceedingly rapid oscillations, and so has no effect on them. Unfortunately we possess no experimental knowledge as to how the oscillatory discharge of Leyden jars is affected by the presence of iron.

Vibration Nodes

The oscillations which we excited in the secondary circuit, and which we measured by means of the sparks in the micrometer, are not the only ones, but are the simplest possible ones in that circuit. While the potential at the ends oscillates backwards and forwards continually between two limits, it always retains the same mean value in the middle of the circuit. This middle point is therefore a node of the electric oscillation, and the oscillation has only this one node. Its existence can also be shown experimentally, and this in two ways. In the first place, it can be done by bringing a small insulated sphere near the wire. The mean value of the potential of the small sphere cannot differ appreciably from that of the neighboring bit of wire. Sparking between the knob and the wire can therefore only arise through the potential of the neighboring point of the system experiencing sufficiently large oscillations about the mean value. Hence there should be vigorous sparking at the ends of the system and none at all near the node. And this is in fact so, except that when the nodal point is touched, the sparks indeed do not entirely disappear, but are only reduced to a minimum. A second way of showing the nodal point is clearer. Adjust the secondary circuit for resonance and draw the knobs of the micrometer so far apart that sparks can only pass by the assistance of the action of resonance. If any point of the system is now touched with a conductor of some capacity, we should in general expect that the resonance would be disturbed and that the sparks would disappear; only at the node would there be no interference with the period of the oscillation. Experiment confirms this. The middle of the wire can be touched with an insulated sphere, or with the hand, or can even be placed in metallic connection with the gas pipes without affecting the sparks; similar interference at the side-branches or the poles causes the sparks to disappear.

After the possibility of fixing a nodal point was thus proved, it appeared to me to be worthwhile to experiment on the production of a vibration with two nodes. I proceeded as follows: The straight primary conductor CC' and the rectilinear secondary circuit $abcd$ were set up as in the earlier experiments and brought to resonance. An exactly similar rectangle $efgh$ was then placed opposite to $abcd$ as shown in Fig. 11, and the neighboring poles of both were joined (1 with 3 and 2 with 4). The whole system forms a closed metallic circuit, and the lowest or fundamental vibration possible in it has two nodes. Since the period of this vibration must very nearly agree with the period of either half, and

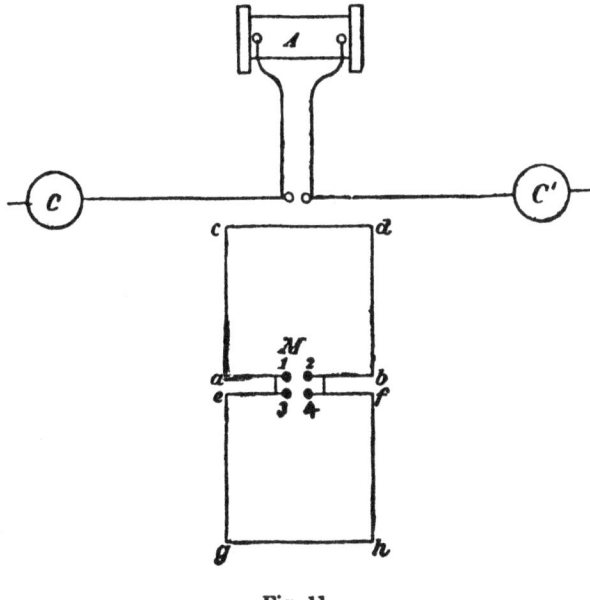

Fig. 11.

therefore with the period of the primary conductor, it was supposed that vibrations would develop having two antinodes at the junctions 1–3 and 2–4, and two nodes at the middle points of cd and gh. These vibrations were always measured by the sparking distance between the knobs of the micrometer that formed the poles 1 and 2. The results of the experiment were as follows: Contrary to what was expected, it was found that the sparking distance between 1 and 2 was considerably diminished by the addition of the rectangle $efgh$. From about 3 mm it fell to 1 mm. Nevertheless there was still resonance between the primary circuit and the secondary. For every alteration of $efgh$ reduced the sparking distance still further, and this whether the alteration was in the direction of lengthening or shortening the rectangle. Further, it was found that the two nodes that were expected were actually present. By holding a sphere near cd and gh, only very weak sparks could be obtained as compared with those from ae and bf. And it could also be shown that these nodes belonged to the same vibration that, when strengthened by resonance, produced the sparks 1–2. For the sparking distance between 1 and 2 was not diminished by touching along *cd* or *gh*, but it was by touching at every other place.

The experiment may be modified by breaking one of the connections 1–3 or 2–4; let us say the latter. As the current strength of the induced oscillation is always zero at these points, this cannot interfere much with the oscillation. And, in fact, after the connection has been broken, it can be shown as before that resonance takes place, and that the vibrations corresponding to this resonance have two nodes at the same places. Of course, there was this difference that the vibration with two nodes was no longer the lowest possible vibration; the vibration of longest period would be one with a single node between *a* and *e* and having the highest potentials at the poles 2 and 4. And if we bring the knobs at these poles nearer together we find that there is feeble sparking between them. We may attribute these sparks to an excitation, even if only feeble, of the fundamental vibration; and this supposition is made almost a certainty by the following extension of the experiment: We stop the sparks between 1 and 2 and direct our attention to the length of the sparks between 2 and 4, which measures the intensity of the fundamental vibration. We now increase the period of oscillation of the primary circuit by extending it to the full length and adding to its capacity. We observe that the sparks then increase to a maximum length of several millimeters and then again become shorter. Clearly they are longest when the oscillation of the primary current is in step with the fundamental oscillation. And, while the sparks between 2 and 4 are longest, it can be easily shown that this time only a single nodal point corresponds to these sparks. For only between *a* and *e* can the conductor be touched without interfering with the sparks, whereas touching the previous nodal points interrupts the stream of sparks. Hence it is in this way possible, in any given conductor, to make either the fundamental vibration or the first overtone predominant.

Meanwhile, there are several further problems that I have not solved; among others, whether it is possible to establish the existence of oscillations with several nodes. The results already described were obtained only by careful attention to insignificant details; and so it appeared probable that the answers to further questions would turn out to be more or less ambiguous. The difficulties that present themselves arise partly from the nature of the methods of observation,

On Very Rapid Electric Oscillations

and partly from the nature of the electric disturbances observed. Although these latter manifest themselves as undoubted oscillations, they do not exhibit the characteristics of perfectly regular oscillations. Their intensity varies considerably from one discharge to another, and from the comparative unimportance of the resonance effects we conclude that the damping must be fast; many secondary phenomena point to the superposition of irregular disturbances upon the regular oscillations, as indeed was to be expected from the complex nature of the system of conductors. If we wish to compare, with respect to their mathematical relations, our oscillations with any particular kind of acoustic oscillations, we must not choose the long-lasting harmonic oscillations of uniform strength that are characteristic of tuning-forks and [vibrating] strings, but rather such as are produced by striking a wooden rod with a hammer—oscillations which rapidly die away, and with which are mingled irregular disturbances.[13] And when we are dealing with oscillations of the latter class we are obliged, even in acoustics, to content ourselves with mere indications of resonance, formation of nodes, and similar phenomena.

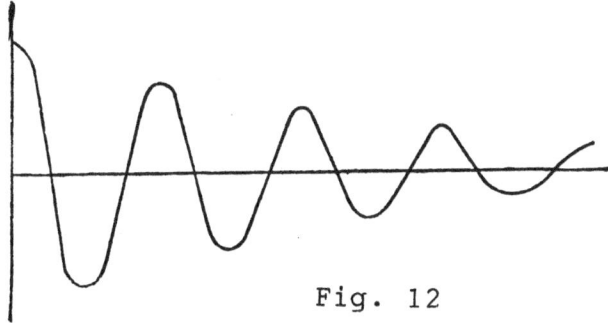

Fig. 12

This graph shows, as a function of time, the variations in the electric field emitted by Hertz's spark oscillator.

---

[13]This remark in my first paper shows clearly that I never conceived the oscillations of my primary conductor as perfectly regular and long-lasting sinusoidal oscillations. The value of the damping has recently been carefully determined by Herr V. Bjerknes (*Annalen* 44 [1891], p. 74). Fig. 12 shows, in accordance with the results of his experiments, the form of oscillation produced by a conductor similar to our primary conductor.

For the sake of those who may wish to repeat these experiments and obtain the same results, I must add one remark the exact significance of which may not be clear at first. In all the experiments described, the apparatus was set up in such a way that the spark of the induction coil was visible from the place where the spark in the micrometer took place. When this is not the case the phenomena are qualitatively the same, but the spark lengths appear to be diminished. I have undertaken a special investigation of this phenomenon and intend to publish the results in a separate paper.[14]

Theoretical

It is highly desirable that quantitative data respecting the oscillations should be obtained by experiment. But as there is at present no obvious way of doing this, we are obliged to have recourse to theory to obtain at least some approximate values of these data. The theory of electric oscillations that has been developed by Sir W. Thomson, v. Helmholtz, and Kirchhoff has been verified as far as the oscillations of open induction coils and oscillatory Leyden-jar discharges are concerned.[15] We may therefore feel certain that the application of this theory to the present phenomena will give results that are correct, at least as far as the order of magnitude is concerned.

To begin with, the period of oscillation is the most important element. As an example that may be used for a calculation, let us determine the (simple or half) period of oscillation T of the primary conductor which we used in the resonance experiments. Let L denote the inductance of this conductor in magnetic units, expressed in centimeters; C the capacitance of either of its ends in electrostatic units (and therefore also expressed in centimeters); and finally c the

---

[14]See *El.Waves*, pp. 63–79. [This is Hertz's famous paper: "On an Effect of Ultraviolet Light upon the Electric Discharge," the first recorded report of the photoelectric effect. This paper is included here as Paper No. 7.]

[15]L. Lorenz, *Annalen* 7 (1879), p. 161.

velocity of light in cm/s.[16] Then, assuming that the resistance is small,

$$T = \left(\frac{\pi}{c}\right)\sqrt{LC}$$

In our experiments the capacitance of the ends of the conductor consisted mainly of the spheres attached to them. We shall therefore not be far wrong if we take C as being the radius of either of these spheres, and so put C = 15 cm.[17] As regards the inductance L, it was that of a straight wire, of diameter d = 0.5 cm, and of length w = 150 cm. when resonance occurred. Calculated by Neumann's formula

$$L = \iint \frac{\beta}{r} ds\, ds'\,^{18}$$

the value of L for such a wire is $2w\{\ln(4w/d) - 0.75\}$ and therefore in our experiments it is L = 1902 cm.

At the same time we know that it is not certain whether Neumann's formula is applicable to open circuits. The most general

---

[16] [In this section we have changed Hertz's notation to bring it into closer conformity with today's accepted notation. Note that his periods and wavelengths are really half-periods and half-wavelengths. Another possible source of confusion is that in this paper Hertz uses the same symbol (A) for the velocity of light that he had used for its reciprocal in Paper No. 3.]

[17] Just at this point there has crept into the calculation a fatal mistake, the unfortunate effects of which extend even to some of the subsequent papers.

The capacitance C in the formula

$$T = \left(\frac{\pi}{c}\right)\sqrt{LC}$$

determines the amount of electricity that exists at one end of an oscillating conductor when the difference of potential between the two ends is equal to unity [since C = Q/v]. Now if these two ends consist of two spheres that are far away from each other, and if their difference of potential is equal to unity, then the difference of potential between each of them and the surrounding space is equal to + or − $1/2$. Therefore the charge on each of the spheres, measured in absolute units, is found by dividing its capacity, i.e., its radius measured in centimeters, by 2. Hence we should here put C = $15/2$ cm, and not C = 15 cm. The [half] period of oscillation, T, now becomes smaller in the proportion of 1 to 1.41, so that T is now equal to 1.26 hundred-millionths of a second.

M.H. Poincaré, as already stated in the introduction, first drew attention to this error (*Comptes Rendus* 111 [1890], p. 322.)

[18] [Here ds and ds' are two length elements of the conductor, r is the distance between their centers, and β is the angle between the vectors marking the direction of ds and ds'.]

formula, as given by v. Helmholtz, contains an undetermined constant k, and this formula is in accordance with the known experimental data. Calculated according to the general formula, we get for a straight cylindrical wire of length w and diameter d the value

$$L = 2w\left[\ln\left(\frac{4w}{d}\right) - 0.75 + \left(\tfrac{1}{2}\right)(1-k)\right].$$

If in this we put $k = 1$, we arrive at Neumann's value; if we put $k = 0$, or $k = -1$, we obtain values that correspond to Maxwell's theory, or Weber's theory, respectively. If we assume, at any rate, that one of these values is the correct one, and therefore exclude the assumption that it may have a very large negative or positive value, then the true value of k is not of much moment. Thus the values of the inductance calculated with these various values of k differ from each other by less than one-sixth of their value; and so if an inductance of 1902 cm does not exactly correspond to a length of wire of 150 cm, it does correspond to a length of our primary conductor not differing greatly from this. From the values of L and C it follows that the length $\pi\sqrt{(LC)}$ is 531 cm. This is the distance through which light travels in the time of an oscillation, and is at the same time the wavelength of the electromagnetic waves which, according to Maxwell's view, are supposed to be the external effect of the oscillations. From this length it follows that the period of oscillation (T) is 1.77 hundred-millionths of a second; thus the statement that we made in the beginning as to the order of magnitude of the period is justified.[19]

Let us now turn our attention to what the theory can tell us about the damping ratio of the oscillations. In order that oscillations may be possible in the open circuit, its resistance must be less than $2c\sqrt{L/C}$. For our primary conductor, $\sqrt{L/C}$ is 11.25; now, since the velocity c is equal to 30 earth-quadrant/sec, or to 30 ohms, it follows that the limit for r admissible in our experiment is 676 ohms. It is very probable that the true resistance of a powerful discharge lies below this limit, and thus from the theoretical point of view there is no

---

[19][It should be noted that, according to footnote 17 above, the corrected value of the half period T is $1.77/1.41 = 1.26$ hundred millionths of a second. In the above paragraph Hertz is referring to half periods and half wavelengths, and so $c = \lambda/T$ remains unchanged when both wavelengths and periods are replaced by half their values.]

contradiction of our assumption of oscillatory motion. If the actual value of the resistance lies somewhat below this limit, the amplitude of any one oscillation would bear to the amplitude of the one immediately following it the ratio of 1 to $\exp(-rT/2L)$. The number of oscillations required to reduce the amplitude in the ratio of 2.71 to 1 is therefore equal to $2L/rT$ or to

$$\frac{2c\sqrt{L/C}}{\pi r}.$$

It therefore bears to 1 the same ratio that $1/\pi$ of the calculated limiting value bears to the actual value of the resistance, or the same ratio as 215 ohms to r. Unfortunately we have no means of even approximately estimating the resistance of a spark-gap. Perhaps we may regard it as certain that this resistance amounts to at least a few ohms, for even the resistance of strong electric arcs does not fall below this. It would follow from this that the number of oscillations we have to consider should be counted by tens and not by hundreds or thousands.[20] This is in complete agreement with the character of the phenomenon, as has already been pointed out at the end of the preceding section. It is also in accordance with the behavior of the very similar oscillatory discharges of Leyden jars, in which case the oscillations of perceptible strength are similarly limited to a very small number.

In the case of purely metallic secondary circuits the conditions are quite different from those of the primary currents to which we have confined our attention. In the former a disturbance would, according to theory, only come to rest after thousands of oscillations. There is no good reason for doubting the correctness of this result; but a more complete theory would certainly have to take into consideration the reaction upon the primary conductor, and would thus probably arrive at higher values for the damping of the secondary conductor as well.

Finally, we may raise the question whether the induction effects of the oscillations that we have observed were of the same order as those that theory would lead us to expect, or whether there is here any

---

[20]The result is about right, but the way in which it is deduced is not sound. We have just referred [in footnote 17] to an error in the calculation that would have to be corrected; and, in addition, no account has been taken of damping due to radiation. Indeed, I had not thought of this when writing the paper.

appearance of contradiction between the phenomena themselves and our interpretation of them. We may answer this question by the following considerations: We observe, in the first place, that the maximum value of the electromotive force which the oscillation induces in its own circuit must be very nearly equal to the maximum difference of potential at the ends, for, if the oscillations were not damped, there would exist complete equality between the two magnitudes; inasmuch as the potential difference of the ends and the electromotive force of induction would in that case be in equilibrium at every instant. Now in our experiments the potential difference between the ends was of a magnitude corresponding to a sparking distance of 7 or 8 mm, and any such sparking distance fixes the value of the greatest inductive effect of the oscillation in its own path. We observe, in the second place, that at every instant the induced electromotive force in the secondary circuit bears to that induced in the primary circuit the same ratio as the mutual inductance M between the primary and secondary circuits bears to the self-inductance L of the primary circuit. There is no difficulty in calculating according to known formulas the approximate value of M for our resonance experiments. It was found to vary in the different experiments between one-ninth and one-twelfth of L. From this we may conclude that the maximum electromotive force which our oscillation excites in the secondary circuit should be of such strength as to give rise to sparks of $\frac{1}{2}$ to $\frac{2}{3}$ mm in length. And accordingly the theory allows us, on the one hand, to expect visible sparks in the secondary circuit under all circumstances, and, on the other hand, we see that we can only explain sparks of several millimeters in length by assuming that several successive inductive effects strengthen one other. Thus from the theoretical side as well we are compelled to regard the phenomena that we have observed as being the results of resonance.[21]

Further applications of theory to these phenomena can only be of service when we shall have succeeded by some means in determining the period of oscillation directly.[22] Such a measurement

---

[21][Note the effective use of order-of-magnitude calculations made by Hertz here and in his other papers. Other great experimental physicists like Fermi possessed this same impressive ability.]

[22][Hertz's experiments would obviously have been very much easier if he had a convenient instrument for measuring the frequency of his spark

On Very Rapid Electric Oscillations

would not only confirm the theory but would also lead to an extension of it. The purpose of the present research is simply to show that, even in short metallic conductors, oscillations can be induced, and to indicate in what manner their characteristic oscillations can be excited.

---

oscillator, but no such device existed for the 100 MHz (10^8 cycles per second) frequencies he was generating. It was only with the advent of radar at the beginning of World War II that frequency meters capable of measuring such high frequencies came into use.]

PAPER NO. 7

On an Effect of Ultraviolet Light upon the Electric Discharge

(Heinrich Hertz, "Über einen Einfluss des ultravioletten Lichtes auf die elektrische Entladung," *Sitzungsberichte der Berliner Akademie der Wissenschaften*, 9 June 1887; Wiedemann's *Annalen der Physik und Chemie* 31 [1887], pp. 983–1000.)[1]

In a series of experiments on the effects of resonance between very rapid electric oscillations that I have carried out and recently published,[2] two electric sparks were produced by the same discharge of an induction coil, and were therefore produced simultaneously. One of these, the spark A, was the discharge-spark of the induction coil and served to excite the primary oscillation. The second, the spark B, belonged to the induced or secondary oscillation. The latter was not very luminous; in the experiments its maximum had to be accurately measured. I occasionally enclosed the spark B in a dark case so as more easily to make the observations; and in so doing I observed that the maximum spark length became decidedly smaller inside the case than it was before. On removing in succession the various parts of the case, it was seen that the only portion of it that exercised this prejudicial effect was that which screened the spark B from the spark A. The screen on that side exhibited this effect, not only when it was in the immediate neighborhood of the spark B, but also when it was interposed at greater distances from B between A and B. A phenomenon so remarkable called for closer investigation. The following communication contains the results that I have been able to establish in the course of the investigation.

1. The phenomenon could not be traced to any screening effect of an electrostatic or electromagnetic nature. For the effect was not only exhibited by good conductors interposed between A and B, but

---

[1] [*El.Waves*, pp. 63–79. On this paper, see pp. 37–39 in the Introductory Biography, and Stuewer (1971).]

[2] See the paper "On Very Rapid Electric Oscillations." [This is the preceding paper (No. 6) in this book. Most of Hertz's references to his previous work, and the terminology he uses in the present paper, refer back to this paper.]

also by perfect non-conductors, in particular by glass, paraffin, and ebonite, which cannot possibly exert any [electrical] screening effect. In addition, metal gratings of coarse texture showed no effect, although they act as effective [electrical] screens.

2. The fact that both sparks A and B corresponded to very rapid, synchronous oscillations was not important. For the same effect could be exhibited by exciting two simultaneous sparks in any other way. It also appeared when, instead of the induced spark, I used a side-spark (this term having the same significance as in my earlier paper). It also appeared when I used as the spark B a side-discharge (according to Riess's terminology), such as is obtained by connecting one pole of an induction coil with an insulated conductor and introducing a spark-gap. But it can best and most conveniently be exhibited by inserting in the same circuit two induction coils with a common interrupter, the one coil giving the spark A and the other the spark B. This arrangement was almost exclusively used in the subsequent experiments. As I found that the experiment succeeded with a number of different induction coils, it could be carried out with any combination of induction apparatus that suited me. At the same time it will be convenient to

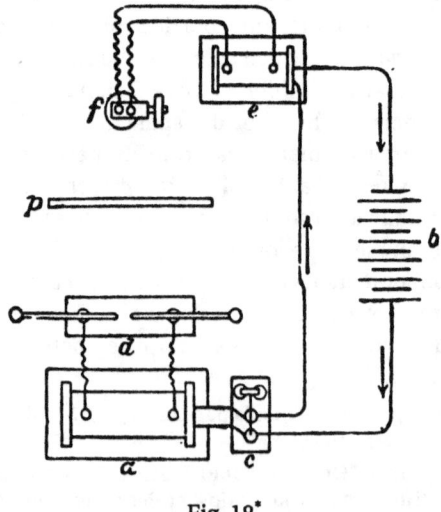

Fig. 18*

---

*As noted on p. xvii, gaps occur in the numbering of figures in this book. The missing numbers belong to figures that appear in articles from *Electric Waves* not included here.

On an Effect of Ultraviolet Light

describe the particular experimental arrangement that gave the best results and was most frequently used.

The spark A was produced by a large Ruhmkorff coil (a in Fig. 18), 52 cm long and 20 cm in diameter, fed by six large Bunsen cells (b) and provided with a separate mercury switch (c). With the current used it could give sparks up to 10 cm long between point and plate, and up to about 3 cm between two spheres. The spark generally used was one of 1 cm length between the points of a common discharger (d). The spark B was produced by a smaller coil (originally intended for medical use) of relatively greater current strength, but having a maximum spark length of only 0.5 to 1.0 cm. As it was here introduced into the circuit of the larger coil, its condenser did not come into play, and thus it produced sparks of only 1 to 2 mm length. The sparks used were ones about 1 mm long between the nickel-plated knobs of a Riess spark-micrometer (f), or between brass knobs of 5 to 10 cm diameter. When the apparatus was thus arranged with both spark gaps parallel and not too far apart, the interruptor set going and the spark-micrometer drawn out just far enough to still permit sparks to pass regularly, then on placing a plate (p) of metal, glass, etc., between the two spark-gaps d and f, the sparks disappeared immediately and completely. On removing the plate, they immediately reappeared.

3. The effect becomes more marked as the spark A is brought nearer to the spark B. The distance between the two sparks when I first observed the phenomenon was 1.5 m, and the effect is, therefore, easily observed at this distance. I have been able to detect indications of it up to a distance of 3 m between the sparks. But at such distances the phenomenon manifests itself only in the greater or less regularity of the stream of sparks at B; at distances less than a meter its strength can be measured by the difference between the maximum spark length before and after the interposition of the plate. In order to indicate the magnitude of the effect I give the following, understandably rough, observations that were obtained with the experimental arrangement shown in Fig. 18.

It will be seen [from the table on page 226] that, under certain conditions, the length of the spark is doubled by removing the plate.

4. The observations given in the table may also be adduced as proof of the following statement, which the reader will probably have assumed from the first: the phenomenon does not depend on any

Distance between Sparks in cm	Length of Spark B in mm *before* and *after* Inserting the Plate		Difference
great	0.8	0.8	0
50	0.9	0.8	0.1
40	1.0	0.8	0.2
30	1.1	0.8	0.3
20	1.3	0.8	0.5
10	1.5	0.8	0.7
5	1.6	0.8	0.8
2	1.8	0.8	1.0

inhibiting effect of the plate on the spark B, but on its annulling a certain action of the spark A that tends to increase the sparking distance. When the distance between the sparks A and B is great, if we so adjust the spark-micrometer that sparks no longer pass at B, and then bring the spark-micrometer nearer to A, the stream of sparks in B reappears; this is the action. If we now introduce the plate, the sparks are extinguished; this is the cessation of the action. Thus the plate only forms a means of exhibiting conveniently and plainly the action of the spark A. I shall in what follows call A the active spark and B the passive spark.

5. The efficiency of the active spark is not confined to any special form of this spark. Sparks between knobs, as well as sparks between points, proved to be effective. Short, straight sparks, as well as long, jagged ones, exhibited the effect. There was no difference of any imporance between faintly luminous bluish sparks and brilliant white ones. Even sparks 2 mm long made their influence felt to considerable distances. Nor does the action proceed from any special part of the spark; every part is effective. This statement can be verified by drawing a glass tube over the spark-gap. The glass does not allow the effect to pass through, and so the spark under these conditions is inactive. But the effect reappears as soon as a short bit of the spark is exposed at one pole or the other, or in the middle. I have not observed any influence due to the metal of the pole, and in arranging the experiment it is not of any importance that the active spark should be parallel to the passive one.

6. On the other hand, the susceptibility of the passive spark to the action is to a certain extent dependent on its form. I could detect no susceptibility with long, jagged sparks between points, and but little with short sparks between points. The effect was best displayed by sparks between knobs, and of these most strikingly by short sparks. It is advisable to use for the experiments sparks 1 mm long between knobs of 5 to 10 mm diameter. Still I have distinctly recognized the effect with sparks 2 cm long. Perhaps the absolute lengthening that such sparks experience is really as great as in the case of shorter sparks, but still the relative increase in length is much smaller, and hence the effect disappears in the differences that occur between the single discharges of the coil. I have not discovered any perceptible influence due to the material of the poles. I examined sparks between poles of copper, brass, iron, aluminum, tin, zinc, and lead. If there was any difference between the metals with respect to the susceptibility of the spark, it appeared to be slightly in favor of the iron. The poles must be clean and smooth; if they are dirty or corroded by long use, the effect is not produced.

7. The relationship between the two sparks is reciprocal. That is to say, not only does the larger and stronger spark increase the spark-length of the smaller one, but conversely the smaller spark has the same effect on the spark length of the larger one. For example, using the same apparatus as before, let us adjust the spark-micrometer so that the discharge in it jumps the gap, but let the discharger be so adjusted that the discharges of the large coil just miss fire. On bringing the spark-micrometer nearer, we find that these discharges are again produced; but that on introducing a plate the action ceases. For this purpose the spark of the large coil must naturally be fairly sensitive; and, inasmuch as long sparks are less sensitive, the effect is not so striking. If both coils are just at the limit of their distance for sparking, complications arise that probably have no connection with the matter now under discussion.[3] One frequently has occasion to notice a long

---

[3] The complications here mentioned, and the starting of long sparks by other much shorter ones, refer to the following phenomena. Let the primary coils of two induction coils be placed in the same circuit, and let their spark-gaps be so adjusted as to be just on the point of sparking. Any cause that starts sparking in one of them will now make the other begin to spark as well; and this quite independently of the mutual action of the light emitted by the two sparks—which, indeed, can easily be excluded. Sparking begins either in both of them, or else in neither. Again, let a Topler-Holtz induction machine, with a

spark being started by other ones that are much smaller, and in part this may certainly be ascribed to the action that we are investigating. When the discharge of a coil is made to take place between knobs, and the knobs are drawn apart until the sparks cease, then it is found that the sparking begins again when an insulated conductor is brought near one of the knobs so as to draw small side-sparks from it. I have proved to my entire satisfaction that the side-discharges here perform the function of an active spark in the sense of the present investigation. It is even sufficient to touch one of the knobs with a non-conductor, or to bring the nib [of a pen] somewhat near it, in order to give rise to the same action. It appears at least possible that the function of an active spark is here performed by the scarcely visible side-discharges over the surface of the non-conductor and the pen nib.

8. The effect of the active spark spreads out on all sides in straight lines and forms rays exactly in accordance with the laws of the propagation of light. Suppose the axes of both the sparks in use are placed vertically, and let a plate with a vertical edge be pushed gradually from the side in between the sparks. It is then found that the effect of the active spark is stopped, not gradually, but suddenly, and in a definite position of the plate. If we now look along the edge of the plate from the position of the passive spark, we find that the active

---

disc 40 cm in diameter, be turned rapidly so as to give sparks having a maximum length of about 15 cm. Now draw the poles 20 to 25 cm apart, so that the sparking stops entirely; it will now be found that a long crackling spark can again be consistently obtained every time a small spark is drawn from the negative conductor, either with a knuckle of the hand or with the knob of a Leyden jar; or the negative pole may be connected to a long conductor, and sparks may be drawn from this with the same result. The "releasing" spark may be quite short and weak; if it is drawn with the knob of a Leyden jar, the jar only appears slightly charged. The same effect cannot be obtained by drawing sparks from the positive pole. The phenomenon must have often been observed before; but I have not found any mention of it in the literature on the subject.

I can give no explanation of these phenomena. They clearly have the same origin as the phenomena that Herr G. Jaumann has described in his paper entitled "Einfluss rascher Potentialveränderungen auf den Entladungsvorgang," *Sitzungsberichte d. Akad. d. Wissensch. zu Wien* 97, Abth. IIa (July, 1888). Herr Jaumann arrives at the conclusion that "not only the form, condition and potential difference of the discharge-field," but also "the manner in which the potential difference varies, and probably its rate of variation, materially influence the discharge." It is to be hoped that these phenomena will be further explained.

spark is just hidden by the plate. If we adjust the plate with its edge vertical between the two sparks and slowly move it sideways, the action begins again in a definite position, and we now find that, viewed from the position of the passive spark, the active spark has just become visible beyond the edge of the plate. If we place between the sparks a plate with a small vertical slit and move it backwards and forwards, we find that the action is only transmitted in one perfectly definite position, namely, when the active spark is visible through the slit from the position of the passive spark. If several plates with such slits are interposed behind one another, we find that in one particular position the action passes through the whole lot. If we seek these positions by trial, we end by finding (most easily, of course, by looking through) that all the slits lie in the vertical plane that passes through the two sparks. If at any distance from the active spark we place a plate with an aperture of any shape, and by moving the active spark about, fix the limits of the space within which the action is exerted, we obtain as this limit a conical surface determined by the active spark as apex and by the limits of the aperture. If we place a small plate in any position in front of the active spark, we find by moving the passive spark about, that the plate stops the action of the active spark within exactly the space that it shelters from its light. It scarcely needs explanation that the action is not only annulled in the shadows cast by external bodies, but also in the shadows of the knobs of the passive spark. In fact, if we turn the latter so that its axis remains in the plane of the active spark, but is perpendicular to it instead of being parallel, the action immediately ceases.

9. Most solid bodies hinder the action of the active spark, but not all; a few solid bodies are transparent to it. All the metals that I have tried proved to be opaque, even in thin sheets, as also did paraffin, shellac, resin, ebonite, and india-rubber; all kind of colored and uncolored, polished and unpolished, thick and thin glass, porcelain, and earthenware; wood, pasteboard, and paper; ivory, horn, animal hides, and feathers; lastly, agate and, in a very remarkable manner, mica, even in the thinnest possible flakes. Further investigation of crystals showed variations from this behavior. Some indeed were equally opaque, e.g., copper sulphate, topaz, and amethyst; but others, such as crystallized sugar, alum, calcspar, and rock salt, transmitted the action, although with diminished intensity; finally, some proved to be completely transparent, such as gypsum

(selenite)[4], and above all rock-crystal,[5] which scarcely interfered with the action even when in layers several centimeters thick.

This research can be conveniently carried out in the following fashion. The passive spark is placed a few centimeters away from the active spark, and is brought to its maximum length. The substance to be examined is now interposed. If this does not stop the sparking, the substance is very transparent. But if the sparking is stopped, the spark-gap must be shortened until it comes again into action. An opaque plate is now interposed in addition to the substance under investigation. If this stops the sparking once more, or weakens it, then the [original] substance must have been at least partially transparent; but if the plate produces no further effect, the substance must have been quite opaque. The influence of the interposed bodies increases with their thickness, and it may properly be described as an absorption of the action of the active spark; in general, however, even those substances that only act as partial absorbers exert this influence even in very thin layers.

10. Liquids also proved to be partly transparent and partly opaque to the action. In order to experiment upon them the active spark was brought about 10 cm vertically above the passive one, and between both was placed a glass vessel, of which the bottom consisted of a circular plate of rock-crystal 4 mm thick. Into this vessel a layer, more or less deep, of the liquid was poured, and its influence was then estimated in the manner described above for solid bodies. Water proved to be remarkably transparent; even a depth of 5 cm scarcely hindered the action. In thinner layers pure concentrated sulphuric acid, alcohol, and ether were also transparent. Pure hydrochloric acid, pure nitric acid, and liquid ammonia proved to be partially transparent. Molten paraffin, benzene, petroleum, carbon bisulphide, a solution of ammonium sulphide, and strongly colored liquids like solutions of fuchsine and potassium permanganate were completely, or nearly completely, opaque. The experiments with salt solutions proved to be interesting. A layer of water 1 cm deep was introduced into the rock-crystal vessel; the concentrated salt solution was added

---

[4][Selenite is the name given to gypsum in the form of clear, colorless crystals.]

[5][Rock-crystal is another name for transparent, colorless quartz. Such quartz is transparent both to visible light and to ultraviolet radiation down to wavelengths of about 200 nm.]

On an Effect of Ultraviolet Light

to this drop-by-drop, stirred, and the effect observed. With many salts the addition of a few drops, or even a single drop, was sufficient to extinguish the passive spark; this was the case for nitrate of mercury, sodium hyposulphite, potassium bromide and potassium iodide. When iron and copper salts were added, the extinction of the passive spark occurred before any distinct coloring of the water could be perceived. Solutions of sal-ammoniac, zinc sulphate, and common salt[6] exercised an absorption when added in larger quantities. On the other hand, the sulphates of potassium, sodium, and magnesium were very transparent even in concentrated solution.

11. It is clear from the experiments made in air that some gases permit the transmission of the action even to considerable distances. Some gases, however, are very opaque to it. In experimenting on gases a tube 20 cm long and 2.5 cm in diameter was interposed between the active and passive sparks; the ends of this tube were closed by thin quartz plates, and by means of two side-tubes any gas could at will be led through it. A diaphragm prevented the transmission of any action excepting through the glass tube. Between hydrogen and air there was no noticeable difference. Nor could any falling off in the action be perceived when the tube was filled with carbonic acid. But when coal-gas was introduced, the sparking at the passive spark-gap immediately ceased. When the coal-gas was driven out by air, the sparking began again; and this experiment could be repeated with perfect regularity. Even introduction of air with which some coal-gas had been mixed hindered the transmission of the action. Hence a much shorter stratum of coal-gas was sufficient to stop the action. If a current of coal-gas 1 cm in diameter is allowed to flow freely into the air between the two sparks, a shadow of it can be plainly perceived on the side remote from the active spark, i.e., the action of this is more or less completely annulled. A powerful absorption like that of coal-gas is exhibited by the brown vapors of nitrous oxide. With these, again, it is not necessary to use the tube with quartz-plates in order to show the action. On the other hand, although chlorine and the vapors of bromine and iodine do exercise absorption, it is not at all in proportion to their opacity. No absorptive action could be recognized when bromine vapor had been introduced into the tube in sufficient

---

[6]According to my experiments a concentrated solution of common salt is a more powerful absorber than crystallized rock-salt. This result is so remarkable as to require confirmation.

quantity to produce a distinct coloration; and there was a partial transmission of the action even when the bromine vapor was so dense that the active spark (colored a deep red) was only just visible through the tube.

12. The intensity of the action increases when the air around the passive spark is rarefied, at any rate up to a certain point. The plan was to measure this increase in terms of the difference between the lengths of the protected and the unprotected sparks. In these experiments the passive spark was produced under the bell-jar of an air-pump, between adjustable poles that passed through the sides of the bell-jar.[7] A window of rock-crystal was inserted in the bell-jar, and through this the action of the other spark had to pass. The maximum spark length was now observed, first with the window open, and then with the window closed; varying air pressures being used, but a constant current. The following table may be regarded as typical of the results.

Air Pressure in mm of Mercury	Length in mm of Spark with Window		Difference
—	Closed	Open	—
760	0.8	1.5	0.7
500	0.9	2.3	1.4
300	1.0	3.7	2.7
100	2.0	6.2	4.2
80	very great	very great	undetermined

It will be seen that as the pressure diminishes, the length of the spark that is not influenced increases only slowly; the length of the spark that is influenced increases more rapidly, and so the difference between the two becomes greater. But at a certain pressure the blue glow-light (*Glimmlicht*) spread over a considerable portion of the cathode, the sparking distance became very great, the discharge altered its character, and it was no longer possible to perceive any influence due to the active spark.

---

[7][For a photograph of this apparatus see Bryant (1988), p. 24, or Mulligan (1989a), p. 55.]

13. The phenomenon is also exhibited when the sparking takes place in other gases than air; and also when the two sparks are produced in two different gases. In these experiments the two sparks were produced in two small tubular glass vessels that were closed by plates of rock-crystal and could be filled with different gases. The experiments were tried mainly because certain circumstances led to the supposition that a spark in any given gas would only act upon another spark in the same gas, and on this account the four gases— hydrogen, air, carbonic acid, and coal-gas—were tried in the sixteen possible combinations.[8] The main conclusion arrived at was that the above supposition was erroneous. It should, however, be added that, although there is no great difference in the efficiency of sparks when employed as active sparks in different gases, there is, on the other hand, a notable difference in their susceptibility when employed as passive sparks. Other things being equal, sparks in hydrogen experienced a perceptibly greater increase in length than sparks in air, and these again about double the increase of sparks in carbonic acid and coal-gas. It is true that no allowance was made for absorption in these experiments, for its effect was not known when they were carried out; but it could only have been perceptible in the case of coal-gas.

14. All parts of the passive spark do not share equally in the action; it takes place near the poles, more especially near the negative pole.[9] In order to show this, the passive spark is made from 1 to 2 cm long, so that the various parts of it can be shaded separately. Shading the anode has but a slight effect; shading the cathode stops the greater part of the action. But the verification of this fact is somewhat difficult, because with long sparks there is a want of distinctness about the phenomenon. In the case of short sparks (the parts of which cannot be separately shaded) the statement can be illustrated as follows. The passive spark is placed parallel to the active one and is turned to right

---

[8][Hertz's thoroughness in these experiments refutes the criticism sometimes proffered that his discovery of the photoelectric effect was accidental, that he never pushed it very far, and then relinquished the field to others. A similar criticism might be levelled with more justice against Roentgen for his discovery of x-rays, and Roentgen received the first Nobel Prize for his work.]

[9]Soon afterwards Herren E. Wiedemann and H. Ebert showed that the action of the light only affects the negative pole, and only the surface of it (*Annalen* 33 [1888], p. 241.).

and left from the parallel into the perpendicular position until the action stops. It is found that there is more play in one direction than in the other, the advantage being in favor of that direction in which the cathode is turned toward the active spark. Whether the effect is produced entirely at the cathode, or only chiefly at the cathode, I have not been able to decide with certainty.

15. The action of the active spark is reflected from most surfaces. From polished surfaces the reflection takes place according to the laws of regular reflection of light. In the preliminary experiments on reflection a glass tube, 50 cm long and 1 cm in diameter, was used. This tube was open at both ends, and was pushed through a large sheet of cardboard. The active spark was placed at one end so that its action could only pass the sheet by way of the tube. If the passive spark was now moved about beyond the other end of the tube, it was affected when in the continuation of the tubular space, and only then; but in this case a far more powerful action was exhibited than when the tube was removed and only the diaphragm was retained. It was this latter phenomenon that suggested the use of the tube; of itself it indicates a reflection from the walls of the tube. The spark-micrometer was now placed to one side of the beam proceeding out of the tube, and so disposed that the axis of the spark was parallel to the direction of the beam. The micrometer was now adjusted so that the sparking just ceased; it was found to begin again if a plane surface inclined at an angle of 45° to the beam was held in it so as to direct the beam, according to the usual law of reflection, upon the passive spark. Reflection took place more or less from glass, crystals, and metals, even when these were not particularly smooth; also from such substances as porcelain, polished wood, and white paper. I obtained no reflection from a well-smoked glass plate.

In the more accurate experiments the active spark was placed in a vertical straight line; at a little distance from it was a large plate with a vertical slit, behind which could be placed polished plane mirrors of glass, rock-crystal, and various metals. The limits of the space within which the action was exerted behind the slit were then determined by moving the passive spark about. These limits were quite sharp and always coincided with the limits of the space within which the image of the active spark was visible in the mirror. On account of the feebleness of the action these experiments could not be carried out with

On an Effect of Ultraviolet Light

unpolished objects; such objects may be supposed to give rise to diffuse reflection.

16. In passing from air into a solid transparent medium, the action of the active spark exhibits a refraction, like that of light; but it is more strongly refracted than visible light.[10] The glass tube used in the reflection experiments served here again for the rougher experiments. The passive spark was placed in the beam proceeding out of the tube and at a distance of about 30 cm from the end farthest from the active spark; immediately behind the opening a quartz prism was pushed sideways into the beam with its refracting edge toward the front. In spite of the transparency of quartz, the effect on the passive spark ceased as soon as the prism covered the end of the tube. If the spark was then moved in a circle about the prism in the direction in which light would be refracted by the prism, it was soon found that there were places at which the effect was again produced. Now let the passive spark be fixed in the position in which the effect is most powerfully exhibited; on looking from this point toward the tube through the prism, the inside of the tube and the active spark at the end of it cannot be perceived; in order to see the active spark through the tube, the eye must be shifted backwards through a considerable distance toward the original position of the spark. The same result is obtained when a rock-salt prism is used. In the more accurate experiments the active spark was again fixed vertically; at some distance from it was placed a vertical slit, and behind this a prism. By inserting a Leyden jar the active spark could be made luminous, and the space thus illuminated behind the prism could easily be determined. With the aid of the passive spark it was possible to mark out the limits of the space within which was exerted the action here under investigation. Fig. 19 gives the result thus obtained by direct experiment. The space $a\,b\,c\,d$ is filled with light; the space $a'\,b'\,c'\,d'$ is filled with the action we are considering. Since the limits of this latter space were not sharp, the rays $a'\,b'$ and $c'\,d'$ were fixed in the following way. The passive spark was placed in a somewhat distant position, near c', at the outer edge of the region over which the action was exerted. A screen $m\,n$ (in Fig. 19) with vertical edge was then pushed in sideways until it stopped the action. The position m of its edge then

---

[10][This seems a clear indication that ultraviolet light is involved in the process Hertz is exploring in these experiments.]

gave one point of the ray $c'd'$. In another experiment a prism of small refracting angle was used, and the width of the slit was made as small as possible, and the spark placed as far from it as would still allow the perception of the action. The visible light was then spread out into a short spectrum, and the influence of the active spark was found to be exerted within a comparatively limited region, which corresponded to a deviation decidedly greater than that of the visible violet. Fig. 20 shows the positions of the rays as they were directly drawn with the prism as a base, r being the direction of the red, v of the violet, and w the direction in which the influence of the active spark was most powerfully exerted.

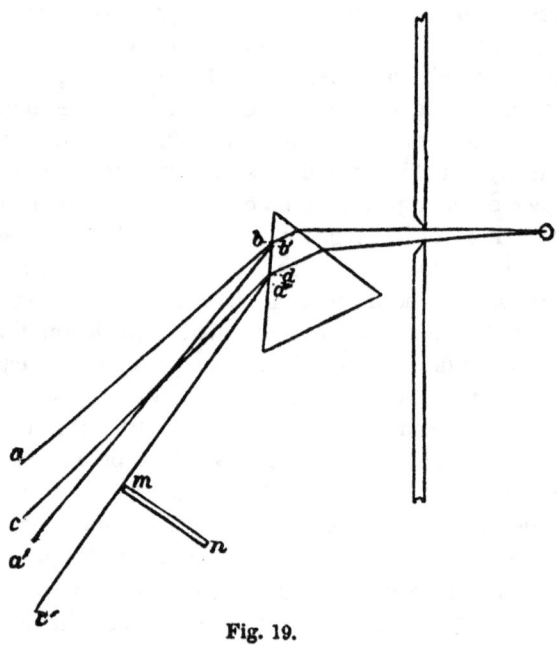

Fig. 19.

I have not been able to decide whether any double refraction of the action takes place. My quartz prisms would not produce a sufficient separation of the beams, and the pieces of calcspar that I possessed proved to be too opaque.[11]

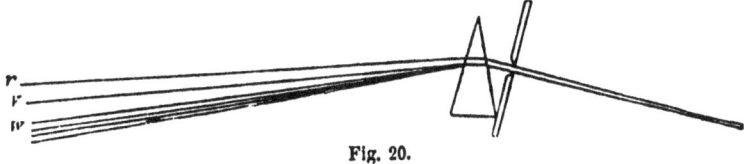

Fig. 20.

17. After what has now been stated, it will be agreed (at any rate until the contrary is proved) that the light of the active spark must be regarded as the prime cause of the action that proceeds from it. Every other conjecture that is based on known facts is contradicted by one or other of the experiments. And if the observed phenomenon is an effect of light at all, it must, according to the results of the refraction experiments, be solely the effect of the ultraviolet light. That it is not an effect of the visible parts of the light is shown by the fact that glass and mica are opaque to this [ultraviolet light], while they are transparent to these [visible parts of the light]. On the other hand, the absorption experiments of themselves make it probable that the effect is due to ultraviolet light. Water, rock-crystal, and the sulphates of the alkalies are remarkably transparent to ultraviolet light and to the action here investigated; benzene and allied substances are strikingly opaque to both. Again, the active rays in our experiments appear to lie at the outermost limits of the known spectrum. The spectrum of the spark when received on a sensitive dry-plate scarcely extended to the place at which the most powerful effect upon the passive spark was produced. And, photographically, there was scarcely any difference between light that had, and light that had not, passed through coal-gas, whereas the difference in the effect on the spark was very marked. Fig. 21 shows the extent of some of the spectra taken. In a the position of the visible red is indicated by r, that of the visible violet by v, and that of the strongest effect on the passive spark by w. The rest of the series give the photographic impressions produced—b after simply

---

[11]Somewhat later I succeeded in this. I had hoped to observe an influence of the state of polarization of the light upon the action, but was not able to detect anything of the kind.

passing through air and quartz, c after passing through coal-gas, d after passing through a thin plate of mica, and e after passing through glass.

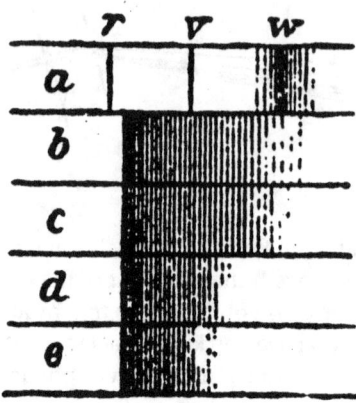

Fig. 21.

18. Our supposition that this effect is to be attributed to light is confirmed by the fact that the same effect can be produced by a number of common sources of light. It is true that the intensity of the light [*Lichtstärke*], in the ordinary sense of the word, forms no measure of its activity as here considered; and for the purpose of our experiments the faintly visible light of the spark of the induction coil remains the most powerful source of light. Let sparks from any induction coil pass between knobs, and let the knobs be drawn so far apart that the sparks fail to pass; if now the flame of a candle be brought near (about 8 cm away), the sparking begins again. The effect might at first be attributed to the hot air from the flame; but when it is observed that the insertion of a thin small plate of mica stops the action, whereas a much larger plate of quartz does not stop it, we are compelled to recognize here again the same effect. The flames of gas, wood, benzene, etc. all act in the same way. The non-luminous flames of alcohol and of the Bunsen burner exhibit the same effect, and in the case of the candle flame the action seems to proceed more from the lower, non-luminous part than from the upper, luminous part. From a small hydrogen flame scarcely any effect could be obtained. The light from platinum glowing at a white heat in a flame, or heated by an

electric current, a powerful phosphorus flame burning quite near the spark, and burning sodium and potassium—all proved to be inactive. So also does burning sulphur; but this can only have been on account of the feebleness of the flame, for the flame of burning carbon bisulphide produced some effect. Magnesium light produced a far more powerful effect than any of the above sources; its action extended to a distance of about a meter. The limelight, produced by means of coal-gas and oxygen, was somewhat weaker, and acted up to a distance of half a meter; the action was mainly due to the jet itself; it made no great difference whether the lime cylinder was brought into the flame or not.

On no occasion did I obtain a decisive effect from sunlight at any time of the day or year when I was able to test it.[12] When the sunlight was concentrated by means of a quartz lens on the spark, there was a slight action; but this was obtained equally well when a glass lens was used, and must therefore be attributed to the heating. But of all sources of light the electric arc is by far the most effective; it is the only one that can compete with the spark. If the knobs of the induction coil are drawn so far apart that sparks no longer pass, and if an arc light is started at a distance of 1, 2, 3, or even 4 meters, the sparking begins again at the same time, and stops again when the arc light goes out. By means of a narrow opening held in front of the arc light, we can separate the violet light of the feebly-luminous arc proper from the light of the glowing carbons; and we then find that the action proceeds chiefly from the former. With the light of the electric arc I have repeated most of the experiments already described, e.g., the experiments on the rectilinear propagation, reflection, and refraction of the action, as well as its absorption by glass, mica, coal-gas and other substances.

According to the results of our experiments, ultraviolet light has the property of increasing the sparking distance of the discharge of an induction coil, and of other discharges. The conditions under which it exerts its effect on such discharges are certainly very complicated, and it is desirable that the action should be studied under simpler conditions, and especially without using an induction coil. In endeavoring to make progress in this direction I have met with

---

[12][This was, of course, because most of the ultraviolet radiation from the sun was absorbed in the Earth's upper atmosphere by a series of photochemical reactions involving ozone and molecular oxygen.]

difficulties.[13] Hence I confine myself at present to communicating the results obtained, without attempting any theory respecting the manner in which the observed phenomena are brought about.[14]

---

[13]By this I did not mean to say that I had not succeeded in observing the action of light upon discharges other than those of induction coils; but only that I had not succeeded in replacing spark discharges—the nature of which is so little understood—by simpler means. This was first done by Herr Hallwachs (*Annalen* 33 [1888], p. 301–312). The simplest effect that I obtained was with the glow-discharge from 1000 small Planté accumulators between brass knobs in free air; by the action of light I was able to make the glow-discharge pass when the knobs were so far apart that it could not pass without the aid of the light.

[14][We know today that light falling on the pole pieces of the secondary spark gap produced electron emission from the negative pole and thus facilitated sparking across the gap. Since the metals Hertz used for his pole pieces all had work functions close to 4 eV, their threshold wavelength was below 310 nm, and ultraviolet light was needed to produce electron emission. After he had completed this paper, Hertz was probably wise to abandon further work on the photoelectric effect and return to his electromagnetic wave experiments, since any satisfactory explanation of this effect would turn out to require the discovery of the electron and the insights of quantum theory.]

PAPER NO. 8

On Electromagnetic Waves in Air and Their Reflection

(Heinrich Hertz, "Über elektrodynamische Wellen im Luftraume und deren Reflexion," Wiedemann's *Annalen der Physik und Chemie* 34 [1888], pp. 610–623)[1]

I have recently endeavored to prove by experiment that electromagnetic actions are propagated through air with finite velocity.[2] The inferences upon which that proof rested appear to me to be perfectly valid; but they are deduced in a complicated manner from complicated facts, and perhaps for this reason will not quite carry conviction to any one who is not already well-disposed to the views adopted therein. In this respect the demonstration there given may be fitly supplemented by a consideration of the phenomena now to be described, for these exhibit the propagation of induction through the air by wave motion in a visible and almost tangible form. These new phenomena also admit of a direct measurement of the wavelength in air. The fact that the wavelength thus obtained by direct measurement only differs slightly from the previous indirect determinations (using the same apparatus), may be regarded as an indication that the earlier demonstration was essentially correct.

In experimenting upon the action between a rectilinear oscillation and a secondary conductor I had often observed phenomena that seemed to point to a reflection of the induction action from the walls of the building. For example, feeble sparks

---

[1] [*El.Waves*, pp. 124–136.]

[2] [Heinrich Hertz, "On the Finite Velocity of Propagation of Electromagnetic Actions," *Annalen* 34 (1888), pp. 551–569. [*El.Waves*, pp. 107–123. On reevaluating his data from this paper in 1891 for the publication of the original German edition of *Electric Waves*, Hertz concluded (*El.Waves*, Supplementary Note 16 on p. 273):

"The correctly calculated period of oscillation is one hundred millionth of a second. This, with a wavelength of 2.8 meters, gives a velocity of 280,000 km per second, or approximately the velocity of light."

Both the period and the wavelength in this quote are *half* the similar quantities in use today.]

frequently appeared when the secondary conductor was so situated that any direct action was quite impossible, as was evident from simple geometrical considerations of symmetry; and this most frequently occurred in the neighborhood of solid walls. In particular, I continually encountered the following phenomenon: In examining the sparks in the secondary conductor at great distances from the primary conductor, where the sparks were exceedingly feeble, I observed that in most positions of the secondary conductor the sparks became appreciably stronger when I approached a solid wall, but again disappeared almost suddenly close to the wall. It seemed to me that the simplest way of explaining this was to assume that the electromagnetic action, spreading outwards in the form of waves, was reflected from the walls, and that the reflected waves reinforced the advancing waves at certain distances, and weakened them at other distances, since standing waves were produced by the interference of the two systems. As I made the conditions more and more favorable for reflection, the phenomenon appeared more and more distinct, and the explanation of it given above more probable. But without dwelling upon these preliminary trials I proceed at once to describe the principal experiments.

The physics lecture-room in which these experiments were carried out is about 15 m long, 14 m wide, and 6 m high. Parallel to the two longer walls there are two rows of iron pillars, each row of which behaves much like a solid wall with respect to the electromagnetic action, so that the parts of the room that lie outside these pillars cannot be taken into account. Thus only the central space, 15 m long, 8.5 m wide, and 6 m high, remained for the experiment. From this space I had the hanging parts of the gas-pipes and the chandeliers removed, so that it contained nothing except wooden tables and benches that could not be easily removed.[3] No objectionable effects were to be feared from these, and none were observed. The front wall of the room, from which the reflection was to take place, was a massive sandstone wall in which there were two doorways; a good many gas-pipes also extended into the wall. In order to give the wall more of the

---

[3] [It is worth mentioning that Hertz carried out these experiments during the 1887 Christmas holidays, for that was the only time during the winter semester when the lecture-room was not needed for classes. This put an enormous pressure on him to complete the experiments in a few weeks time.]

nature of a conducting surface a sheet of zinc 4 m high and 2 m broad was attached to it; this metal sheet was connected by wires with the gas-pipes and with a neighboring water-pipe, and special care was taken that any electricity that might accumulate at the upper and lower ends of the sheet would be able to flow away as freely as possible.

The primary conductor was set up opposite the middle of this wall at a distance of 13 m from it, and was therefore 2 m away from the opposite wall. It was the same conductor that had already been used in the experiments on the rate of propagation. The direction of the conducting wire was now vertical; hence the forces which have here to be considered oscillate up and down in a vertical direction. The middle point of the primary conductor was 2.5 m above the level floor; the observations were also carried out at the same distance above the floor, a gangway for the observer being built using tables and boards at a suitable height. We shall denote as the normal a straight line drawn from the center of the primary conductor perpendicular to the reflecting surface. Our experiments are restricted to the neighborhood of this normal; experiments at greater angles of incidence would be complicated by having to take into consideration the varying polarization of the waves. Any vertical plane parallel to the normal will be called a plane of oscillation, and any plane perpendicular to the normal will be called a wave-plane.

The secondary conductor was the circle of 35 cm radius that had also been used previously. It was mounted so as to revolve about an axis passing through its center and perpendicular to its plane. In the experiments the axis was horizontal; it was mounted in a wooden frame, so that both circle and axis could be rotated about a vertical axis. For the most part it does well enough for the observer to hold the circle, mounted in an insulating wooden frame, in his hand, and then to bring it as may be most convenient into the various positions required. But, inasmuch as the body of the observer always exerts a slight influence, the observations thus obtained must be controlled by other observations obtained from greater distances. The sparks are strong enough to be seen in the dark from a distance of several meters; but in a well-lighted room, even at close quarters, practically nothing can be seen of the phenomena we are about to describe.

After we completed these preparations, the most striking phenomenon that arose is the following: we place the secondary circle with its center on the normal and its plane in the plane of oscillation, and turn the spark-gap first toward the wall and then away from it. Generally the sparks differ greatly in the two positions. If the experiment is arranged at a distance of about 0.8 m from the wall, the sparks are much stronger when the spark-gap is turned toward the wall. The length of the sparks can be so regulated that a continuous stream of sparks passes over when the spark-gap is turned toward the wall, whereas no sparks whatever pass over in the opposite position. If we repeat the experiment at a distance of 3 m from the wall, we find, on the contrary, a continuous stream of sparks when the spark-gap is turned away from the wall, whereas the sparks disappear when the spark-gap is turned toward the wall. If we proceed further to a distance of 5.5 m, a fresh reversal takes place; the sparks on the side toward the wall are stronger than the sparks on the opposite side. Finally, at a distance of 8 m from the wall, we find that another reversal occurs; the sparking is stronger on the side remote from the wall, but the difference is no longer so noticeable. Nor does any further reversal occur; for it is prevented by the greater strength of the direct action and by the complicated forces that exist in the neighborhood of the primary oscillation. Our figure (Fig. 26, the scale on which indicates the distances from the wall) shows at I, II, III, IV the secondary circle in those positions in which the sparks were most strongly developed. The alternating character of the conditions of the space is clearly exhibited.

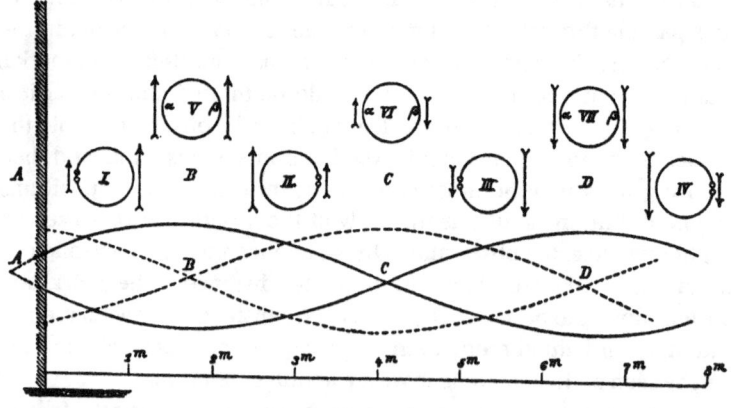

Fig. 26.

At distances lying between those mentioned both sets of sparks under consideration were of equal strength, and also in the immediate neighborhood of the wall the distinction between them diminishes. We may therefore denote these points—namely the points A,B,C,D in the figure—as being nodal points in a certain sense. Still we must not consider the distance between any one of these points and the next as being the half wavelength. For if *all* the electrical disturbances change their direction in passing through one of these points, then the phenomena in the secondary circle should repeat themselves without reversal; for in the spark length there is nothing that corresponds to a change of direction in the oscillation. We should rather conclude from these experiments that in passing through any one of these points one part of the action undergoes reversal, while another part does not. On the other hand, it is allowable to assume that double the distance between any two of the points corresponds to the half wavelength, so that these points each indicate the end of a quarter wavelength. And, indeed, on the basis of this assumption and of the fundamental view just expressed, we shall arrive at a complete explanation of the phenomenon.

For let us suppose that a vertical wave of electric force proceeds toward the wall, is reflected with slightly diminished intensity, and so gives rise to standing waves. If the wall were a perfect conductor a node would form at its very surface. For inside a conductor or at its boundary the electric force[4] must always be vanishingly small. Now our wall cannot be regarded as a perfect conductor. For, in the first place, it is only partly metallic, and the part that is metallic is not very extensive. Hence at its surface the force will still have a certain value, and this in the sense of the advancing wave. The node, which would be formed at the wall itself if it were perfectly conducting, must therefore lie in actuality somewhat behind the surface of the wall, say at the point A in the figure. If double the distance AB, that is the distance AC, corresponds to the half wavelength, then the geometrical relations of the standing wave are of the kind represented in the usual symbolic

---

[4] [As noted by Hertz in footnote 52 to Paper No. 5 in this collection, here and in all the papers reprinted from *Electric Waves*, by "electric force" Hertz really means what we would now call the "electric field." He is therefore arguing from the fact that the electric field must be zero inside a metal, since the free electric charges inside the metal will arrange themselves to cancel out any external voltage that is applied.]

fashion by the continuous line in the figure. The forces acting on both sides of the circle in the positions I, II, III, and IV are correctly represented at any given instant, both in magnitude and direction, by the arrows at the sides. If, then, in the neighborhood of a node the spark-gap is turned toward the node, we have in the circle a stronger force acting under favorable conditions against a weaker force, which acts under unfavorable conditions. But if the spark-gap is turned away from the node, the stronger force now acts under unfavorable conditions against a weaker force, which in this case is acting under favorable conditions. And, whether in this latter case the one predominates or the other, the sparks must necessarily be weaker than in the former case. Thus the change of sign of our phenomenon every quarter wavelength is explained.

Our explanation carries with it a means of further testing its correctness. If it is correct, then the change of sign at the points B and D should occur in a manner quite different from the change in sign at C. At V, VI, and VII in the figure the circle and the acting forces in these positions are represented, and it is easily seen that if at B or D we transfer the spark-gap from one position to the other by rotating the circle within itself, the oscillation changes its direction relative to a fixed direction within the circle; during this rotation the sparks must therefore become zero either once or an uneven number of times. On the other hand, if the same operation is performed at C, the direction of the oscillation does not change; and therefore the sparks must either not disappear at all, or else they must disappear an even number of times. Now when we actually perform the experiment, what we actually observe is this: At B the intensity of the sparks diminishes as soon as we move the spark-gap from α; it becomes zero at the highest point, and again increases to its original value when we come to position β. Similarly at D. At C, on the other hand, the sparks persist without change during the rotation or, if anything, are somewhat stronger at the highest and lowest points than at those that we have been considering. Furthermore, it strikes the observer that the change of sign ensues after a much smaller displacement at C than at B and D, so that in this respect there is a contrast between the change at C and that at B and D.

The picture of the electric wave that we have thus sketched can be verified in yet another way, and that a very direct one. Instead of placing the plane of our circle in the plane of oscillation, let us place it in the wave-plane; the electric force is now equally strong at all parts of the circle, and for similar positions of the sparks their intensity is simply proportional to this electric force. As might be expected, the sparks are now zero at the highest and lowest points of the circle at all distances,[5] and are strongest at the points that lie in the same horizontal plane as the normal. Let us then bring the spark-gap into one of these latter positions, and move slowly away from the wall. This is what we observe: Just at the conducting metallic surface there are no sparks, but they make their appearance at a very small distance from it; they increase rapidly, are comparatively strong at B, and then again diminish. At C again they are exceedingly feeble, but become stronger as we proceed further. They do not, however, again diminish, but continue to increase in strength, for we are now approaching the primary oscillation. If we were to illustrate the strength of the sparks along the interval AD by a curve carrying positive and negative signs, we should obtain almost exactly the curve that has been sketched. And perhaps it would have been better to start with this experiment. But it is not really so striking as the first one described; and furthermore, a periodic change of sign seems to be a clearer proof of wave motion than a periodic waxing and waning.

We are now quite certain that we have recognized nodes of the electric wave at A and C, and antinodes at B and D. We might, however, in another sense call B and D nodes, for these points are nodes of a stationary wave of magnetic force, which, according to theory, accompanies the electric wave and is displaced a quarter wavelength relative to it.[6] This statement can be illustrated experimentally as follows. We again place our circle in the plane of oscillation, but now bring the spark-gap to the highest point. In this position the electric force, if it were homogeneous over the whole extent of the secondary circle, could induce no sparks. It only produces an effect insofar as its magnitude varies in various parts of the circle, and its integral taken around the circle is not zero. This integral is proportional to the number of lines of magnetic force that flow backwards and forwards

---

[5] [The reason is that the primary oscillator vibrated in a vertical direction, and at the highest and lowest points of the circle the secondary spark-gap was horizontal.]

[6] [In a plane electromagnetic wave, the electric and magnetic field vectors are also perpendicular to each other, and both are perpendicular to the direction of propagation.]

through the circle. In this sense, we may say that in this position, the sparks measure the magnetic force, which is perpendicular to the plane of the circle.[7] But now we find that in this position near the wall there is vigorous sparking, which rapidly diminishes, disappears at B, increases again up to C, and then again decreases to a marked minimum at D, after which it continuously increases as we approach the primary oscillation. Representing the strength of these sparks as ordinates with positive and negative signs, we obtain approximately the dotted line in our figure, which thus represents correctly the magnetic wave.[8] The phenomenon that we first described can also be explained as resulting from the cooperation of the electric and the magnetic force. The former changes sign at the points A and C, the latter at the points B and D; thus one part of the action changes sign at each of these points while the other retains its sign; hence the resulting action (which is the product) changes sign at each of the points. Clearly this explanation only differs in mode of expression, and not in meaning, from the one first given.

Hitherto we have only considered the phenomena in some of the more important positions of the circle and of the spark-gap.[9] The

---

[7] It should be observed that we are here only able to determine the position of the magnetic force by the aid of theory. From the experiments we cannot conclude that a second kind of force is present together with the electric force. If we confine ourselves to the experiments, we can only regard the expression "magnetic force" as a short name for a certain mode of distribution of the electric force. That this magnetic force produces effects that cannot be explained by the electric force is first verified by experiments [in a later paper], and, of course, only for waves on wires. [Hertz's reference here is to his paper, "On the Mechanical Action of Electric Waves on Wires," in *El.Waves*, pp. 186–194. This paper is not included in the present collection.]

[8] [Hertz's tremendous experimental ingenuity thus enabled him to use a simple circle of wire containing a spark-gap to detect both the electric and magnetic fields, and to measure the relative strength of each field at various distances from the reflecting wall.]

[9] [In the translation of *Electric Waves* by D.E. Jones the phrase "and of the spark-gap" is omitted. This destroys the meaning of the next sentence: "*Die Zahl der Übergänge zwischen denselben ist in dreifachem Sinne unendlich*," in which Hertz seems to be referring to the fact that there are three variables to be selected in determining a unique experimental situation: the location of the circular detector and the orientation of the plane of the detector with respect to the reflecting wall, plus the location of the spark gap on the circumference of

number of changes in these positions is in a threefold sense infinite. We shall therefore content ourselves with describing the changes for the case in which the plane of the circle lies in the plane of oscillation. Near the wall the sparking is greatest on the side toward the wall, and least on the opposite side; on rotating the circle within itself the sparking changes from the one value to the other, attaining only intermediate values; there are no zero-points on the circle. As we move away from the wall the sparking on the side remote from it gradually diminishes and becomes zero when the center of the circle is 1.08 m distant from the wall; this distance can be ascertained within a few centimeters. As we proceed further, the sparks on the side remote from the wall reappear and at first are still weaker than on the side toward the wall; but the strength of the sparks does not change from the one value to the other simply by passing through intermediate values; on rotating the circle within itself the sparking becomes zero once in the upper and once in the lower half of the circle. The two zero-points develop out of the one that was first formed and gradually separate more from each other, until at B they lie at the highest and lowest points of the circle. By this indication the point B can be determined with fair accuracy, but more exactly still by a further observation of the zero-points. On proceeding further, these zero-points slide over toward the side of the circle facing the wall, approach each other, and again coincide in a single zero-point at a certain distance from the wall that can be sharply determined. In this case the distance of the center [of the circle] from the wall is 2.35 m. The point B must lie exactly between this and the analogous point first observed, i.e., at a distance of 1.72 m from the wall; this agrees within a few centimeters with the direct observation. If we proceed further toward C, the sparks at all points of the circle tend to become of equal strength, and do become so at C. Beyond C the same performance begins over again. In this region there are no zero-points in the circle. In spite of this, the position of the point C can be determined with fair accuracy, inasmuch as in its neighborhood the phenomena first described alter very rapidly. In my experiments C was 4.10 to 4.15 m or, let us say, 4.12 m from the wall. The point D could not be accurately determined for the phenomena had here become very feeble; only this much could

---

the detector. Each of these variables had a very large number of possible settings, making for a threefold infinity of choices.]

be asserted, that its distance from the wall was between 6 and 7.5 m. For an explanation of the details I may refer to a previous paper.[10] The mathematical developments therein indicated admit of being carried much further; but the experiments seem to be sufficiently intelligible without calculation.

According to our measurement, the distance between B and C is 2.4 m. If we assume this to be the correct value, the nodal point A lies 0.68 m behind the wall, the point D 6.52 m in front of it, which agrees sufficiently well with the experiments. According to this, the half wavelength is 4.8 m. By an indirect method I had obtained 4.5 m as the [half] wavelength for the same apparatus. The difference is not so great as to prevent us from regarding the new measurement as confirming the earlier one.[11] If in our earlier measurements we substitute 2.9 for 2.8 m as the wavelength in the wire, and 7.1 for 7.5 as the length of the coincidence (which will be found to agree with the observations), we can deduce the new value from the earlier observations. Perhaps, indeed, a mean value would be nearest to the truth; and I scarcely think it likely that the nodal point A should lie nearly 0.7 m behind the metallic wall. Assuming a mean value for the wavelength, and a velocity of propagation equal to that of light, we get for the period of oscillation of our apparatus about 1.55 hundred millionths of a second, instead of the 1.4 hundred-millionths obtained by calculation.[12]

I have repeated the experiments with some alterations. Changing the distance of the primary oscillation from the reflecting wall did not result in much fresh information. If this distance could

---

[10]H. Hertz, "On the Action of a Rectilinear Electric Oscillation upon a Neighboring Circuit," *Annalen* 34 (1888), 155–170. [*El.Waves*, pp. 80–94.]

[11]The wavelength measured depends, therefore, very much on the distance from B to C; and hence on the assumption that C is quite accurately measured. If we assume that the position of C is altered by general conditions of the surrounding space, the first node should be placed nearer to the wall, and we might obtain much smaller values for the wavelength. But the experiments give no reason for believing that the position of C is uncertain.

[12][Note that Hertz uses *half* wavelengths and *half* periods throughout this discussion. Since the distance BC is one-quarter wavelength, the measured wavelength is $4 \times 2.4 = 9.6$ m. Using this value and the known speed of light, the frequency of Hertz's primary coil was approximately 31 million vibrations per second, and the full period of oscillation was 3.2 hundred millionths of a second, in reasonable agreement with Hertz's stated value of $2(1.55) = 3.10$ hundred millionths of a second.]

have been considerably extended, we might certainly have expected a distinct formation of a second and third wavelength; but there was not sufficient space for such an extension. When the distance was diminished the phenomena simply became less interesting, since in the direction of the primary oscillator they were more and more indistinct, and in the opposite direction the reversal of sign was lost. The experiments with an oscillation of a different period are more worthy of description, for they show that the points which have attracted our attention are determined, not by the form of the wall or of the room, but only by the dimensions of the primary and secondary oscillation. I therefore used for some experiments a secondary circle of 17.5 cm radius, and a primary oscillation of the same periodic time as this circle. The primary oscillator was placed at a distance of 8 or 9 m from the wall. It is, however, difficult to work with apparatus of such small dimensions. Not only are the sparks exceedingly minute but the phenomena of resonance, etc., are very feebly developed. I suspect that oscillations of such rapidity are very rapidly damped. Thus it was not possible here to make out as much detail as in the case of the larger circle; but the main features, such as those first described above, could be plainly recognized. Near the wall, and at distances of 2.5 and 4.5 m from it, the stronger sparks were on the side nearer to the wall; at the intermediate positions (1.5 and 3.5 m from the wall) the stronger sparks were on the side near the primary oscillator. A change of sign occurred about every meter; accordingly the half wavelength was here only 2 m, and the oscillation was more than twice as rapid as that first used.

Finally, I may remark that as far as the above experiments are concerned, no great preparations are essential if one is content with more or less complete indications of the phenomena. After some practice one can find indications of reflection at any wall. Indeed, the action of the reflected waves can be quite well recognized between any one of the iron pillars referred to above and the primary oscillator; and similarly on the opposite side the electromagnetic shadow can be observed.

Let us now extend our experiments in a new direction. Hitherto the secondary conductor has been placed between the reflecting wall and the primary oscillator—that is to say, in a space in which the direct and reflected waves travel in opposite directions and by interference produce stationary waves. If, instead, we place the primary oscillator between the wall and the secondary conductor, the latter is situated in a space in which the direct and reflected waves travel in the same direction. Hence these must combine to produce a progressive wave, the intensity of which will, however, depend on the difference of phase between the two interfering waves. If the phenomena are to be at all striking, the two waves must be of similar intensity; hence the distance of the primary oscillator from the wall must not be large compared with the dimensions of the latter, and must be small compared with its distance from the secondary circuit.

In order to test whether the corresponding phenomena could be observed in an actual working experiment, I arranged my apparatus as follows. The secondary circle was now set up at a distance of 14 m from the reflecting wall, and therefore 1 m away from the opposite wall. Its plane was parallel to what we have called the plane of oscillation, and its spark-gap was turned towards the nearer wall so that the conditions were particularly favorable for the appearance of sparks in it. The primary conductor was set up parallel to its original position in front of the conducting wall, and initially at a very short distance—about 30 cm—from it. The sparks in the secondary circle were extremely feeble. The spark-gap was now adjusted so that no sparks whatever passed over. The primary conductor was next shifted step by step away from the wall. Single sparks soon appeared in the secondary conductor, and these ran into an unbroken stream of sparks when the primary conductor arrived at a distance of 1.5 to 2 m from the wall—that is to say, at the point B. This might be referred to the decrease in the distance between the two conductors. But when I now removed the primary conductor further away from the wall, and therefore nearer to the secondary, the sparks again diminished and disappeared when the primary conductor arrived at C. On proceeding still further, the sparks began to increase and now did so continuously. No exact measurement of the wavelength can be deduced from these experiments, but from what has been stated above it will be seen that the wavelengths already obtained are in accordance with the phenomena. The experiments could be very well carried out with the smaller apparatus. The primary conductor was set up at a distance of 1 m from the wall, and the corresponding secondary conductor 9 m from the wall. The sparks in the latter were certainly small, but could be quite easily observed. They disappeared when the primary conductor was moved out of its position, whether it was moved toward the wall or toward the secondary conductor. The sparks only reappeared when the distance from the wall was increased to 3 m, and from there on they did not again disappear on approaching nearer to the secondary conductor. It is worthy of notice that at the same distance of 2 m the presence of the wall proved to be of assistance in propagating the induction in the case of the slower oscillation, whereas it was a hindrance in the case of the more rapid one. This shows plainly that the position of the points to which we have drawn attention is determined by the dimensions of the oscillator, and not by those of the wall or room.

In acoustics there is an experiment analogous to those last described, in which it is shown that when a tuning-fork is brought near a wall the sound is strengthened at certain distances and weakened at others. The analogous experiment in optics is Lloyd's form of Fresnel's mirror experiment.[13] In optics and acoustics these experiments count

---

[13]Lloyd's experiment is the optical analogue of the experiments in which the primary conductor is gradually moved away from the wall. The experiments of the first kind, in which we moved the secondary conductor away from a reflecting wall, have also found an optical analogue in the beautiful experiment that Herr O. Wiener has published in his paper on "Stationary Light-Waves and the Direction of Vibration of Polarized Light" (*Annalen* 40 [1890], p. 303).

As to the acoustic analogues, I find that the phenomenon that forms the analogue to the experiments of the first kind was discovered by N. Savart many years ago (see *Annalen* 46 {1839}, p. 458; also a number of Seebeck's papers in the subsequent volumes). If a steady source of sound is placed at a distance of 15 to 20 m in front of a plane wall, and if we listen near the wall (best with the aid of a resonator), we find that the sound swells out at certain points—the antinodes—and becomes weak at other points—the nodes. A correct analogue to the experiments of the second kind—in which the primary conductor is moved—has already been given in the text. Another analogue—in itself interesting—is the following. Take a glass tube about 60 cm long and 2 cm in diameter and lower it gradually over a Bunsen burner, of which the flame is not too large. At a given depth the Bunsen flame will begin, but not without some difficulty, to make the tube sing loudly. Now bring the system near to a wall. Quite near the wall the sound disappears; it reappears at a distance of a quarter wavelength, and again vanishes at a distance of half a wavelength. By very careful adjustment, which up to the present I have not been able to secure at will, I have been able to observe two further positions of sound and silence at distances of half a wavelength. I do not know of any complete explanation of this phenomenon. Probably it has some connection with the fact that such a tube becomes silent if a resonator, tuned to the same note, is brought near its

as arguments in favor of the wave nature of light and sound; and so the phenomena here described may be regarded as arguments in favor of the propagation of the inductive action of an electric oscillation by means of waves.

I have described the present set of experiments, as also the first set on the propagation of induction, without paying special regard to any particular theory; and, indeed, the demonstrative power of the experiments is independent of any particular theory. Nevertheless, it is clear that the experiments amount to so many reasons in favor of that theory of electromagnetic phenomena that was first developed by Maxwell from Faraday's views. It also appears to me that the hypothesis about the nature of light which is connected with that theory now forces itself upon our minds with even stronger reason than heretofore. Certainly it is a fascinating idea that the processes in air, which we have been investigating, represent to us on a million-fold larger scale the same processes that go on in the neighborhood of a Fresnel mirror or between the glass plates used for exhibiting Newton's rings.

That Maxwell's theory, notwithstanding its great intrinsic probability, cannot dispense with such confirmation as it has already received and may yet receive, is proven—if indeed proof be needed—by the fact that the electric action is not propagated along wires of good conductivity with approximately the same velocity as through air. Hitherto it has been inferred from all theories, Maxwell's included, that the velocity along wires should be the same as that of light. I hope in time to be able to investigate and report on the causes of this conflict between theory and experiment.[14]

---

end. This last experiment is due—as far as I am aware—to Professor A. Christiani (*Verhandl.d.phys.Gesellsch.zu Berlin*, 15 Dec. 1882; at end of *Fortschritte der Physik*, vol. 36).

[14]This remark refers to the experiments with wires, which I was arranging at the time this paper was written. It has already been stated in the introduction [to *Electric Waves*] that the hope here expressed has not been fulfilled.

[Hertz's point here is that he had hoped to prove that electromagnetic waves on wires and in air travel at the same speed, that of light, as predicted by Maxwell's theory. His measurements of the speed of such waves in air were disturbed by diffraction effects and spurious reflections from objects and the walls of the laboratory. It was the experiments of Sarasin and de la Rive, carried out in a much larger space, that finally convinced Hertz that Maxwell's theory was correct in this prediction (On this see especially footnote 26 to Paper No. 5 in the present collection). Hertz's later work with much shorter wavelengths led to an additional confirmation of the equality of the velocities of electromagnetic waves on wires and in air.]

PAPER NO. 9

On Electric Radiation

(Heinrich Hertz, "Über Strahlen elektrischer Kraft," *Sitzungsber. d. Berl. Akad. d. Wiss.*, 13 December 1888; Wiedemann's *Annalen der Physik und Chemie* 36 [1889], 769–783.)[1]

As soon as I had succeeded in proving that the action of an electric oscillation spreads out as a wave in space, I planned experiments with the object of concentrating this action and making it perceptible at greater distances by putting the primary conductor in the focal line of a large concave parabolic mirror. These experiments did not lead to the desired result, and I felt certain that the lack of success was a necessary consequence of the disproportion between the length (4 to 5 m) of the waves used and the dimensions which I was able, under the most favorable circumstances, to give to the mirror. Recently I have observed that the experiments that I have described can be carried out quite well with oscillations of more than ten times the frequency, and with waves less than one-tenth the wavelength of those that I first discovered. I have, therefore, returned to the use of concave mirrors, and have obtained better results than I had hoped for. I have succeeded in producing distinct beams of electric force and in carrying out with them the elementary experiments that are commonly performed with light and radiant heat. The following is an account of these experiments.

The Apparatus

The short waves were excited by the same method that we used for producing the longer waves. The primary conductor used may be most

---

[1] [*El. Waves*, pp. 172–185. This is the simplest and yet the most important of Hertz's experimental papers. On the basis of easily-understood experiments he was able to demonstrate conclusively that light and electromagnetic waves with wavelengths of 66 cm had the same properties. His experiments startled the scientific world and eventually convinced physicists of the validity of Maxwell's theory.]

simply described as follows: Imagine a cylindrical brass body,[2] 3 cm in diameter and 26 cm long, interrupted midway along its length by a spark-gap whose poles on either side are formed by spheres of 2 cm radius. The length of the conductor is approximately equal to the half wavelength of the corresponding oscillation in straight wires; from this we are at once able to estimate approximately the period of oscillation. It is essential that the pole-surfaces of the spark-gap should be frequently repolished, and also that during the experiments they should be carefully protected from illumination by simultaneous side-discharges; otherwise the oscillations are not excited. Whether the spark-gap is in a satisfactory state can always be recognized by the appearance and sound of the sparks. The discharge is led to the two halves of the conductor by means of two wires covered with gutta-percha,[3] which are connected near the spark-gap on either side. I no longer made use of the large Ruhmkorff [induction-coil], but found it better to use a small induction-coil by Keiser and Schmidt; the longest sparks between points produced by this coil were 4.5 cm long. It was supplied with current from three accumulators, and gave sparks 1 to 2 cm long between the spherical knobs of the primary conductor. For the purpose of the experiments the spark-gap was reduced to 3 mm.

Here, again, the small sparks induced in a secondary conductor were the means used for detecting the electric forces in space. As before,[4] I used as a component a circle that could be rotated within itself and that had about the same period of oscillation as the primary conductor. It was made of copper wire 1 mm thick and had in the present instance a diameter of only 7.5 cm. One end of the wire carried a polished brass sphere a few millimeters in diameter; the other end was pointed and could be brought up, by means of a fine screw insulated from the wire, to within an exceedingly short distance from the brass sphere. As will be readily understood, we have here to deal only with minute sparks of a few hundredths of a millimeter in length; and after a little practice one judges more according to the brilliancy than the length of the sparks.

---

[2]See Figs. 35 and 36 and the description of them at the end of this paper.

[3][Gutta-percha was a rubbery substance much used for electrical insulation in Hertz's day.]

[4][See Paper No. 8.]

On Electric Radiation

The circular conductor gives only a differential effect, and is not adapted for use in the focal line of a concave mirror. Most of the work was therefore carried out with another conductor arranged as follows: Two straight pieces of wire, each 50 cm long and 5 mm in diameter, were arranged in a straight line so that their near ends were 5 cm apart. From these ends two wires, 15 cm long and 1 mm in diameter, were carried parallel to one another and perpendicular to the wires first mentioned to a spark-gap arranged just as in the circular conductor. In this conductor the resonance action was given up, and indeed it only comes slightly into play in this case. It would have been simpler to put the spark-gap directly in the middle of the straight wire, but the observer could not then have handled and observed the spark-gap in the focus of the mirror without obstructing the aperture. For this reason the arrangment above described was chosen in preference to the other, which would in itself have been more advantageous.

The Production of the Beam

If the primary oscillator is now set up in a fairly large free space, one can, with the aid of the circular conductor, detect in its neighborhood all those phenomena that I have already observed and described as occurring in the neighborhood of a larger oscillation,[5] but now on a smaller scale. The greatest distance at which sparks could be observed in the secondary conductor was 1.5 m, or, when the primary spark-gap was in very good order, as much as 2 m. When a plane reflecting plate is set up at a suitable distance on one side of the primary oscillator and parallel to it, the action on the opposite side is strengthened. If the distance chosen is either very small, or somewhat greater than 30 cm, the plate weakens the effect; it strengthens the effect greatly at distances of 8 to 15 cm, slightly at a distance of 45 cm, and exerts no influence at greater distances. We have drawn attention to this phenomenon in an earlier paper, and we conclude from it that the wave in air corresponding to the primary oscillation has a half wavelength of about 30 cm. We may expect to find a still further reinforcement if we replace the plane surface by a concave mirror

---

[5]See *El.Waves*, pp. 80–94, 107–136. [One of the three papers Hertz refers to here is Paper No. 8 in this book.]

having the form of a parabolic cylinder, in the focal line of which the axis of the primary oscillation lies.[6] The focal length of the mirror should be chosen as small as possible, if it is properly to concentrate the action. But if the direct wave is not to annul immediately the action of the reflected wave, the focal length must not be much smaller than a quarter wavelength. I therefore fixed on 12.5 cm as the focal length, and constructed the mirror by bending a zinc sheet 2 m long, 2 m wide, and 0.5 mm thick into the desired shape over a wooden frame of the exact curvature. The height of the mirror was thus 2 m, the width of its aperture 1.2 m, and its depth 0.7 m. The primary oscillator was fixed in the middle of the focal line. The wires that conducted the discharge were led through the mirror; the induction-coil and the cells were accordingly placed behind the mirror so as to be out of the way.

If we now investigate the neighborhood of the oscillator with our conductors, we find that there is no action behind the mirror or at either side of it; but in the direction of the optic axis of the mirror the sparks can be perceived up to a distance of 5 to 6 m. When a plane conducting surface was set up so as to oppose the advancing waves at right angles, the sparks could be detected in its neighborhood at even greater distances—up to about 9 or 10 m. The waves reflected from the conducting surface reinforce the advancing waves at certain points; at other points the two sets of waves weaken one another. In front of the plane reflector one can recognize very distinct maxima and minima with the rectilinear conductor, and with the circular conductor the characteristic interference phenomena of stationary waves that I have described in an earlier paper. I was able to distinguish four nodal points, which were situated at the wall and at distances of 33, 65, and 98 cm from it. We thus get 33 cm as a closer approximation to the half wavelength of the waves used, and 1.1 thousand-millionth of a second as their period of oscillation, assuming that they travel at the speed of light. In wires the oscillation gave a wavelength of 29 cm. Hence it appears that these short waves also have a somewhat lower velocity on wires than in air; but the ratio of the two velocities comes very near to the theoretical value—unity—and does not differ from it as much as

---

[6][The parabolic "dishes" used today for microwave transmission and radio astronomy are direct descendants of the parabolic mirrors used by Hertz in these experiments.]

appeared to be probable from our experiments on longer waves.[7] This remarkable phenomenon still needs elucidation. Inasmuch as the phenomena are only exhibited in the neighborhood of the optic axis of the mirror, we may describe the result as a beam of electric radiation proceeding from the concave mirror.

I now constructed a second mirror, exactly similar to the first, and attached the rectilinear secondary conductor to it in such a way that the two wires of 50 cm length lay in the focal line, and the two wires connected to the spark-gap passed directly through the walls of the mirror without touching it. The spark-gap was thus situated directly behind the mirror, and the observer could adjust and examine it without obstructing the course of the waves. I expected to find that, on intercepting the beam with this apparatus, I should be able to observe it at even greater distances; the event proved that I was not mistaken. In the rooms at my disposal I could now perceive the sparks from one end to the other. The greatest distance to which I was able, by availing myself of a doorway, to follow the beam, was 16 m; but according to the results of the reflection experiments (to be presently described), there can be no doubt that sparks could be obtained at least up to 20 m in open spaces. For the remaining experiments such great distances are not necessary, and it is convenient to have the sparking in the secondary conductor not too feeble; for most of the experiments a distance of 6 to 10 m is most suitable. We shall now describe the simple phenomena that can be exhibited with the beam without any difficulty. When the contrary is not expressly stated, it is to be assumed that the focal lines of both mirrors are vertical.

Rectilinear Propagation

If a screen of sheet zinc 2 m high and 1 m wide is placed on the straight line joining the two mirrors, and at right angles to the direction of the

---

[7][This is what was to be expected if the source of this discrepancy was diffraction. The shorter the wavelength used, the less the diffraction, and therefore the more closely the measured values of the velocity of the radiation in free space should agree with that on wires. In his papers on *Electric Waves* Hertz did not make any allowance for diffraction effects, which could have washed out or modified his results. This was first pointed out by Poincaré. See footnote 74 in the Introductory Biography.]

beam, the secondary sparks disappear completely. An equally complete shadow is thrown by a screen of tinfoil or gilt paper. If an assistant walks across the path of the beam, the secondary spark-gap becomes dark as soon as he intercepts the beam, and again lights up when he leaves the path clear. Insulators do not stop the beam; it passes right through a wooden partition or door; and it is with amazement that one sees the sparks appear inside a closed room. If two conducting screens, 2 m high and 1 m wide, are set up symmetrically on the right and left of the beam, and perpendicular to it, they do not interfere at all with the secondary spark so long as the width of the opening between them is not less than the aperture of the mirrors, namely, 1.2 m. If the opening is made narrower, the sparks become weaker and disappear when the width of the opening is reduced below 0.5 m. The sparks also disappear if the opening is left with a width of 1.2 m, but is shifted to one side of the straight line joining the mirrors. If the optic axis of the mirror containing the oscillator is rotated to the right or left about 10 degrees out of its proper position, the secondary sparks become weak, and a rotation of 15 degrees causes them to disappear.

There is no sharp geometrical limit to either the beam or the shadows; it is easy to produce phenomena corresponding to diffraction.[8] As yet, however, I have not succeeded in observing maxima and minima at the edge of the shadows.

Polarization

From the manner in which our beam was produced we can have no doubt whatever that it consists of transverse vibrations and is plane-polarized in the optical sense. We can also prove by experiment that this is the case. If the receiving mirror be rotated about the beam as axis until its focal line, and therefore the secondary conductor also, lies in a horizontal plane, the secondary sparks become more and more

---

[8] In connection with these phenomena we may refer to the observation that Herren Hagenbach and Zehnder have brought forward as an objection to my interpretation of the experiment (*Annalen* 43 [1891], p. 611). My meaning is that light behaves just as the electric waves here behave; but we must imagine the dimensions of everything concerned in the experiment to be reduced in the same proportion, not only the length of the waves.

feeble, and when the two focal lines are at right angles, no sparks whatever are obtained even if the mirrors are moved close to one another. The two mirrors behave like the polarizer and analyzer of a polarization apparatus. I next had made an octagonal frame, 2 m high and 2 m wide; across this were stretched copper wires 1 mm thick, the wires being parallel to one another and 3 cm apart. If the two mirrors were now set up with their focal lines parallel, and the wire screen was interposed perpendicularly to the ray and so that the direction of the wires was perpendicular to the direction of the focal lines, the screen practically did not interfere at all with the secondary sparks. But if the screen was set up in such a way that its wires were parallel to the focal lines, it stopped the ray completely. With regard, then, to transmitted energy the screen behaves toward our beam just as a tourmaline plate behaves toward a plane-polarized beam of light. The receiving mirror was now placed once more so that its focal line was horizontal; under these circumstances, as already mentioned, no sparks appeared. Nor were any sparks produced when the screen was interposed in the path of the ray, so long as the wires in the screen were either horizontal or vertical. But if the frame was set up in such a position that the wires were inclined at 45 degrees to the horizontal on either side, then the interposition of the screen immediately produced sparks in the secondary spark-gap. Clearly the screen resolves the advancing oscillation into two components and transmits only that component which is perpendicular to the direction of its wires. This component is inclined at 45 degrees to the focal line of the second mirror and may thus, after being again resolved [into components] by the mirror, act upon the secondary conductor. The phenomenon is exactly analogous to the brightening of the dark field of two crossed Nicols[9] by the interposition of a crystalline plate in a suitable position.

With regard to the polarization it may be further observed that, with the means employed in the present investigation, we are only able

---

[9] [A Nicol prism is a device for producing plane polarized light. It consists of a crystal of calcite cut at a 68 degree angle, cleaved along the optic axis, and stuck together with a thin layer of Canada balsam between the two pieces of calcite. The calcite is birefringent and breaks an incoming ray into two rays polarized at right angles to each other. The refractive index of the material for one ray is such that when it strikes the Canada balsam, it is totally internally reflected. The other ray, however, passes through the Canada balsam without reflection and emerges as a ray of plane polarized light. Two crossed Nicol prisms act in the same way as do two crossed pieces of polaroid.]

to recognize the electric force. When the primary oscillator is in a vertical position the oscillations of this force undoubtedly take place in the vertical plane through the beam, and are absent in the horizontal plane. But the results of experiments with slowly alternating currents leave no room for doubt that the electric oscillations are accompanied by oscillations of magnetic force that take place in the horizontal plane through the beam and are zero in the vertical plane. Hence the polarization of the beam does not so much consist in the occurrence of oscillations in the vertical plane, but rather in the fact that the oscillations in the vertical plane are of an electrical nature, while those in the horizontal plane are of a magnetic nature. Obviously, then, the question in which of the two planes the oscillations in our beam occurs, cannot be answered unless one specifies whether the question relates to the electric or the magnetic oscillation. It was Herr Kolacek[10] who first pointed out clearly that this consideration is the reason why an old optical dispute has never been settled.

Reflection

We have already demonstrated the reflection of the waves from conducting surfaces by the interference between the reflected and the advancing waves, and have also made use of the reflection in the construction of our concave mirrors. But now we are able to go further and to separate the two systems of waves from one another. I first placed both mirrors in a large room side by side, with their apertures facing in the same direction and their axes converging to a point about 3 m away. The spark-gap of the receiving mirror naturally remained dark. I next set up a plane vertical screen made of thin sheet zinc, 2 m high and 2 m wide, at the point of intersection of the axes, and adjusted it so that it was equally inclined to both. I obtained a vigorous stream of sparks arising from the reflection of the beam by the screen. The sparking ceased as soon as the screen was rotated around a vertical axis through about 15 degrees on either side of the correct position; from this it follows that the reflection is regular, not diffuse. When the screen was moved away from the mirrors, the axes of the latter still being kept converging toward the wall, the sparking

---

[10]F. Kolacek, *Annalen* 34 (1888), p. 676.

On Electric Radiation

degrees. The refracting edge was placed vertical, and the height of the whole prism was 1.5 m. But since the prism weighed about 12 hundredweight [1,200 pounds], and would have been too heavy to move in one piece, it was constructed of three pieces, each 0.5 m high, placed one above the other. The material was cast in wooden boxes, which were left around it, since they did not appear to interfere with its use.[12] The prism was mounted on a support of such height that the middle of its refracting edge was at the same height as the primary and secondary spark-gaps. When I was satisfied that refraction did take place, and had obtained some idea of its amount, I arranged the experiment in the following manner: The source mirror was set up at a distance of 2.6 m from the prism and facing one of the refracting surfaces, so that the axis of the beam was directed as nearly as possible toward the center of mass of the prism, and met the refracting surface at an angle of incidence of 65 degrees with respect to the side meeting the rear face.[13] Near the refracting edge and also at the opposite side of the prism were placed two conducting screens that prevented the ray from passing by any other path than that through the prism. On the side of the emerging ray there was marked on the floor a circle of 2.5 m radius, having as its center the center of mass of the lower end of the prism. The receiving mirror was now moved around the circle, its aperture always being directed toward the center of the circle. No sparks were obtained when the mirror was lined up with the direction of the incident beam; in this direction the prism cast a complete shadow. But sparks appeared when the mirror was moved toward the base of the prism, starting when the angular deviation from its initial position was about 11 degrees. The sparking increased in intensity until the deviation amounted to about 22 degrees, and then again decreased. The last sparks were observed at a deviation of about 34 degrees. When the mirror was placed in a position of maximum effect,

---

[12][For a photograph of this prism and of the wire grating used in Hertz's polarization experiments, see the first page of the present section (IV) in this book.]

[13][This translation is more faithful to Hertz's German than is Jones' translation in *Electric Waves*. The German reads "und die brechende Flache von der Seite der Hinterflache her unter einem Winkel von 65 Grad traf." Jones converts this into "and met the refracting surface at an angle of incidence of 25 degrees (on the side of the normal towards the base)." The meaning is, in any case, the same.]

and then moved away from the prism along the radius of the circle, the sparks could be traced up to a distance of 5 or 6 meters. When an assistant stood either in front of the prism or behind it, the sparking invariably ceased, which shows that the action reaches the secondary conductor through the prism and not in any other way. The experiments were repeated after placing both mirrors with their focal lines horizontal, but without altering the position of the prism. This made no difference in the phenomena observed. A refracting angle of 30 degrees and a deviation of 22 degrees in the neighborhood of the minimum deviation corresponds to a refractive index of 1.69. The refractive index of pitch-like materials for light is given as being between 1.5 and 1.6. We must not attribute any importance to the magnitude or even the direction of this difference,[14] seeing that our method was not an accurate one, and that the material used was impure.

We have applied the term "beams of electric force" to the phenomena we have investigated. We may perhaps further designate them as "beams of light of very great wavelength." The experiments described appear to me, at any rate, eminently adapted to remove any doubt as to the identity of light, radiant heat, and electromagnetic wave motion. I believe that from now on we shall have greater confidence in making use of the advantages which the acceptance of this identity provides for us both in the field of optics and of electricity.[15]

Explanation of the Figures

In order to facilitate the repetition and extension of these experiments, I append in the accompanying Figs. 35, 36a and 36b, illustrations of the apparatus I used, although these were constructed simply for the purpose of experimenting at the time and without any regard to durability. Fig. 35 shows in ground-plan and vertical projection (cross-

---

[14] Messrs. Oliver Lodge and Howard have actually succeeded in showing the refraction and concentration of electric beams by means of large lenses (*Phil.Mag.* 27 [1889], p. 48).

[15] [There is a copy of this paper of Hertz, with handwritten annotations by Oliver Heaviside, in the Burndy Library. At this point in the paper the following annotation appears: "This is a splendid paper. O.H."]

On Electric Radiation

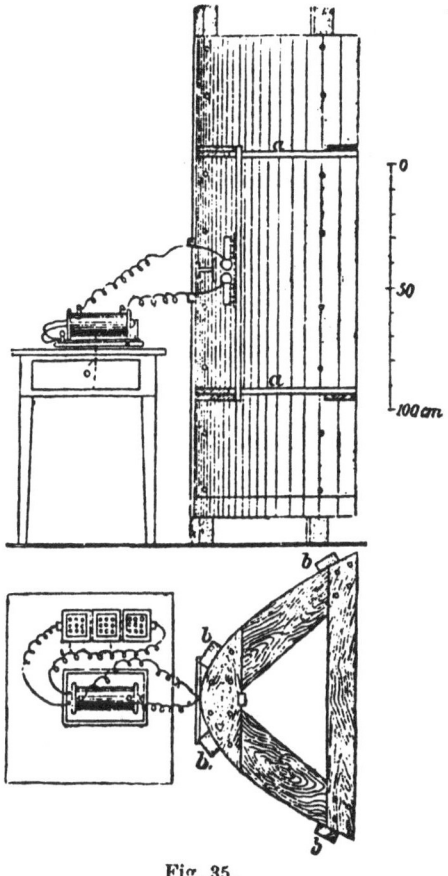

Fig. 35.

section) the source mirror. It will be seen that its framework consists of two horizontal frames (*a,a*) of parabolic form, and four vertical supports (*b,b*), which are screwed to each of the frames so as to support and connect them. The sheet metal reflector is clamped between the frames and the supports, and fastened to both by numerous screws. The supports project above and below the sheet metal so that they can be used as handles in handling the mirror. Fig. 36a represents the primary conductor on a somewhat larger scale. The

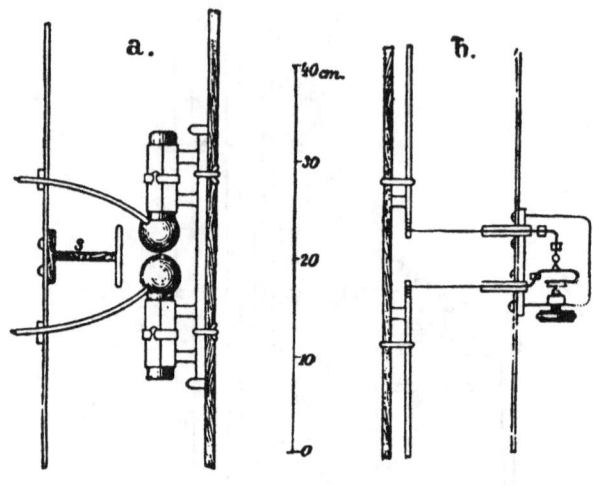

Fig. 36.

two metal parts slide with friction in two sleeves of strong paper that are held together by bands of India rubber. The sleeves themselves are fastened by four rods of sealing wax to a board that again is tied by India-rubber bands to a strip of wood forming part of the frame that can be seen in Fig. 35. The two leading wires (covered with gutta-percha) terminate in two holes bored in the knobs of the primary conductor. This arrangement permits all necessary motions and adjustments of the conductor; it can be taken to pieces and put together again in a few minutes, and this is essential in order that the knobs may be frequently repolished. Just at the points where the leading wires pass through the mirror, they are surrounded during the discharge by a bluish light. The smooth wooden screen s is introduced [in Fig. 36a] for the purpose of shielding the spark-gap from this light, which otherwise would interfere seriously with the production of the oscillations.

Lastly, Fig. 36b represents the secondary spark-gap. Both parts of the secondary conductor are again attached by sealing-wax rods and rubberbands to a moulding that is part of the wooden framework. From the inner ends of these parts the leading wires, enclosed in glass tubes, can be seen proceeding through the mirror and bending toward each other. The upper wire carries at its pole a small brass knob. To the lower wire is soldered a piece of watch-spring that carries the second pole, consisting of a fine copper point. The point is intentionally chosen of softer metal than the knob; unless this precaution is taken, the point easily penetrates into the knob, and the minute sparks disappear from sight in the small hole thus produced. The figure shows how the point is adjusted by a screw that presses against the spring, which is insulated from it by a glass plate. The spring is bent in a particular way in order to secure finer motion of the point than would be possible if the screw alone were used.

No doubt the apparatus here described can be considerably modified without interfering with the success of the experiments. Acting upon friendly advice, I have tried to replace the spark-gap in the secondary conductor by a frog's leg prepared for detecting currents; but this arrangement, which is so sensitive under other conditions, does not seem to be useful for these purposes.[16]

---

[16]Since then many have succeeded in a variety of ways in demonstrating the objectivity of these experiments. Herr R. Ritter has employed successfully a frog's leg (*Annalen* 40 [1890], p. 53). Mr. Dragoumis has used Geissler tubes (*Nature* 39 [1888–89], p. 548). Herr [Ludwig] Boltzmann has given a very convenient method in which a gold-leaf electroscope is used (*Annalen* 40 [1890], p. 399). Herr Klemencic has used a thermoelement (*Annalen* 42 [1891], p. 416). The method that is most elegant and best adapted for demonstration, although it is far from being an easy one, is the bolometer method, which Herren H. Rubens and R. Ritter have employed for exhibiting the experiments and for further useful researches (*Annalen* 40 [1890], p. 55, and subsequent volumes).

[This note was written by Hertz near the end of 1891.]

PAPER NO. 10

On the Relations between Light and Electricity

(An invited address delivered by Heinrich Hertz in Heidelberg on 20 September 1889 at the Sixty-Second Conference of German Scientists and Physicians.)[1]

When one speaks of the relations between light and electricity, the lay mind at once thinks of the electric light. With this the present lecture is not concerned. To the mind of the physicist there occurs a series of delicate mutual interactions between these two forces, such as the rotation of the plane of polarization [of light] by a current, or the change of the resistance of a conductor by the action of light. In these, however, light and electricity do not directly meet; between the two there intrudes an intermediate agent—ponderable matter. With this group of phenomena also we shall not concern ourselves. But between these two entities there are still other relations—relations in a closer and stricter sense than those already mentioned. I am here today to support the assertion that light of every kind is itself an electrical phenomenon—the light of the sun, the light of a candle, the light of a glow-worm. Take away from the world electricity, and light disappears; remove from the world the luminiferous ether, and electric and magnetic actions can no longer traverse space. This is our assertion. It does not date from today or yesterday; it has already a long history behind it. In this history its foundations lie. Such researches as I have made upon this subject form but a link in a long chain. And it is of the chain, and not only of the single link, that I would speak to you. I must confess that it is not easy to speak of these matters in a way at once intelligible and accurate. It is in empty space, in the free ether,[2] that the processes we have to describe take place. They cannot be felt with the hand, heard by the ear, or seen by the eye. They are accessible to our intuition and inner perception, but only with difficulty to our senses. Hence we shall try to make use, as far as possible, of the insights and perceptions that we already possess. Let us, therefore,

---

[1][*Misc. Pprs.*, pp. 313–327. For Hertz's difficulties in preparing this lecture, see Part V of the Introductory Biography.]

[2][As this lecture clearly indicates, Hertz retained throughout his life a firm commitment to the luminiferous ether.]

273

stop to inquire what we know with certainty about light and electricity before we proceed to connect the one with the other.

What, then, is light? Since the time of Young and Fresnel we know that it is a wave motion. We know the velocity of the waves, we know their wavelength, we know that they are transverse waves; in short, we know completely the geometrical aspects of their motion. To the physicist it is inconceivable that this view should be refuted; we can no longer entertain any doubt about the matter. The wave theory of light is, from a human point of view, certain, and the conclusions that follow from it with logical necessity are equally certain. It is therefore certain that all space known to us is not empty, but is filled with a substance, the ether, which is able to support waves. But whereas our knowledge of the geometrical aspects of the processes occurring in this substance is clear and definite, our conception of the physical nature of these processes is vague, and our assumptions as to the properties of the substance itself are not altogether consistent. At first, following the analogy of sound, waves of light were freely regarded as elastic waves, and treated as such. But elastic waves in fluids are only known in the form of longitudinal waves. Transverse elastic waves in fluids are unknown. They are not even possible; they contradict the nature of the fluid state. Hence men were forced to assert that the ether that fills space behaves like a solid body. But when they considered and tried to explain the unhindered course of the stars in the heavens, they found themselves forced to admit that the ether behaves like a perfect fluid. These two statements taken together land us in a painful and unintelligible contradiction, which disfigures the otherwise beautiful development of optics. Instead of trying to conceal this defect let us turn to electricity; in investigating it we may perhaps make some progress in removing this difficulty.

What, then, is electricity? This is at once an important and a difficult question. It interests the lay as well as the scientific mind. Most people who ask it never doubt about the existence of electricity. They expect a description of it—an enumeration of the properties and the forces of this wonderful thing [*dieses wunderbaren Stoffes*]. To the scientific mind the question rather presents itself in the form: Is there such a thing as electricity? Cannot electrical phenomena be traced back, like all others, to the properties of the ether and of ponderable matter? We are far from being able to answer this question definitely in the affirmative. In our presentation the thing conceived of as electricity

On the Relations between Light and Electricity

plays a large role. The traditional conceptions of electricities that attract and repel each other, and that are endowed with actions-at-a-distance as with almost spiritual properties—with these we are all familiar, and in a way we are fond of them; they hold undisputed sway as common modes of expression at the present time. The period at which these conceptions were formed was the period in which Newton's law of gravitation won its most glorious successes, and in which the idea of direct action-at-a-distance was familiar. Electric and magnetic attractions followed the same law as gravitational attraction; no wonder men thought the simple assumption of action-at-a-distance sufficient to explain these phenomena, and to trace them back to their ultimate intelligible cause. Of course, things changed in the present century, when the reactions between electric currents and magnets became known; for these have an infinity of possible interactions, in which both motion and time play an important part. It became necessary to increase the number of actions-at-a-distance, and to improve their form. Thus the conception [of action-at-a-distance] gradually lost its simplicity and physical probability. Men tried to make up for this loss by seeking for more comprehensive and simple laws, so-called elementary laws. Of these the celebrated law of Weber is the most important example. Whatever we may think of its correctness, the completion of this endeavor produced a comprehensive system full of scientific charm; those who were once attracted into its magic circle remained prisoners there.

And if the path indicated was a false one, warning could only come from an intellect of great freshness, from a man who looked at phenomena with an open mind and without preconceived opinions, who started from what he saw, not from what he had heard, learned, or read. Such a man was Faraday.[3] Faraday undoubtedly heard it said that when a body was electrified something was introduced into it; but he saw that the changes that occurred only made themselves felt outside and not inside the body. Faraday had learned that forces simply acted across space; but he saw that an important part was played by the particular kind of matter filling the space across which the forces were supposed to act. Faraday read that electricities certainly existed, whereas there was much contention as to the forces exerted by them;

---

[3][This tribute to Faraday is one of the most frequently-quoted parts of Hertz's eloquent address.]

but he saw that the effects of these forces were clearly displayed, whereas he could perceive nothing of the electricities themselves. And so he formed a quite different, opposite conception of the matter. To him the electric and magnetic forces became the actually present, tangible realities; to him electricity and magnetism were the things whose existence might be disputed. The lines of force, as he called the forces independently considered, stood before the eye of his intellect as states of space, as tensions, vortices, currents, whatever they might be—this he himself was unable to determine—but there they were, acting on each other, pushing and pulling bodies about, spreading themselves around and carrying the disturbance from point to point. To the objection that complete rest is the only condition possible in empty space, he could answer: Is space really empty? Do not the phenomena of light compel us to regard it as filled with something? Might not the ether that transmits light waves also be capable of transmitting the changes that we call electric and magnetic forces? Might there not conceivably be some connection between these changes and light waves? Might not the latter be due to something like a vibration of the lines of force?

Faraday had advanced as far as this in his ideas and conjectures. He could not prove them, although he eagerly sought such a proof. He delighted in investigating the connection between light, electricity, and magnetism. The beautiful connection that he did discover[4] was not the one that he sought. So he tried again and again, and his search only ended with his life. Among the many questions that he raised, the one question that continually presented itself to him was whether the electric and magnetic forces require time for their propagation. When we suddenly excite an electromagnet by a current, is the effect perceived simultaneously at all distances? Or does it first affect magnets close at hand, then more distant ones, and lastly those that are quite far away? When we electrify and discharge a body in rapid succession, does the force vary at all distances simultaneously? Or do the oscillations arrive later the farther we move away from the body? In the latter case the oscillation would propagate itself as a wave through space. Are there such waves? To those questions Faraday could obtain no answer. And yet the answer is most closely connected with his own

---

[4] [The reference here appears to be to the Faraday Effect, the rotation by a magnetic field of the plane of polarization of polarized light.]

fundamental assumptions. If such waves of electric force exist, travelling freely from their origin through space, they plainly exhibit to us the independent existence of the forces that produce them. There can be no better way to prove that these forces do not act [instantaneously] across space, but are propagated from point to point, than by actually following their progress from instant to instant. These questions are not unanswerable; indeed they can be attacked by very simple methods. If Faraday had had the good fortune to hit upon these methods, his views would forthwith have gained recognition. The connection between light and electricity would at once have become so clear that it could not have escaped notice even by eyes with less clear vision than his own.

But a path so short and straight as this was not vouchsafed to science. For a while experiments did not point to any solution, nor did the prevailing theory tend in the direction of Faraday's ideas. The assertion that electric forces could exist independently of their electricities was in direct opposition to the accepted electrical theories. Similarly the prevailing theory of optics refused to accept the idea that light waves could be other than elastic waves. Any attempt at a thorough discussion of the one or the other of these assertions seemed almost to be idle speculation. All the more must we admire, then, the happy genius of the man who could connect together these apparently disconnected conjectures in such a way that they mutually supported each other and formed a theory that every one had immediately to agree was at least plausible. The man of whom I speak was the Englishman Maxwell. You know the paper he published in 1865 on the electromagnetic theory of light. It is impossible to study this wonderful theory without feeling as if the mathematical equations had an independent life and an intelligence of their own, as if they were wiser than ourselves, indeed wiser than their discoverer, as if they gave forth more than he had put into them.[5] And this is not altogether impossible; it may happen when the equations prove to be more correct than their discoverer could have known with certainty. It is true that such comprehensive and accurate equations only reveal themselves to those who with keen insight pick out every indication of the truth that is only faintly visible in nature.

---

[5][This sentence on Maxwell's theory is also frequently quoted.]

The clue that Maxwell followed is well known to the initiated. It had attracted the attention of other investigators; it had suggested to Riemann and Lorenz speculations of a similar nature,[6] although not so fruitful in their results. This was the following circumstance. Electricity in motion produces magnetic force, and magnetism in motion produces electric force; but both of these effects are only observable at high velocities. As a consequence velocities appear in the mutual relations between electricity and magnetism, and the constant that governs these relations and continually recurs in them is itself a velocity of exceedingly great magnitude. This constant was determined in various ways, first by Kohlrausch and Weber,[7] who employed purely electrical experiments. When proper allowance was made for the experimental errors that are unavoidable in such a difficult experiment, this constant proved to be identical with another important velocity, the velocity of light. This might be an accident, but a pupil of Faraday could scarcely regard it as such. To him [Maxwell] it appeared as an indication that the same ether must be the medium for the transmission of both electric force and light. The two velocities that were found to be nearly equal must really be identical. But in that case the most important optical constants must occur in the electrical equations. This was the bond that Maxwell set himself to strengthen. He developed the electrical equations to such an extent that they embraced all the known phenomena, and in addition to these a class of phenomena hitherto unknown—electric waves. These waves would be transverse waves, which might have any wavelength, but would always be propagated in the ether with the same velocity, that of light. And now Maxwell was able to point out that waves having just these geometrical properties do actually occur in nature, although we are accustomed to denote them not as electrical phenomena, but by the special name of light. If Maxwell's electrical theory were regarded as false, there was no reason for accepting his views as to the nature of light. And if light waves were held to be purely elastic waves, his

---

[6][On the theories of Riemann and Lorenz, see footnote 10 to Paper No. 3.]

[7][In 1855 Rudolf Kohlrausch (1809–1858) and Wilhelm Weber (1804–1891) measured the ratio of the electrostatic unit of charge to the electromagnetic unit of charge. This ratio is dimensionally a velocity. In this way they obtained the value 3.1×10^8 m/s, which was very close to the measured velocity of light at that time.]

electrical theory lost its whole significance. But if one approached the structure [of Maxwell's theory] with no prejudices arising from commonly-accepted views, one saw that its parts supported each other like the stones of an arch, and the entire structure stretched over the deep abyss of the unknown to tie together those parts that were known.

On account of the difficulty of the theory, the number of its disciples was necessarily small. But everyone who studied it thoroughly became an adherent, and forthwith sought diligently to test its original assumptions and its ultimate conclusions. Naturally the test of experiment could for a long time only be applied to separate propositions, limited to the superficial aspects of the theory. I have just compared Maxwell's theory to an arch stretching across an abyss of the unknown. If I may carry the analogy a bit further, I would say that for a long time the only additional support that was given to this arch was by strengthening its two abutments. The arch was thus enabled to carry its own weight safely; but still its span was so great that we could not venture to add additional weight to it as if it were on a secure foundation. For this purpose it was necessary to have special pillars built up from the solid ground and serving to support the center of the arch. One such pillar would consist in proving that electric or magnetic effects can be directly produced by light. This pillar would support the optical side of the structure directly and the electrical side indirectly. Another pillar would consist in proving the existence of waves of electric or magnetic force capable of being propagated after the manner of light waves. This pillar again would directly support the electrical side, and indirectly the optical side. In order to complete the structure symmetrically, both pillars would have to be built; but it would suffice to begin with one of them. With the former we have not as yet been able to make a start;[8] but fortunately, after a protracted search, a safe point of support for the latter has been found. A sufficiently extensive foundation has been laid down; a part of the pillar has already been built up; with the help of many willing hands it

---

[8][We might think that Hertz's own work on the photoelectric effect (Paper No. 7) was a start "in proving that electric or magnetic effects can be directly produced by light." But Hertz did not consider the photoelectric effect to be a direct effect of light, but rather an indirect effect requiring ponderable matter as an intermediary, as discussed on the first page of the present paper. Hence he does not include it here.]

will soon reach the height of the arch, and so enable this arch to bear the weight of the additional structure to be erected on top of it. At this stage I was so fortunate as to be able to take part in the work. To this circumstance I owe the honor of speaking to you today; and you will therefore pardon me if I now try to direct your attention solely to this part of the structure. Lack of time compels me, against my will, to pass by the researches of many other investigators; so that I am not able to show you in how many ways the path was prepared for my experiments, and how close several investigators came to performing these experiments themselves.

Was it then so difficult to prove that electric and magnetic forces need time for their propagation? Would it not have been easy to charge a Leyden jar and to observe directly whether the corresponding disturbance in a distant electroscope took place somewhat later? Would it not have sufficed to watch the behavior of a magnetic needle while someone at a distance suddenly excited an electromagnet? As a matter of fact these and similar experiments had already been performed without indicating that any interval of time elapsed between the cause and the effect. To an adherent of Maxwell's theory this is simply a necessary result of the enormous velocity of propagation. We can only perceive the effect of charging a Leyden jar or exciting a magnet at moderate distances, say up to ten meters. To traverse such a distance light, and therefore according to the theory likewise the electric force, takes only one thirty-millionth part of a second. Such a small fraction of time we cannot directly measure or even observe. It is still more unfortunate that there are no adequate means at our disposal for indicating with sufficient precision the beginning and end of such a short interval. If we wish to measure a length correctly to the tenth part of a millimeter, it would be absurd to indicate the beginning of the length with a broad chalk line. If we wish to measure a time correctly to the thousandth part of a second, it would be absurd to denote its beginning by the stroke of a large clock. Now the time of discharge of a Leyden jar is, according to our ordinary ideas, inconceivably short. It would certainly be that short if it took about one thirty-thousandth part of a second. And yet for our present purpose even that would be a thousand times too long. Fortunately nature in this case does provide us with a more accurate method. It has long been known that the discharge of a Leyden jar is not a continuous process, but that, like the striking of a clock, it consists of a large number of oscillations, of discharges in opposite directions that follow each other at exactly equal intervals.[9] Electricity is able to simulate the phenomena of elasticity.[10] The period of a single oscillation is much shorter than the total duration of the discharge, and this suggests that we might use a single oscillation as an indicator. But, unfortunately, the shortest oscillation yet observed takes fully a millionth of a second. While such an oscillation is actually in progress, its effects spread out over a distance of three hundred meters; within the modest dimensions of a room they would be perceived almost at the instant the oscillation commences. Thus no progress was possible with the known methods; some fresh ideas were required. This came in the form of the discovery that not only the discharge of Leyden jars but, under suitable conditions, the discharge of every kind of conductor gives rise to oscillations. These oscillations may be much shorter than those of the jars. When you excite the conductor of an electrical machine, you excite oscillations whose period lies between a hundred-millionth and a thousand-millionth of a second. It is true that these oscillations do not follow each other in a long continuous series; they are few in number and rapidly die out. It would suit our experiments much better if this were not the case. But there is still the possibility of success if we can only obtain two or three such sharply-defined indicators. Thus, in the realm of acoustics, if we were denied the continuous tones of pipes and strings, we could get a poor kind of music by striking strips of wood.

We thus have indicators for which the thirty-millionth[11] part of a second is no longer short. But these would be of little use to us if we were not in a position to actually observe their action up to the distance under consideration, namely, about ten meters. This can be

---

[9][The oscillatory nature of the discharge of a Leyden jar had been demonstrated by Joseph Henry, William Thomson (Lord Kelvin), and Hermann von Helmholtz, among others. See, for example, Helmholtz, in Kahl (1971), p. 31, and his "Über elektrische Oscillationen" in Helmholtz (1882–1895), vol. 1, pp. 531–536.]

[10][For example, the expression for the resonant frequency of an electric circuit containing only capacitance and inductance is completely analogous to that for the resonant frequency of a mechanical oscillator consisting of a mass vibrating at the end of a spring.]

[11][There is a slip on the part of the translators here: *dreissigmillionte* means "thirty-millionth," not "thirty-thousandth."]

done by very simple means.[12] Just at the spot where we wish to detect the electric force we place a conductor, say a straight wire, which is interrupted in the middle by a small spark-gap. The rapidly oscillating force sets the electricity of the conductor in motion and gives rise to a spark at the gap. This method had to be found by experiment, for no amount of thought could have enabled one to predict that it would work satisfactorily. For the sparks are microscopically short, scarcely a hundredth of a millimeter long; they last only about a millionth of a second. It almost seems absurd and impossible that they should be visible; but in a perfectly dark room they *are* visible to an eye that has been well rested in the dark. Upon this thin thread hangs the success of our undertaking. In beginning it we are met by a number of questions. Under what conditions can we get the most powerful oscillations? These conditions we must carefully investigate and use to our advantage. What is the best form we can give to the receiver? We may choose straight wires or circular wires, or conductors of other forms; in each case the choice will have some effect upon the phenomena observed. When we have settled on the form, what size shall we select? We soon find that this is a matter of some importance, that a given conductor is not suitable for the investigation of all kinds of oscillations, that there are relations between the two which remind us of the phenomena of resonance in acoustics. And lastly, are there not an endless number of positions in which we can expose a given conductor to the oscillations? In some of these the sparks are strong, in others weaker, and in others they entirely disappear. I dare not venture to take up your time with these details, for in the large picture they are of minor importance. But for the researcher in this field they are certainly not of minor importance; they are the characteristics of his research tools; and the success of a workman depends on how well he understands his tools.[13] The thorough study of these tools, and of the questions posed above, formed a very important part of the task to be accomplished. After this was achieved, the method of attacking the main problem became obvious. If you give a physicist a number of tuning-forks and resonators and ask him to demonstrate to you the

---

[12][The means of producing and detecting these high-frequency oscillations are discussed in Paper No. 6.]

[13][One of the secrets of Hertz's success as an experimental physicist was that he understood his research instruments remarkably well because he had designed them and had often built them with his own hands.]

propagation in time of sound waves, he will find no difficulty in doing so even within the narrow confines of a room. He places a tuning fork anywhere in the room, listens with the resonator at various points, and observes the intensity of the sound. He shows how at certain points this is very small, and how this arises from the fact that at these points every oscillation is annulled by another one that started subsequently but travelled to the point along a shorter path. When a shorter path requires less time that a longer one, the propagation is a propagation in time. Thus the problem is solved. But the physicist now shows us additionally that the positions of silence follow each other at regular and equal distances; from this he determines the wavelength and, if he knows the time of vibration of the tuning fork, he can deduce the velocity of the wave.

In exactly the same way we proceed with our electric waves. In place of the tuning fork we use an oscillating conductor. In place of the resonator we use our interrupted wire, which may also be called an electric resonator. We observe that in certain places [in the room] there are sparks at the gap, in other places none; we see that the dead places follow each other periodically in ordered succession. Thus the propagation in time is proved and the wavelength can be measured.[14] Next comes the question whether the waves thus demonstrated are longitudinal or transverse. At a given place we hold our wire in two different positions with reference to the wave; in one position it responds, in the other it does not.[15] This is sufficient; the question is decided; our waves are transverse waves. Their velocity must now be found. We multiply the measured wavelength by the calculated period[16] of oscillation, and find a velocity that is approximately that of light waves. If doubts are raised as to whether the calculation is trustworthy, there is still another method available to us. In wires, as well as in air, the velocity of electric waves is enormously great, so that we can make a direct comparison between the two. Now the velocity of

---

[14][This technique is discussed at length in Paper No. 8.]

[15][Hertz means that sparks are produced when the receiving spark-gap is parallel to the transmitting conductor, and no sparks are produced when the receiving spark-gap is perpendicular to the transmitter. The waves are therefore polarized and must be transverse, since longitudinal waves cannot be polarized. On this see the section "Polarization" in Paper No. 9.]

[16][The period can be calculated from the values of the capacitance and the inductance of the transmitting circuit.]

electric waves in wires has long since been directly measured. This was an easier problem to solve, since such waves can be followed for several kilometers. Thus we obtain another measurement, purely experimental, of our velocity, and if the result is only an approximate one, at any rate it certainly does not contradict our first result.

All these experiments in themselves are very simple, but they lead to conclusions of the highest importance. They are fatal to any and all theories which assume that electric force acts across space independently of time. They mark a brilliant victory for Maxwell's theory. No longer does this theory unexpectedly tie together greatly diverse phenomena of nature. Even those who once felt that this conception of the nature of light had but a faint air of probability about it, now find it hard to reject it. In this sense we have reached our goal.

But perhaps the intervention of theory may in this case be completely unnecessary. Our experiments were carried out at the summit of the arch which, according to the theory, connects the domain of optics with that of electricity. It was only natural to move a few steps further and to attempt the descent into the known domain of optics. There may be some advantage in putting theory aside. There are many lovers of science who are curious as to the nature of light and are interested in simple experiments, but to whom Maxwell's theory is nevertheless a book with seven seals. The economy of science, too, requires of us that we should avoid roundabout ways when a straight road is available. If with the aid of our electric waves we can directly exhibit the phenomena of light, we shall have no need for theory as an interpreter; the experiments themselves will clearly demonstrate the relationship between the two fields. As a matter of fact, such experiments can be performed. We set up the conductor in which the oscillations are excited in the focal line of a very large concave mirror.[17] The waves are thus kept together and proceed from the mirror as a powerful parallel beam. We indeed cannot see this beam directly, or feel its effects; its effects are manifested by the sparks excited in the conductors on which it falls. It only becomes visible to our eyes when they are provided with one of our resonators. But in other respects it is really a beam of light. By rotating the mirror we can send it in various directions, and by examining the path it follows we

---

[17][All the experiments discussed in this paragraph are described in more detail in Paper No. 9.]

can prove that it travels in a straight line. If we place a conducting body in its path, we find that the beam does not pass through; the body casts a shadow. This does not extinguish the beam, but only reflect it back on itself; we can follow the reflected beam and convince ourselves that the laws of its reflection are the same as those of light. We can also refract the beam in the same way as light. In order to refract a beam of light we send it through a prism, and it then suffers a deviation from its straight path. In the present case we proceed in the same way and obtain the same result; excepting that the dimensions of the waves and of the beam make it necessary for us to use a very large prism. For this reason we make our prism of a cheap material, such as pitch or asphalt. Lastly, we can with our beam observe those phenomena that hitherto have only been observed with beams of light, the phenomena of polarization. By interposing a suitable wire grating in the path of the beam we can extinguish or excite sparks in our resonator in accordance with just the same laws as those that govern the brightening or darkening of the field of view in a polarizing apparatus when we interpose a crystalline plate.

Thus far the experiments. In carrying them out we are decidedly working in the region of optics. In planning the experiments, in describing them, we no longer think electrically, but optically. We no longer see currents flowing in conductors and electricities accumulating on them; we see only waves in the air, how they intersect, how they decay, how they combine, how they strengthen and weaken each other. Starting with purely electrical phenomena, we have gone step by step until we find ourselves in the realm of purely optical phenomena. We have crossed the summit of the arch; our path is downwards and soon begins to become level again. The connection between light and electricity, of which there were hints and suspicions and even predictions in the theory, is now established; it is accessible to our senses and intelligible to our understanding. From the highest point to which we have climbed, from the very top of the arch, we can better survey both regions. They are more expansive than we had ever realized. Optics is no longer restricted to minute ether-waves a small fraction of a millimeter in length; its domain is extended to waves that are measured in decimeters, meters and kilometers. And in spite of this extension it appears, when examined from this point of view, as a mere tiny appendage to the great domain of electricity. We see that this latter has become a mighty kingdom. We perceive electricity in a thousand places where we had no proof of its existence before. In every flame, in every luminous particle we see an electrical process. Even if a body is not luminous, if it only radiates heat it is a center of electrical disturbances. Thus the domain of electricity extends over the whole of nature. It even affects our own selves closely: we perceive that we actually possess an electrical organ, the eye. These are the things that we see when we look downward from our high standpoint.

No less attractive is the view when we look upward towards the lofty peaks, the highest pinnacles of science. We are at once confronted with the question of direct actions-at-a-distance. Are there such? Of the many in which we once believed there now remains but one, gravitation. Is this too a deception? The law according to which it acts makes us suspicious.[18]

In another direction looms the question of the nature of electricity. Viewed from this standpoint, the question conceals itself behind the more definite question of the nature of electric and magnetic forces in space. Directly connected with these is the great problem of the nature and properties of the ether that fills space, of its structure, of its rest or motion, of its finite or infinite extent. More and more we feel that this is the all-important problem, and that the solution of it will reveal to us not only the nature of what used to be called imponderables, but also the nature of matter itself and of its most essential properties, weight and inertia. The quintessence of ancient systems of physical science is preserved for us in the assertion that all things have been made out of fire and water. At the present time physics is more inclined to ask whether all things have not been fashioned out of the ether.[19] These are the ultimate problems of physical science, the icy summits of its loftiest range. Shall we ever be permitted to set foot upon one of these summits? Will it be soon? Or

---

[18][Newton's law of universal gravitation, Coulomb's law in electrostatics, and the equivalent law for magnetic poles all take the same mathematical form. Since Maxwell's theory and Hertz's experiments have excluded the possibility of action-at-a-distance for electric and magnetic forces, the same conclusion about gravitational forces is strongly suggested.]

[19][Here Hertz's use of the word *Quintessenz* seems to be intended as a play on words. In addition to Aristotle's four elements—earth, air, fire, and water—the ancients believed that there was a fifth element, a *quinta essentia*, out of which all the celestial bodies were made. The *ether* later came to be regarded as filling the role of this fifth element. Clearly the German *Quintessenz* and the English *quintessence* are both derived from the Latin *quinta essentia*.]

have we long to wait? We know not; but we have found a starting-point for further attempts that is at a higher point than any available before. Here the path does not end abruptly in a rocky wall; the first steps that we can see form a gentle ascent, and among the rocks there are paths that lead upwards. There is no lack of eager and experienced investigators;—how can we feel otherwise than hopeful that future undertakings will achieve success?

PAPER NO. 11

Review of Hertz's *Electric Waves* by George Francis FitzGerald

(*Nature* 48 [Oct. 5, 1893], pp. 538–539.[1] Reprinted with permission from *Nature* 48. Copyright © 1893 Macmillan Magazines Limited.)

A discoverer's own account of his work is always of interest, and when it is an epoch-making work and the account so clear and well described as to be intelligible to all, it deserves the most careful attention, and should be studied by all who feel any interest in the subject. Dr. Hertz's account of his discovery of the propagation of electric energy is eminently a work of this kind. The subject is of immense importance; the work described is of the highest order of experimental investigation; the results attained have contributed more than any other recent results to revolutionize the view taken by the majority of scientific workers as to the nature of electromagnetic waves. It is to be hoped that a translation[2] of this account of one of the greatest advances in our knowledge of nature will soon be in the hands of all who care to learn how the functions of the ether have been raised from obscurity into light, from being in the opinion of many a pious belief to be the momentous question of the hour.

Prof. Hertz gives in his introduction an interesting account of the steps by which Maxwell's theory may be connected with the older theories. These latter supposed action-at-a-distance pure and simple, and postulated two fluids, etc., etc. They neglected the intervening medium. The second step was to introduce the medium as performing some function when it was a material medium, but still to retain the positive and negative electricities acting across the space from molecule to molecule. This was practically Mossotti's theory as to the

---

[1] [FitzGerald is not indicated as the author of this review in *Nature*, nor is the review included in his collected works (FitzGerald [1902]). Bruce J. Hunt has concluded, however, on the basis of strong internal evidence, that the review is "almost certainly by FitzGerald." (On this see Hunt [1991], pp. 199–200, especially footnote 84.) An inquiry to *Nature* by this editor brought a reply on 9 February 1993, that their records clearly indicate that FitzGerald was indeed the author of this review of the 1892 German edition of *Electric Waves*.]

[2] [Such a translation, by D.E. Jones, appeared as *Electric Waves* in 1893.]

properties of the dielectric founded on Poisson's theory of magnetic induction. M. Poincaré seems to have got to about this stage, or perhaps a little further. The third stage was to transfer the molecular action to the ether, but still to consider it as due to electrical fluids attracting and repelling one another, producing the etherial stresses. The fourth stage was to see that these attractions and repulsions of electrical fluids are quite superfluous, and to attribute the whole phenomenon to stresses in the ether set up by straining it. In this last stage there is no room for an electrical fluid with attracting and repelling properties, and accordingly it is suppressed. What the structure of the ether may be which is strained, and thereby electromagnetic stresses produced, is still unknown, and consequently the nature of the strain is unknown. It certainly differs from the ordinary straining of a solid in two important respects. In the first place, the mechanical stresses are proportional to the squares of the quantities that represent the strains; and in the second place, they depend on the absolute strain, and not on the relative displacement of the parts of the medium. Solid structures can be invented that have laws of this kind. The change of longitudinal stress in a stretched string is proportional to the square of the transverse displacement, and, if the ends of the string are fixed, this stress depends on the absolute value of the displacement.

Upon a foundation of a somewhat similar kind a theory as to the structure of the ether being like a solid in tension may be founded, which gets over many of the difficulties of the simple elastic-solid theory of the ether. We are, however, still a good way off any really satisfactory theory as to the structure of the ether, but the leading idea of Maxwell's theory, that electromagnetic attractions and repulsions are due to some sort of strain in the ether, is the direction in which scientific men are at present seeking for a dynamical explanation of electromagnetism and for a structure of the ether. Prof. Hertz, however, seems content to look upon Maxwell's theory as the series of Maxwell's equations.[3] This is hardly fair. Maxwell has done much more than produce a series of equations that represent electromagnetic actions. Weber and Clausius went very close to that without

---

[3] [The reference here is to Hertz's famous statement in his Introduction to *Electric Waves*: "To the question, 'What is Maxwell's theory?,' I know of no shorter or more definite answer than the following: Maxwell's theory is Maxwell's system of equations." (See Paper No. 5, footnote 47.)]

revolutionizing our ideas as to the nature of these actions. Any exposition of Maxwell's theory which does not clearly put before the reader that energy is stored in the ether by stresses working on strains, is a very incomplete representation of Maxwell's theory.

The bulk of Prof. Hertz's work is, however, not concerned with any theory, but with the practical study of electromagnetic propagation along conducting wires and throughout space. This is the work for which Prof. Hertz is so justly famous, and on account of which Hertzian oscillators, Hertzian receivers, Hertzian waves have become in the few years since 1888 the objects of universal attention.[4] No physical experiments since those by which Joule founded the theory of the conservation of energy have produced as great an effect on science as these experiments here described by their author. The subject is brought down to last year, and the experiments of others are mentioned and discussed. In this connection it may be worthwhile remarking that the observation that the waves emitted by a Hertzian oscillator are of all sorts of wavelengths was clearly stated by Prof. Hertz himself when he explained how rapidly they die out. For what is a rapidly dying-out oscillation except a Fourier series of all sorts of waves? There is consequently no essential difference between these two statements. The first states more than the second, for it explains the character of what in the other statement is described by the vague term, "all sorts of waves."

The whole work is most interesting, and well deserves the best attention of all interested in the greatest scientific advance of the last quarter of the nineteenth century, a century that has seen

---

[4] [FitzGerald also wrote a "very popular" account of Hertz's experiments on electromagnetic waves. It was entitled simply "Hertz's Experiments," and appeared in a series of anonymous articles in *Nature*. See FitzGerald (1891). Bruce Hunt has informed me that there are two letters from *Nature* (12 Dec. and 17 Dec. 1890) in the FitzGerald papers in Dublin, the first asking FitzGerald to write "three or four very popular articles on the Hertz experiments," and the second offering him a premium rate of 15 s. per column, as against the usual rate of 10 s. per column. FitzGerald agreed, but was not happy with the results, later asking Oliver Lodge in a letter (21 April 1891): "Are those Hertz articles in *Nature* long-winded rot?" He was on to something here, since the articles are overly simple and lack the depth and sharpness usually found in FitzGerald's writing. His overall impression of Hertz's work is reflected in a phrase from the first of these articles: "some of the most beautifully conceived, ingeniously devised and laboriously executed of experiments" (FitzGerald [1891], p. 537). I am grateful to Bruce Hunt for the above information.]

thermodynamics founded by Carnot and Clausius, conservation of energy by Joule, bacteriology by Pasteur, the origin of species by Darwin, and the functions of the ether by Faraday, Maxwell, and Hertz.

PAPER NO. 12

Nomination of Heinrich Hertz as Corresponding Member of the Berlin Academy of Sciences[1]

(Presented by Hermann von Helmholtz at the 31 January 1889 session of the Physics-Mathematics section of the Berlin Academy of Sciences; translated by Joseph F. Mulligan. Originally published in *Physiker über Physiker* [edited by Christa Kirsten and Hans-Günther Körber], Vol. 1 [1975], on pages 114–115: "Wahlvorschlag für Heinrich Hertz [1857–1894]." Reprinted with permission from Akademie-Verlag, Berlin.)

The undersigned wish to introduce a motion to elect as Corresponding Member of the Academy Professor Heinrich Hertz, at the present time still Professor of Physics at the *Technische Hochschule* in Karlsruhe, but already appointed as Ordinary Professor of the same subject at the University of Bonn.

Hertz is still a relatively young man. He has, however, already made himself well known by a series of very ingenious and unusually significant pieces of research. Born in Hamburg, he first studied mechanical engineering but soon switched to physics. While still a student he received a faculty prize for a paper[2] (published in 1880) exploring, for the rapid motion of electricity in wires, whether alongside the apparent inertia of electricity produced by electromagnetic induction there is an additional true inertia that is independent of the neighboring currents. Even then he had pinned down a source of error that had up to that time plagued all previous attempts, in which strong induction currents were produced in large spirals. With great perspicacity he later found the way to use a much more rapid, but more accurately calculable, oscillating current in straight stretched wires, and thus succeeded in overcoming the great experimental difficulties that stood in the way of such an undertaking.

Throughout this work his attention was continually called to the fundamental questions of the theory of electrodynamics; the most

---

[1][Kirsten and Körber (1975), pp. 114–115. The footnotes in Kirsten and Körber have been considerably modified to make them more useful to readers of this book.]

[2][H. Hertz, "Experiment to Determine an Upper Limit to the Kinetic Energy of an Electric Current" (1880); *Misc.Pprs.*, pp. 1–34.]

important research accomplished by him thus far is in this field. This area of physics was in no sense uncultivated. On the contrary, because at stake here was the uncovering of a new system of forces that must differ essentially in their fundamental behavior from that of Newton's force of gravitation, the attention of the most prominent theoreticians and experimentalists had been unceasingly directed to this problem.

But as soon as one moves from the easily observable actions of a constant closed current, and of the related partially open current that can be interrupted by the separated plates of a capacitor, to the transient motions that arise from the discharge of electricity by conductors at rest, the experimental difficulties appear to be insuperable. Here we have to deal with actions that spread through space at the speed of light and with currents that last only millionths of a second.

But Hertz first wrote a number of relevant, extremely penetrating and highly original theoretical articles (1884,1885).[3] He also employed the large scale of the rotating magnetic earth to exclude certain possible hypotheses.[4] Finally he found a practical way to make visible the above-mentioned fluctuating movements of electricity and their effects, to determine their mode of propagation and velocity, and even to measure this velocity approximately. This was made possible by employing tiny sparks that jumped across very narrow gaps between the conductors involved. He proved in this way that the electrodynamic and magnetic actions need time to propagate through space, and that the nature of this method of propagation is, in the highest degree, similar to that of light, an insight that had been made theoretically probable by Maxwell, but that left other explanations as still possible.[5]

---

[3][H. Hertz, "On the Relations between Maxwell's Fundamental Electromagnetic Equations and the Fundamental Equations of the Opposing Electromagnetics" (1884), which is Paper No. 3; H. Hertz, "On the Dimensions of Magnetic Pole in Different Systems of Units" (1885), *Misc.Pprs.*, pp. 291–295.]

[4][H. Hertz, "On Induction in Rotating Spheres" (1880), *Misc.Pprs.*, pp. 35–126. This was Hertz's doctoral dissertation. Here Helmholtz is apparently referring to one of the examples Hertz considers at the end of this paper (pp. 106–107), in which he discusses the rotation of the magnetic Earth in dielectric space. This topic seems out-of-place in the context, and Helmholtz makes no effort to indicate the relevance or importance of Hertz's work on this problem.]

[5][See Papers No. 5,6,8,9, and 10 in this collection.]

We need not explore these researches of Hertz any further, since in the last two years they have been presented to the Academy.[6] This research is the most significant but not the only work of the above-mentioned author. In addition, he has written about the rate of evaporation of liquids, in particular mercury in a space free of air; also, he has studied the equilibrium of floating elastic plates, and the method of formation of residual electric charge on insulators.[7] These researches are further demonstrations of a deeply penetrating mind and great skill in translating theoretical problems into experimental tests.

We therefore believe that the motion proposed above should be enthusiastically recommended to the Academy.[8]

---

[6][The researches referred to here, which were published in *Sitzungsberichte d. Berl. Akad. d. Wiss.* are: 1. "On the Effect of Ultraviolet Light on the Electric Discharge" (9 June 1887); 2. "On Electromagnetic Effects Produced by Electrical Disturbances in Insulators" (10 Nov. 1887); 3. "On the Finite Velocity of Propagation of Electromagnetic Actions" (2 Feb. 1888); 4. "On Electric Radiation" (13 Dec. 1888). All four were later published in the *Annalen*, and are collected in Hertz's *Electric Waves*. The first and last of these are included here as Papers No. 7 and 9. In the years 1887-1889 Hertz published nine articles in the *Annalen*, but had only these four presented first to the Berlin Academy by Helmholtz.]

[7][These researches are all contained in Hertz's *Miscellaneous Papers*, and are discussed by Lenard, Helmholtz and Planck in Papers No. 1, 14 and 18 of the present collection.]

[8][This nomination was signed by Helmholtz, August Kundt (1839-1894), and Wilhelm von Bezold (1837-1907). Hertz was formally elected a Corresponding Member of the Berlin Academy on 7 March 1889.]

PART V
Bonn University (1889–1894): Professor of Physics

> As to the details I have nothing to bring forward which is new or which could not have been gleaned from many books. What I hope is new, and to this alone I attach value, is the arrangement and collocation of the whole—the logical or philosophical aspect of the matter. According as it marks an advance in this direction or not, my work will attain or fail of its object.
>
> *Heinrich Hertz, Preface to his* Mechanics *(1894)*

Heinrich Hertz, probably just a few months before his move to Bonn as Professor of Physics in April 1889. This photograph was taken by Hertz himself in his Karlsruhe laboratory. Courtesy of Dr. Klaus-Peter Hoepke, archivist of the University of Karlsruhe.

Introduction

Hertz's research output during his brief tenure in Bonn was limited by his inability to find a worthwhile experimental project and by increasingly bad health. He did complete one piece of experimental research on electric waves on wires that had been left unfinished when he moved from Karlsruhe (*Electric Waves*, pp. 186–194). The only other experimental research of any consequence that he did in Bonn was on cathode rays, which demonstrated that cathode rays could pass through thin metallic foils. This paper (No. 13) was probably stimulated by contact with Philipp Lenard, who, like Hertz, had done previous work on cathode rays, and who later extended Hertz's work by developing the "Lenard window" to obtain cathode rays outside the vacuum chamber in which they had been produced.

The rest of Hertz's research publications in Bonn consisted of two important theoretical papers on the fundamental equations of electromagnetism (*Electric Waves*, pp. 195–268), and his well-known book on mechanics.

Hertz's 271-page book, *The Principles of Mechanics Presented in a New Form*, is strikingly different from anything else Hertz wrote. He makes no attempt to obtain predictions that can be tested by experiment but is only interested in codifying well-known facts of mechanics in a consistent, logical system. His interest seems to be more in logic and epistemology than in physics, and for this reason his book is now of more interest to philosophers than to physicists. In addition, the new approach to mechanics that Hertz recommends in this book has been overtaken and undermined by the developments in quantum mechanics and relativity theory in this century. For this reason it would be a futile gesture to include here any brief samples of a work that must stand or fall as a whole.

But the introductory material to Hertz's *Mechanics* is, even today, of great interest to both physicists and philosophers. Therefore, the second paper included in this section is the Preface by H. von Helmholtz, in which he reviews the life and research contributions of his most famous student. Helmholtz seems, however, to be strangely reluctant to come to grips with the thrust and significance of Hertz's *Mechanics*. Only on the last few pages of this preface does he discuss the book at all, and then it is in a remarkably detached and somewhat vague way. His comments are all based on Hertz's forty-one-page Introduction, with no judgments at all on the merit or importance of the text proper.

The third paper selected for inclusion in this section is Hertz's Preface to his *Mechanics*. This brief piece is mainly concerned with his intentions in writing this book and with the sources he found useful in carrying out these intentions.

The last of Hertz's papers reprinted in this collection is the Introduction to his *Mechanics*. In it Hertz outlines with great clarity what he is attempting to do in the rest of the book. He compares three different approaches to the foundations of mechanics, the third being his own novel formulation of mechanics on the basis of only three fundamental quantities (time, space, and mass), instead of including either force or energy in addition to these three, as did the two most important schools of mechanics at that time. Because of its philosophical awareness and technical competence, this Introduction still remains one of the most frequently quoted of Hertz's writings.

The present section, devoted to Hertz's years in Bonn, concludes with George Francis FitzGerald's 1895 review of Hertz's *Mechanics* for *Nature*. This is in great part a paraphrase of Helmholtz's Preface and Hertz's own Introduction, but is illuminated by the mathematical sophistication and probing questioning FitzGerald brought to everything he read or heard.

PAPER NO. 13

On the Passage of Cathode Rays through Thin Metallic Layers

(Heinrich Hertz, "Über den Durchgang der Kathodenstrahlen durch dünne Metallschichten," Wiedemann's *Annalen der Physik und Chemie* 45 [1892], pp. 28–32.)[1]

One of the chief differences between light and cathode rays lies in their power to pass through solid bodies. The very substances which are most transparent to all kinds of light offer, even in the thinnest layers that can be made, an insuperable resistance to the passage of cathode rays. I have been all the more surprised, then, to find that metals, which are opaque to light, are slightly transparent to cathode rays. Metallic layers of moderate thickness are of course as opaque to cathode rays as they are to light. But if a metallic layer is so thin as to allow a part of the incident light to pass through, it will also allow a part of the incident cathode rays to pass through; and the fraction transmitted seems to be larger in the latter than in the former case.[2]

This can be demonstrated by a very simple experiment. Take a plane glass plate capable of phosphorescing, preferably a piece of uranium glass; partially cover one side, which we will call the front side, with pure gold leaf, and in front of this fasten a piece of mica. Expose this front side to cathode rays proceeding from a flat circular aluminum cathode of 1 cm diameter, [with the glass plate] at a distance of about 20 cm from the cathode. As long as the vacuum is only moderate, the cathode rays fill the whole of the discharge tube as a powerful cone of light, and the glass only phosphoresces outside the patch covered with gold. At this stage the phosphorescence is caused chiefly by the light of the discharge, and only a very small part of this penetrates through the gold. But as the vacuum improves, there is less

---

[1] [*Misc.Pprs.*, pp. 328–331. Hertz's previous research on cathode rays may be found in his "On a Phenomenon that Accompanies the Electric Discharge" (1883), *Misc.Pprs.*, pp. 216–223; and "Experiments on the Cathode Discharge" (1883), *Misc.Pprs.*, pp. 224–254.]

[2] [Later research by Lenard and others has proved conclusively that cathode rays can also pass through thin metallic foils that are completely opaque to light.]

and less light inside the discharge tube, and the rays that impinge on the glass are more purely cathode rays. The glass now begins to phosphoresce behind the layer of gold leaf, and when the cathode rays have attained their most powerful strength, the gold leaf, when observed from the back, simply looks like a faint veil on the glass plate, chiefly recognizable at its edges and by the slight wrinkles it has. It can scarcely be said to throw a real shadow. On the other hand, the thin mica plate, which we have superimposed on the gold leaf, throws (through the gold leaf) a marked black shadow on the glass. Thus the cathode rays seem to penetrate with but little loss through the layer of gold.

I have tested other metals in the same way with the same result—silver leaf, aluminum leaf, various kinds of impure silver and gold leaf (alloys of tin, zinc, and copper), silver chemically precipitated, and also layers of silver, platinum and copper precipitated by the discharge *in vacuo*. These latter layers were much thinner than the beaten metallic leaves. I have not observed any characteristic differences between the various metals. Commercial aluminum leaf seems to work best. It is almost completely opaque to light but very transparent to the cathode rays; it is easily handled, and is not attacked by the cathode rays, whereas a layer of silver leaf, for example, is soon corroded by them in a peculiar manner.

It might be urged, against the assumption that the cathode rays in these experiments penetrate right through the substance of the metal, that such thin metallic layers are full of small holes, and that the cathode rays might well reach the glass through these without going through the metal itself. It is the behavior of the beaten metallic leaves that is most surprising, and one is bound to admit that these contain many pores; but the aggregate area of the holes scarcely amounts to a few percent of the area of the leaf, and is not sufficient to account for the brilliant luminescence of the covered glass. Furthermore, the covered part of the glass exhibits no luminescence when it is viewed from the front, i.e., from the side on which the cathode is. Hence the cathode rays must have reached the glass by a way which the light excited by them cannot retrace; so that they cannot have reached the glass through openings in the metallic leaf that rests close against it.

Again, if we place two metallic leaves one on top of the other, the number of coincident holes must become vanishingly small. But the cathode rays are able to make glass luminesce brightly under a double layer of metallic leaf; even under a three- or four-fold layer of gold or aluminum leaf we can perceive the phosphorescence of the glass and the shadows of objects in front of the leaf. I have been rather surprised by the extent to which the rays are weakened by passing through a double layer; it is much greater than one would expect from the slight weakening produced by a single layer. I think that the following suggestion provides a satisfactory explanation of this phenomenon. The metallic layer has a reflecting surface by which the phosphorescent light is reflected. This reflecting surface prevents the light from radiating toward the cathode, but it doubles the intensity of the light in the direction away from the cathode. If we assume that the metallic layer allows only $\frac{1}{3}$ of the cathode rays to pass, it will not reduce the luminescence to $\frac{1}{3}$ but only to $\frac{2}{3}$ of its previous value; whereas the second layer will reduce it to $\frac{2}{9}$, and further layers will soon cause the phosphorescence to vanish. If this suggestion is correct, metallic layers capable of transmitting more than half of the cathode rays should not weaken the luminescence at all; behind such metallic layers the glass ought actually to phosphoresce more strongly than in parts where it is not covered. I think I have been able to verify this expectation in the case of layers of silver chemically precipitated and of suitable thickness; but the observation is not yet trustworthy, because in the uncovered parts one cannot avoid seeing the greyish-blue luminescence of the gas through the phosphorescing glass. For this reason it is not easy to separate with any certainty the brightness of this [light from the gas] from that of the green phosphorescent light.

Lastly, if the cathode rays went right through the holes in the metal, they would afterwards continue their rectilinear path. But this is just what they do not do; by their passage through the metal they become diffused, just as light does by passing through a turbid medium such as milk glass. Let part of a cylindrical discharge tube be shut off, say at a distance of 20 cm from the cathode, by a metal plate extending right across it but containing a circular aperture a few mm in diameter; let this aperture be closed by a piece of aluminum leaf. If we now place a suitable glass plate close behind the aperture, we get, as might be expected, a distinct and bright phosphorescent image of the aperture on the glass; but if we move the glass plate even 1 or 2 mm back [from the aperture], the image becomes perceptibly larger and suffers a corresponding loss of brightness, its edge at the same time becoming indistinct. When the glass plate is moved back several mm, the image of the aperture becomes very indistinct, large and faint; and when the plate is shifted still further back, the tube behind the diaphragm appears quite dark.

That this is simply due to the feebleness of the cathode rays that have been diffused from the small aperture can be shown by introducing into the diaphragm several such apertures closed by aluminum leaf. For this purpose the diaphragm is best made of wire gauze hammered flat, upon which is stretched a piece of aluminum leaf. Behind such a diaphragm the whole of the discharge tube becomes filled with a uniform, moderately bright light. The phosphorescence is sufficiently strong to enable us to obtain separate beams by means of additional diaphragms; with these we can convince ourselves that, even after passing through metallic leaf, the cathode rays retain their properties of rectilinear propagation, of being deflected by a magnet, etc.[3]

There must be some connection between the phenomenon of the diffusion of cathode rays on passing through thin layers of bare metal and another phenomenon, namely, that when cathode rays impinge upon such a surface the portion reflected back is diffused, as E. Goldstein has shown.[4]

---

[3][Although the properties here mentioned by Hertz would be those expected of charged material particles, he refused to accept such a hypothesis because he could not understand how such particles could pass through metal foils. Rather Hertz, together with most German physicists, held to the view that cathode rays were some unusual form of electromagnetic radiation or disturbance in the ether.]

[4]E. Goldstein, *Annalen* 15 (1882), p. 246.

PAPER NO. 14

Preface to Hertz's *Principles of Mechanics* by Hermann von Helmholtz[1]

On the first of January, 1894, Heinrich Hertz died. All who regard human progress as consisting in the broadest possible development of the intellectual faculties, and in the victory of the intellect over natural passions as well as over the forces of nature, must have learned with the deepest sorrow of the death of this highly gifted prodigy. Endowed with the rarest gifts of intellect and character, he reaped during his lifetime (alas, so short!) a bounteous harvest which many of the most gifted investigators of the present century have tried in vain to gather. In old classical times it would have been said that he had fallen a victim to the envy of the gods. Here nature and fate appeared to have favored in an exceptional manner the development of a human intellect embracing all that was requisite for the solution of the most difficult problems of science—an intellect capable of the greatest acuteness and clearness in logical thought, as well as of the closest attention in observing apparently insignificant phenomena.[2] The uninitiated readily pass these by without heeding them; but to the practiced eye they point the way by which we can penetrate into the secrets of nature.

Heinrich Hertz seemed to be predestined to open up to mankind many of the secrets that nature had hitherto concealed from us; but all these hopes were frustrated by the malignant disease which, creeping on slowly but surely, robbed us of this precious life and of the achievements that it promised.

To me this has been a deep sorrow; for among all my students I have always regarded Hertz as the one who had penetrated furthest into my own circle of scientific thought; it was to him that I looked with the greatest confidence for the further development and extension of my work.

---

[1] [This Preface was written by Helmholtz during July of 1894. He was in poor health at the time and died on September 8 of the same year.]

[2] [Hertz's discovery of the photoelectric effect was a good example of this. On this see Paper No. 7. Later in Helmholtz's paper, he refers to other examples of the same ability on the part of Hertz.]

Heinrich Rudolf Hertz was born on 22 Feb. 1857 in Hamburg, and was the eldest son of Dr. Hertz, who was then a barrister and subsequently became a senator. Up to the time of his confirmation he was a pupil in one of the municipal primary schools (*Bürgerschulen*). After a year's preparation at home[3] he entered the *Gymnasium Johanneum*, the High School of his native city; here he remained until 1875, when he received his certificate of matriculation. As a boy he won the appreciation of his parents and teachers by his high moral character. Already his pursuits showed his natural inclinations. While still attending school he worked of his own accord at the bench and lathe; on Sundays he attended the *Gewerbeschule* [an industrial trade school] to practice geometrical drawing. There he constructed serviceable optical and mechanical instruments with the simplest possible tools.

At the end of his school course he had to decide on his career, and chose that of an engineer. His modesty, which in later years was such a characteristic feature of his nature, seems to have made him doubtful about his talent for theoretical science. He liked mechanical work and felt surer of success in it, because he already knew well enough what it meant and what it required. Perhaps, too, he was influenced by the prevailing atmosphere in his native city, which was inclined toward the practical life. It is in young men of unusual capacity that one most frequently observes this sort of timid modesty. They have a clear conception of the difficulties that have to be overcome before attaining the high ideal set before their minds; their strength must be tried by some practical test before they can develop the self-reliance requisite for their difficult task. And even in later years, for men of great ability, the higher their abilities and ideals, the less content they are with their own achievements. The most gifted attain the highest and most real success because they are most keenly alive to the presence of imperfection and most tireless in removing it.

For fully two years Heinrich Hertz remained in this state of doubt. Then, in the autumn of 1877, he decided upon an academic career; for as he grew in knowledge, he also grew in the conviction that

---

[3][Helmholtz is mistaken about some of the facts of Hertz's early life. Hertz was enrolled in a private school in 1863 at age six, studied there for nine years, spent the years 1872–1874 at home studying with private tutors, and had only one year at the *Gymnasium* before passing the *Abitur* examination in 1875.]

only in scientific work could he find lasting satisfaction. In the autumn of 1878 he came to Berlin and it was as an university student there, in the physical laboratory under my direction, that I first made his acquaintance. Even while he was going through the elementary course of laboratory work, I saw that I had here to deal with a student of quite unusual talent; and when, toward the end of the summer semester, it fell to me to propose to the students a subject of physical research for a prize, I chose one in electromagnetism, in the belief that Hertz would have an interest in it and would attack it, as he did, with success.[4]

In Germany at that time the laws of electromagnetism were deduced by most physicists from the hypothesis of W. Weber, who sought to trace back electric and magnetic phenomena to a modification of Newton's assumption of direct forces acting at a distance and in a straight line. With increasing distance these forces diminish in accordance with the same law assigned by Newton to the force of gravity, and by Coulomb to the action between pairs of electrified particles. This force was directly proportional to the product of the two quantities of electricity, and inversely proportional to the square of their distance apart; like quantities produced repulsion, unlike quantities attraction. Furthermore, in Weber's hypothesis it was assumed that this force was propagated through infinite space instantaneously, and with infinite velocity. The only difference between the views of W. Weber and Coulomb consisted in Weber's assumption that the magnitude of the force between two quantities of electricity might be affected by the velocity with which the two quantities approached, or receded from, each other, and also by the rate of change of this velocity in time. Side by side with Weber's theory there existed a number of others, all of which had this in common, that they regarded the magnitude of the force expressed by Coulomb's law

---

[4] [Philipp Lenard, the editor of the German edition of Hertz's *Mechanics*, adds the following footnote at this point in the second (1910) edition:

"The problem for the Physics Prize had been proposed to the students at the end of the summer semester in 1878, before Hertz moved to Berlin. The above comment of Helmholtz must therefore relate to the second prize-problem proposed later by the Academy of Science, again on Helmholtz's recommendation. This problem was also solved by Hertz, but only much later, by his electromagnetic discoveries."

Lenard's comment agrees with Hertz's statement that he found the first prize problem on the bulletin-board of the Physics Institute when he arrived in Berlin in October, 1878. On this see Part II of the Introductory Biography.]

as being modified by the influence of some component of the velocity of the electrical quantities. Such theories were advanced by F.E. Neumann, by his son C. Neumann, by Riemann, Grassmann, and subsequently by Clausius. Magnetized molecules were regarded as the axes of circular electric currents, in accordance with an analogy between their external effects previously discovered by Ampère.

This plentiful crop of hypotheses had become very unmanageable, and in dealing with them it was necessary to go through complicated calculations, the resolution of forces into their components in various directions, and so on. So at that time the domain of electromagnetism had become a pathless wilderness. Observed facts and deductions from exceedingly doubtful theories were inextricably mixed together. With the object of clearing up this confusion, I had set myself the task of surveying the domain of electromagnetism and of working out the distinctive consequences of the various theories in order, wherever possible, to decide between them by suitable experiments.

I arrived at the following general result. The phenomena which completely closed currents produce by their circulation through continuous, closed metallic circuits, and which have this common property that, while they flow, there is no considerable variation in the electric charges accumulated on the various parts of the conductor—all these phenomena can be equally well deduced from any of the above-mentioned theories. The deductions that follow from them agree with Ampère's laws of electromagnetic action, with the laws discovered by Faraday and Lenz, and also with the laws of induced electric currents as generalized by F.E. Neumann. On the other hand, the deductions that follow from them in the case of conducting circuits that are not completely closed differ essentially.

The agreement between the various theories and the facts that have been observed in the case of completely closed circuits is easily intelligible when we consider that closed currents of any desired strength can be maintained as long as we please, at any rate long enough to allow the forces exerted by them to exhibit plainly their effects; and on this account the actual effects of such currents and their laws are well known and have been carefully investigated. Thus any divergence between any newly-advanced theory and any one of the known facts in this well-trodden region would soon attract attention and be used to disprove the theory.

But at the open ends of unclosed conductors between which insulating substances have been placed,[5] every motion of electricity along the length of the conductor immediately causes an accumulation of electric charge; this is due to the surging of the electricity, which cannot force its way through the insulator, against the open ends of the conductor. Between the electricity accumulated at the end and the electricity of the same kind surging against it, there is a force of repulsion; and an exceedingly short time suffices for this force to attain such a magnitude that it completely checks the flow of electricity. The surging then ceases, and after an instant of rest there follows a resurging of the accumulated electricity in the opposite direction.

To everyone who was knowledgeable in these matters it was then apparent that a complete understanding of the theory of electromagnetic phenomena could only be attained by a thorough investigation of the processes that occur during these very rapid surgings of unclosed currents. W. Weber had endeavored to remove or lessen certain difficulties in his electromagnetic theory by suggesting that electricity might possess a certain degree of inertia, such as ponderable matter exhibits. In the opening and closing of every electric current, effects are produced that simulate the appearance of such electric inertia. These, however, arise from what is called electromagnetic induction, i.e., from a mutual action of neighboring conductors on each other, according to laws that have been well known since Faraday's time. True inertia should be proportional only to the quantity of the electricity in motion, and independent of the position of the conductor. If anything of the kind existed, we ought to be able to detect it by a retardation in electric oscillations such as those produced by the sudden interruption of an electric current in metallic wires. In this manner it should be possible to find an upper limit to the magnitude of this electric inertia; and so I was led to propose the problem of carrying out experiments on the magnitude of these so-called extra-currents. Extra-currents in double-wound spirals, the currents traversing the branches in opposite directions, were suggested in the statement of the problem as being apparently best adapted for these experiments. Heinrich Hertz's first research of importance consisted in solving this problem. In it he gives a definite

---

[5] [For example, a circuit containing an air gap or a capacitor.]

answer to the question proposed and shows that of the extra-current in a double-wound spiral, at most $1/30$ to $1/20$ of it could be ascribed to the effect of electric inertia. The prize was awarded to him for this investigation.

But Hertz did not confine himself to the experiments that had been suggested. For he recognized that, although the effects of induction are very much weaker in wires stretched out straight, they can be much more accurately calculated than in spirals of many turns, since in the latter he could not measure the geometrical parameters very accurately. Hence he used for further experiments a conductor consisting of two rectangles of straight wire, and found that the extra-current due to inertia could not be more than at most $1/250$ of the magnitude of the induced current.

Investigations on the effect of centrifugal force in a rapidly rotating plate on the motion of electricity passing through it led him to find a still lower value for the upper limit of the electric inertia.

These experiments clearly impressed on his mind the exceedingly great mobility of electricity and pointed out to him the way to his most important discoveries.

Meanwhile in England the ideas introduced by Faraday as to the nature of electricity were spreading. These ideas, expressed as they were in abstract language difficult of comprehension, made but slow progress until they found in Clerk Maxwell a fitting interpreter. In explaining electrical phenomena Faraday was bent on excluding all preconceived notions involving assumptions as to the existence of phenomena or substances that could not be directly observed. Especially did he reject, as did Newton at the beginning of his career, the hypothesis of the existence of action-at-a-distance. What the older theories assumed seemed to him inconceivable—that direct actions could go on between bodies separated in space without any change taking place in the intervening medium. So he first looked for indications of changes in media lying between electrified bodies or between magnetic bodies. He succeeded in detecting magnetism or diamagnetism in nearly all the bodies which up to that time had been regarded as non-magnetic. He also showed that good insulators undergo a change when exposed to the action of electric force; this he denoted as the "dielectric polarization of insulators."

It could not be denied that the attraction between two electrically-charged bodies or between two magnetic poles in the

Helmholtz's Preface to Principles of Mechanics

direction of their lines of force was considerably increased by introducing between them dielectrically or magnetically polarized media. On the other hand there was a repulsion across the lines of force. After these discoveries men were bound to recognize that a part of the magnetic and electric action was produced by the polarization of the intervening medium; another part might still remain, and this might be due to action-at-a-distance.

Faraday and Maxwell inclined toward the simpler view that there was no action-at-a-distance; this hypothesis, which involved a complete reversal of the ideas hitherto accepted, was thrown into mathematical form and developed by Maxwell. According to it, the seat of the changes that produce electrical phenomena must be sought only in the dielectrics; the polarization and depolarization of these are the real causes of the electrical disturbances that apparently take place in conductors. There were no longer any open currents;[6] for the accumulation of electric charges at the ends of a conductor, and the simultaneous dielectric polarization of the medium between them, represented an equivalent electric motion[7] in the intervening dielectric, thus completing the gap in the circuit [and changing the open circuit into a closed one].

Faraday had a very sure and profound insight into geometrical and mechanical questions, and he had already recognized that the distribution of electric action in space according to these new views must exactly agree with that found according to the older theory.

By the aid of mathematical analysis Maxwell confirmed this, and extended it into a complete theory of electromagnetism. For my own part I fully recognized the force of the facts discovered by Faraday and began to investigate the question whether actions-at-a-distance did indeed exist, and whether they must be taken into account. For I felt that scientific prudence required one to keep an open mind at first in

---

[6][Jones and Walley's translation here reads "closed currents," which is clearly a lapse in their otherwise excellent translation. The original German is "Ungeschlossene Ströme."]

[7][The reference here is to Maxwell's famous "displacement current," the essential element he required to derive his set of equations for the electromagnetic field. It is noteworthy that Helmholtz never uses the term "displacement current" in this Preface, since it was unpopular with physicists in Germany at the time he was writing, i.e., in 1894.]

such a complicated matter, and that such an approach might point the way to decisive experiments.

This was the state of the question at the time when Heinrich Hertz attacked it after completing the investigation that we have already described.

It was an essential postulate of Maxwell's theory that the polarization and depolarization of an insulator should produce in its vicinity the same electromagnetic effects as a galvanic current in a conductor. It seemed to me that this should be capable of demonstration, and that it would constitute a problem of sufficient importance for one of the great prizes of the Berlin Academy.

In the Introduction to his interesting book, *Untersuchungen über die Ausbreitung der elektrischen Kraft*,[8] Hertz has described how his own discoveries grew out of the seeds thus sown by his contemporaries, and has done this in such an admirably clear manner that it is impossible for anyone else to improve on it or add anything of importance. His Introduction, a perfectly frank and thorough account of a most important and fruitful discovery, is of exceedingly great value. It is a pity that we do not possess more documents of this kind on the inner psychological history of science. We owe the author a debt of gratitude for allowing us to penetrate into the inmost working of his thoughts, and for recording even his temporary mistakes.

Something may, however, be added as to the consequences which followed from his discoveries.

The views that Hertz subsequently proved to be correct had been propounded, as we have already said, by Faraday and Maxwell before him as being possible, and even highly probable; but as yet they had not been actually verified. Hertz supplied the demonstration. The phenomena that guided him along the path of success were exceedingly insignificant, and could only have attracted the attention of an observer who was unusually acute and able to see immediately the full importance of an unexpected phenomenon that others had overlooked. It would have been a hopeless task to render visible by means of a galvanometer, or by any other experimental method in use at that time, the rapid oscillations of currents having a period as short as one ten-thousandth, or even only one millionth, of a second. For all

---

[8][H. Hertz, *Electric Waves*. London, Macmillan, 1893. We have included Hertz's Introduction as Paper No. 5 in the present volume.]

finite forces require a certain time to produce finite velocities and to displace bodies of any mass, even when they are as light as the magnetic needles of our galvanometers usually are. But electric sparks can become visible between the ends of a conductor even when the potential at its ends only rises for a millionth of a second high enough to cause sparking across a minute air-gap. Through his earlier investigations Hertz was thoroughly familiar with the regularity and enormous velocity of these rapid electric oscillations; and when he attempted in this way to discover and render visible the most transient electric disturbances, success was not long in coming. He very quickly discovered what were the conditions under which he could produce in open circuits oscillations of sufficient regularity. He proceeded to examine their behavior under the most varied circumstances, and thus determined the laws of their development. He next succeeded in measuring their wavelength in air and their velocity. In the whole investigation one scarcely knows which to admire most, his experimental skill or the acuteness of his reasoning, so happily are the two combined.[9]

By these investigations Hertz has enriched physics with new and most interesting views respecting natural phenomena. There can no longer be any doubt that light waves consist of electric vibrations in the all-pervading ether, and that the latter possesses the properties of an [electrical] insulator and a magnetic medium.[10] Electric oscillations in the ether occupy an intermediate position between the exceedingly rapid oscillations of light and the comparatively slow disturbances that are produced by a tuning-fork when thrown into vibration; but as regards their rate of propagation, the transverse nature of their vibrations, the consequent possibility of polarizing them, their reflection and refraction, it can be shown that in all these respects they correspond completely to light and to heat rays.[11] The electric waves

---

[9][This last sentence is an indication of the great pride Helmholtz took in the success of his best and brightest student, Hertz.]

[10][Hertz had determined in the course of his work in Karlsruhe that the crucial question was the one Helmholtz poses here: Does a vacuum—or in terms more familiar to both physicists—Does the *ether* act like any other dielectric with respect to electrical disturbances? Hertz's answer was a resounding "Yes," with, of course, the proviso that the ether had a different electric permittivity and magnetic permeability from other substances.]

[11][On this, see Papers No. 9 and 10.]

only lack the power of affecting the eye, as do also the dark heat rays, whose frequency of oscillation is not high enough for this.

Here we have two great natural agencies—on the one hand light, which is so full of mystery and affects us in so many ways, and on the other hand electricity, which is equally mysterious, and perhaps even more varied in its manifestations. To have furnished a complete demonstration that these two are most closely connected together is to have achieved a great feat. From the standpoint of theoretical science, it is perhaps even more important to be able to understand how apparent actions-at-a-distance really consist in the propagation of an action from one layer of an intervening medium to the next.[12] Gravitation still remains an unsolved puzzle; as yet a satisfactory explanation of it has not been forthcoming, and we are still compelled to treat it as a pure action-at-a-distance.[13]

Among scientific men Heinrich Hertz has secured enduring fame by his researches. But not through his work alone will his memory live; none of those who knew him can ever forget his uniform modesty, his warm recognition of the labors of others, or his genuine gratitude toward his teachers. To him it was enough to seek after truth; and this he did with all zeal and devotion, and without the slightest trace of self-seeking. Even when he had some right to claim discoveries as his own, he preferred to remain quietly in the background. But although naturally quiet, he could be merry enough among friends, and could enliven social intercourse by many an apt remark. He never made an enemy,[14] although he knew how to judge slovenly work, and to appraise at its true value any pretentious claim to scientific recognition.

His career may be briefly sketched as follows. In the year 1880 he was appointed Demonstrator in the Physics Laboratory at the

---

[12][Note that Hertz's research demonstrated that the same is true even when only the "ether" fills the space between the interacting electric charges or magnetic poles.]

[13][Hertz himself had serious doubts that gravitational forces were action-at-a-distance forces. For example, in the second last paragraph of Paper No. 10, Hertz writes: "We are at once confronted with the question of direct actions-at-a-distance. Are there such? Of the many in which we once believed there now remains but one—gravitation. Is this too a deception? The law according to which it acts makes us suspicious."]

[14][This is reminiscent of what was said of Ernest Rutherford (1871–1937) after his death: "He never made an enemy, and never lost a friend."]

Helmholtz's Preface to Principles of Mechanics

University of Berlin. In 1883 he was induced by the Prussian Department of Education to go to Kiel with a view to his promotion to the rank of *Privatdozent* there. In Easter of 1885 he was called to Karlsruhe as ordinary Professor of Physics at the Polytechnic [*Technische Hochschule*]. Here he made his most important discoveries, and it was during his stay in Karlsruhe that he married Miss Elisabeth Doll, the daughter of one of his colleagues. Two years later he received a call to the University of Bonn as ordinary Professor of Physics, and moved there in Easter, 1889.

Few as the remaining years of his life unfortunately were, they brought him ample proof that his work was recognized and honored by his contemporaries. In the year 1888 he was awarded the Matteucci Medal of the Italian Scientific Society, in 1889 the La Caze Prize of the Academy of Sciences in Paris and the Baumgartner Prize of the Imperial Academy of Vienna, in 1890 the Rumford Medal of the Royal Society, and in 1891 the Bressa Prize of the Turin Royal Academy. He was elected a corresponding member of the Academies of Berlin, Munich, Vienna, Göttingen, Rome, Turin, and Bologna, and of many other learned societies; and the Prussian Government awarded him the Order of the Crown.

He was not long spared to enjoy these honors. A painful bone disease developed, and in November, 1892 it became life-threatening. An operation performed at that time appeared to relieve the pain for a while. Hertz was able to carry on his lectures, but only with great effort, up to the 7th of December, 1893. On New Year's Day of 1894 death released him from his sufferings.

In the present treatise on the *Principles of Mechanics*,[15] the last memorial of his labors here below, we again see how strong was his inclination to view scientific principles from the most general viewpoint. In it he has endeavored to give a consistent representation

---

[15][It is surprising that Helmholtz devotes the first 80 percent of this paper, which is intended as a Preface to Hertz's *Mechanics*, to Hertz's life and electromagnetic-wave research. When Helmholtz finally addresses the *Mechanics*, he does not seem as sure of himself as he was on the previous topics. This agrees with what Helmholtz told Lenard, who saw the *Mechanics* through to final publication after Hertz's death. Helmholtz admitted that it was only when he was near the end of Hertz's manuscript that he understood what Hertz was trying to do. For this and for Helmholtz's influence on Hertz's *Mechanics*, see the section "Helmholtz and Hertz's *Mechanics*" in Part V of the Introductory Biography.]

of a complete and connected system of mechanics, and to deduce all the separate special laws of this science from a single fundamental law which, logically considered, can only be regarded as a plausible hypothesis. In doing this he has reverted to the oldest theoretical conceptions, which may also be regarded as the simplest and most natural; and he propounds the question whether these do not suffice to enable us to deduce, by consistent and rigorous methods of proof, all the recently discovered general principles of mechanics, even such as have only made their appearance as inductive generalizations.

The first scientific development of mechanics arose out of investigations on the equilibrium and motion of solid bodies that were directly connected with one another; we have examples of these in simple machines: the lever, the pulley, inclined planes, etc. The law of virtual velocities is the earliest general solution of all the problems that thus arise. Later on Galileo developed the conception of inertia and of the accelerating action of force, although he represented this as consisting of a series of impulses. Newton first conceived the idea of action-at-a-distance, and showed how to determine it by the principle of equal action and reaction. It is well known that Newton, as well as his contemporaries, at first only accepted the idea of direct action-at-a-distance with the greatest reluctance.[16]

From that time on Newton's idea and definition of force served as a basis for the further development of mechanics. Gradually men learned how to handle problems in which conservative forces were

---

[16][Newton's famous statement, "*Hypotheses non fingo,*" occurs in the General Scholium at the end of Book III of the Cajori translation of the *Principia* in the following context:

> But hitherto I have not yet been able to deduce from the phenomena the cause of these properties of gravitation, and *I frame no hypotheses*; for whatever is not deduced from the phenomena should be called an hypothesis; and hypotheses whether metaphysical or physical, whether of occult qualities or mechanical, have no place in experimental philosophy. (1934 University of California edition, vol. 2, p. 547; italics not in original).

Newton was not able to come up with any reasonable explanation of how one mass acted on another at a distance, and was unwilling to speculate further on the matter. He always resisted an action-at-a-distance explanation for gravitation and attempted to explain his inverse-square law by the properties of the ether, but without success.]

Helmholtz's Preface to Principles of Mechanics   *317*

combined with fixed connections; of these the most general solution is given by d'Alembert's Principle. The chief general propositions of mechanics (such as the law of the motion of the center of gravity, the law of areas for rotating systems, the principle of the conservation of *vis viva* [i.e., kinetic energy], the principle of least action) have all been developed from Newton's assumption of constant, and therefore conservative, forces of attraction between material points, and of the existence of fixed connections between them. They were originally discovered and proved only under these assumptions. Subsequently it was discovered by observation that the propositions thus deduced could claim a much more general validity in nature than that which followed from the mode in which they were demonstrated. Hence it was concluded that certain general characteristics of Newton's conservative forces of attraction were common to all the forces of nature; but no proof was forthcoming that this generalization could be deduced from any general principle.

Hertz[17] has now endeavored to furnish mechanics with such a fundamental idea [*Grundanschauung*] from which all the laws of mechanics that have been recognized as of universal validity can be deduced in a perfectly logical manner. He has done this with great acuteness, making use in an admirable manner of new and peculiar generalized kinematical ideas. He has chosen as his starting-point that of the oldest mechanical theories, namely, the conception that all mechanical processes go on as if the connections between the various parts that act on each other were fixed. Of course, he is obliged to make the further hypothesis that there are a large number of imperceptible masses with invisible motions, in order to explain the existence of forces between bodies which are not in direct contact with each other.[18] Unfortunately he has not given examples illustrating the

---

[17][It is only here, at the very end of Helmholtz's Preface, that he begins to come to grips with what Hertz had tried to do in his *Mechanics*. Even then, there seems to be a certain vagueness and lack of enthusiasm about Hertz's new ideas that is never present in Helmholtz's comments on Hertz's other work. This may be due in part to Helmholtz's poor health, since this Preface was written just a few months before he died.]

[18][It seems from other evidence that Hertz was really interested in this treatise on mechanics as an introduction to a subsequent, and in his mind more important, explanation of the workings of the ether, which would, of course, be able to explain the gravitational attraction of two bodies without the

manner in which he supposed such hypothetical mechanisms to act; to explain even the simplest cases of physical forces on these lines will clearly require much scientific insight and imaginative power. In this direction Hertz seems to have relied chiefly on the introduction of cyclical systems with invisible motions.

English physicists—e.g., Lord Kelvin, in his theory of vortex-atoms, and Maxwell, in his hypothesis of systems of cells with rotating contents, on which he bases his attempt at a mechanical explanation of electromagnetic processes—have evidently derived a fuller satisfaction from such explanations than from the simple representation of physical facts and laws in the most general form, as given in systems of differential equations. For my own part, I must admit that I have adhered to the latter mode of representation and have felt safer in so doing; yet I have no essential objections to raise against a method which has been adopted by three physicists of such eminence.[19]

It is true that great difficulties have yet to be overcome before we can succeed in explaining the varied phenomena of physics in accordance with the system developed by Hertz, but in every respect his presentation in the *Principles of Mechanics* must be of the greatest interest to every reader who can appreciate a logical system of dynamics developed with the greatest ingenuity and in the most perfect mathematical form. In the future this book may prove of great heuristic value as a guide to the discovery of new and general attributes of the forces of nature.

---

need for postulating action-at-a-distance forces. On this see Part V of the Introductory Biography.]

[19][There appears to be in this statement of Helmholtz a feeling of regret that Hertz seemed to be moving toward the model-building that was fashionable among physicists in Great Britain at the time, and away from the approach to physics he had learned from Helmholtz. By the "three physicists of such eminence" he meant Maxwell, Kelvin and Hertz. On this see D'Agostino (1971), p. 645.]

PAPER NO. 15

Heinrich Hertz's Preface to His *Principles of Mechanics*

All physicists agree that the task of physics consists in tracing the phenomena of nature back to the simple laws of mechanics. But there is not the same agreement as to what these simple laws are. To most physicists they are simply Newton's laws of motion. But in reality these latter laws only obtain their inner significance and their physical meaning through the tacit assumption that the forces of which they speak are of a simple nature and possess simple properties. But we have in this case no certainty as to what is simple and permissible, and what is not; it is just at this point that we no longer find any general agreement. Hence there arise actual differences of opinion as to whether or not this or that assumption is in accordance with the usual system of mechanics. It is in the treatment of new problems that we recognize the existence of such unanswered questions as a real hindrance to progress. So, for example, it is premature to attempt to base the equations of motion of the ether on the laws of mechanics until we have obtained perfect agreement as to what is understood by this name.[1]

The problem that I have endeavored to solve in the present investigation is the following: To fill up the existing gaps and to give a complete and definitive presentation of the laws of mechanics, which will be consistent with the state of our present knowledge, being neither too restricted nor too extensive in relation to the scope of this knowledge. The presentation must not be too restricted; there must be no natural motion that it does not embrace. On the other hand, it must not be too extensive; it must admit of no motion whose occurrence in nature is excluded by the state of our present knowledge. Whether the presentation here given as the solution to this problem is the only possible one, or whether there are other and perhaps better ones

---

[1][That is, what is meant by the "laws of mechanics." Although Hertz does not say it outright here, his words seem to suggest that one purpose of his *Principles of Mechanics* is to provide a solid foundation for a later treatment of the ether, which was still accepted by German physicists at the time. On this see Part V of the Introductory Biography.]

possible, remains to be seen. But that the presentation given is in every respect a possible one, I prove by developing its consequences, and showing that when fully unfolded it is capable of embracing the whole content of ordinary mechanics, insofar as the latter relates only to the actual forces and connections of nature, and is not regarded as a playground for mathematical exercises.

In the process of this development a theoretical discussion has grown into a treatise that contains a complete survey of all the more important general propositions of dynamics, and that may serve as a systematic textbook of this science. For several reasons it is not well suited for use as a first introduction; but for these very reasons it is the better suited to guide those who have already a fair mastery of mechanics as usually taught. It is to be hoped that it may lead such individuals to a vantage-point from which they can perceive more clearly the physical meaning of mechanical principles, how they are interrelated, and within what limits they remain valid; a vantage-point from which the ideas of force and the other fundamental ideas of mechanics appear stripped of the last remnant of obscurity.

In his papers on the principle of least action and on cyclical systems,[2] von Helmholtz has already treated in an indirect manner the problem that is investigated in this book, and has given one possible solution to it. In the first set of papers he proposes and defends the thesis that a system of mechanics, which regards not only Newton's laws as of universal validity, but also the special assumptions involved (in addition to these laws) in Hamilton's Principle, would still be able to embrace all the processes of nature. In the second set of papers the meaning and importance of concealed motions is for the first time treated in a general way. Both in its broad features and in its details my own investigation owes much to the above-mentioned papers; the chapter on cyclical systems is taken almost directly from them. Apart from matters of form, my own solution differs from that of von Helmholtz chiefly in two respects. Firstly, I endeavor from the start to keep the elements of mechanics free from that[3] which von Helmholtz

---

[2]H. von Helmholtz, "Über die physikalische Bedeutung des Prinzips der kleinsten Wirkung," *Journal für die reine und angewandte Mathematik*, 100 (1887), pp. 137–166, 213–222; "Prinzipien der Statik monocyklischer Systeme," *ibid.* 97 (1884), pp. 111–140, 317–336.

[3][Here Hertz is referring to the concept of "force," which he attempted to eliminate as a fundamental quantity in his *Mechanics*.]

Hertz's Preface to His Principles of Mechanics

only removes by subsequent restrictions, after he has completed the development of his mechanics. Secondly, in a certain sense I eliminate less from mechanics, inasmuch as I do not rely on Hamilton's Principle or any other integral principle.[4] The reasons for this and the consequences that arise from it are made clear in the book itself.

In his important paper on the physical applications of dynamics, J.J. Thomson[5] pursues a train of thought similar to that contained in von Helmholtz's papers. Here again the author develops the consequences of a system of dynamics based on Newton's laws of motion and also on other special assumptions, which are not explicitly stated. I might have derived assistance from this paper as well; but as a matter of fact my own investigation had made considerable progress by the time I became familiar with it. I may say the same of the mathematical papers of Beltrami[6] and Lipschitz,[7] although these are of much older date. Still I found these very suggestive, as also the more recent presentation of their investigations which Darboux[8] has given with additions of his own. I may have missed many mathematical papers that I could and should have consulted. In a general way I owe very much to Mach's splendid book on the *Development of Mechanics*.[9] I have naturally consulted the better-known textbooks of

---

[4][In what follows in his *Mechanics*, however, Hertz does base all of mechanics on the following Fundamental Law:

Every free system persists in its state of rest or of uniform motion along one of its straightest paths.]

[5]J.J. Thomson, "On some Applications of Dynamical Principles to Physical Phenomena," *Philosophical Transactions of the Royal Society*, 176 (1885), II, pp. 307–342.

[6]E. Beltrami, "Sulla teoria generale dei parametri differenziali," *Memorie della Reale Accademia di Bològna* (25 Febbrajo 1869).

[7]R. Lipschitz, "Untersuchungen eines Problems der Variationsrechnung, in welchem das Problem der Mechanik enthalten ist," *Journal für die reine und angewandte Mathematik*, 74 (1872), pp. 116–149; "Bemerkungen zu dem Princip des kleinsten Zwanges," *ibid.* 82 (1877), pp. 316–342.

[8]G. Darboux, *Leçons sur la théorie générale des surfaces*, Livre V, Chapitres 6,7,8 (Paris, 1889).

[9]E. Mach, *Die Mechanik in ihrer Entwicklung historisch-kritisch dargestellt.* Leipzig, 1883. [An English translation may be found in Mach (1960). Although Hertz's *Mechanics* owed much to Mach in its approach to the philosophy of science and in its rejection of "force" as a fundamental concept, Mach always vigorously opposed all unseen and insensible objects (like atoms!) as explanations of physical phenomena. For this reason he did not care much

general mechanics, and especially Thomson and Tait's comprehensive treatise.[10] The notes of a course of lectures on analytical dynamics by Borchardt,[11] which I took down in the winter of 1878-79, have proved useful. These are the sources on which I have drawn; in the text I shall only give such references as are requisite. As to the details I have nothing to bring forward which is new or which could not have been gleaned from many books. What I hope is new, and to this alone I attach value, is the order and arrangement of the whole—the logical or philosophical aspect of the matter. According to whether it marks an advance in this direction or not, my work will succeed or fail in its objectives.

---

for Hertz's approach to mechanics and was led to write: "In the beautiful ideal form which Hertz gave to mechanics, its physical contents have shrunk to an apparently almost imperceptible residue." Mach (1960), pp. 323-324. For a discussion of the philosophical outlooks of Mach and Hertz, see Cohen (1956), Section 3; and Blackmore (1972), pp. 119-120.]

[10]William Thomson and Peter Guthrie Tait, *Treatise on Natural Philosophy*. Two Volumes. Oxford: Clarendon Press, 1867.

[11][Hertz had taken mathematics courses under Borchardt in Berlin, beginning in 1878. Carl Wilhelm Borchardt (1817-1880) *habilitiert* in Berlin in 1851 and later became Professor of Mathematics there. He was also the editor of the *Journal für die reine und angewandte Mathematik.*]

PAPER NO. 16

Heinrich Hertz's Introduction to His *Principles of Mechanics*

The most direct and in a sense the most important problem that our conscious knowledge of nature should enable us to solve is the anticipation of future events, so that we may arrange our present affairs in accordance with such anticipation.[1] As a basis for the solution of this problem we always make use of our knowledge of events that have already occurred, obtained by chance observation or by prearranged experiment. In thus endeavoring to draw inferences about the future from the past, we always adopt the following process. We form for ourselves mental pictures or symbols of external objects[2]; and the form that we give them is such that the necessary consequents in thought of our pictures are always the pictures of the necessary

---

[1] [This opening sentence makes it clear that this paper differs essentially from all of Hertz's previous papers in this collection. In those papers the emphasis was always on experimental observations and theoretical derivations in an attempt to understand the physical world. In this paper the emphasis is much more on philosophy, on how we know and what we are able to know about the physical universe. For further discussion of this paper, see the section, *The Principles of Mechanics*, in Part V of the Introductory Biography.]

[2] [The German here is "Wir machen uns innere Scheinbilder oder Symbole der äusseren Gegenstände." In the rest of Hertz's introduction the word *Scheinbilder* (literally "imitation pictures" or "pseudo-pictures") never recurs, but the word *Bilder* appears again and again. There has been much debate about the proper translation of this word *Bilder*, and what precisely Hertz means by it. Philosophers of science consider the meaning of *Bilder* at the very core of Hertz's philosophy of science. See, for example, Barker (1980), p. 247, and Janik and Toulmin (1973), pp. 139–143. We here translate *Bilder* consistently as "pictures," which is the primary meaning of the German word, and seems to convey Hertz's thought better than the "images" adopted by Jones and Walley, the English translators of Hertz's *Mechanics*. In his *Mechanics* Hertz's "pictures" are deliberately constructed models of real, external objects intended to be used in a mathematical representation of reality. Different individuals may construct different mental pictures of the same object; the only important thing, as Hertz here emphasizes, is that the relations between such pictures in our inner thoughts must correspond to the relations between the actual external objects in the real world.]

consequents in nature of the objects pictured.[3] In order that this requirement may be satisfied, there must be a certain conformity between nature and our thought. Experience teaches us that this requirement can be satisfied, and hence that such a conformity does in fact exist. When on the basis of our accumulated previous experiences we have once succeeded in constructing pictures with the desired properties, we can quickly derive by means of them, as by means of models,[4] the consequents which in the external world occur only after a comparatively long time, or as the result of our own intervention. We are thus able to be in advance of the facts, and to decide as to present affairs in accordance with the insight so obtained. The pictures of which we here speak are our models of things; these models are in conformity with the things themselves in *one* important respect, namely, in satisfying the above-mentioned requirement. For our purpose it is not necessary that they should be in conformity with the things in any other respect whatsoever. As a matter of fact, we do not know, nor have we any means of knowing, whether our models of things are in conformity with the things themselves in any other than this *one* fundamental respect.

The pictures that we may form of things are not determined unambiguously by the requirement that the consequents of the pictures must be the pictures of the consequents. Different pictures of the same objects are possible, and these pictures may differ in various respects. We should at once denote as inadmissible all pictures that implicitly contradict the laws of thought. Hence we postulate in the first place that all our pictures shall be logically permissible—or, briefly, that they shall be permissible. We shall denote as incorrect any permissible pictures if their essential relations contradict the relations of external things, i.e., if they do not satisfy our first fundamental requirement. Hence we postulate in the second place that our pictures shall be correct. But two permissible and correct pictures of the same

---

[3] [This sentence of Hertz is frequently cited by physicists and philosophers of science as a very clear, precise and apt statement of the task of physicists in the twentieth century. In this respect Hertz was well ahead of his times.]

[4] [These words of Hertz indicate that *Bilder* for Hertz was very close to what we would today call "models" of objects in the real world. Note also the second sentence after this one: "The pictures of which we here speak are our models of things."]

external object may yet differ with respect to appropriateness. Of two pictures of the same object, that one is the more appropriate which pictures more of the essential relations of the object—the one that we may call the more distinct. Of two pictures of equal distinctness the more appropriate is the one that contains, in addition to the essential characteristics, the smaller number of superfluous or empty relations, i.e., the simpler of the two. Empty relations cannot be avoided altogether; they enter into the pictures because they are themselves simply pictures, and indeed pictures produced by our own mind and necessarily affected by the characteristics of its mode of picturing them.

The postulates already mentioned are those that we assign to the pictures themselves; entirely different are the requirements we place on a scientific exposition of such pictures. We require of this latter that it should lead us to a clear conception of what properties are to be ascribed to the pictures for the sake of permissibility, what for correctness, and what for appropriateness. Only thus can we attain the possibility of modifying and improving our pictures. What is ascribed to the pictures for the sake of appropriateness is contained in the notations, definitions, abbreviations and, in short, all that we can arbitrarily add or take away. What enters into the pictures for the sake of correctness is contained in the results of experience, from which the pictures are built up. What enters into the pictures, in order that they may be permissible, is given by the nature of our mind. To the question whether a picture is permissible or not, we can without ambiguity answer yes or no, and our decision will hold good for all time. And equally without ambiguity we can decide whether a picture is correct or not, but only according to our present state of experience, and allowing for an appeal to later and more mature experience. But we cannot decide without ambiguity whether a picture is appropriate or not; as to this, differences of opinion may arise. One picture may be more suitable for one purpose, another for another;[5] Only by gradually testing many pictures can we finally succeed in obtaining the most appropriate.

---

[5][A contemporary example of this might be the wave-particle duality in optics. The picture of light as a wave is more appropriate for explaining diffraction and interference; the picture of light as a particle is more appropriate for explaining the Compton Effect and the Photoelectric Effect.]

These are, in my opinion, the standpoints from which we must judge the value of physical theories and the value of the representation [*Darstellung*] of physical theories.[6] They are the standpoints from which we shall here consider the representations [*Darstellungen*] that have been proposed for the Principles of Mechanics. We must first explain clearly what we mean by this name.

Strictly speaking, what was originally called a principle in mechanics was such a statement as could not be traced back to other propositions in mechanics, but was regarded as a direct result obtained from other sources of knowledge. In the course of historical development it inevitably came to pass that propositions, which at one time and under special circumstances were rightly denoted as principles, wrongly retained these names. Since Lagrange's time it has frequently been remarked that the principles pertaining to the center of gravity and to areas are in reality only propositions of a general nature. But we can with equal justice say that other so-called principles cannot bear this name, but must descend to the rank of propositions or corollaries, when the representation of mechanics becomes based upon one or more of the others. Thus the idea of a mechanical principle has not been kept sharply defined. We shall therefore retain for such propositions, when mentioning them separately, their customary names. But these separate concrete propositions are not what we shall have in mind when we speak simply and generally of the principles of mechanics; by this will be meant any selection from among such and similar propositions, which satisfies the requirement that the whole of mechanics can be developed from it by purely deductive reasoning without any further appeal to experience. In this sense the fundamental ideas of mechanics, together with the principles connecting them, represent the simplest picture that physics can produce of things in the sensible world and the processes that occur in it. By varying the choice of the propositions we take as fundamental, we can give various representations [*Darstellungen*] of the principles of mechanics. Hence we can obtain various pictures [*Bilder*] of things; and these pictures we can test and

---

[6][Here Hertz uses the word *Darstellungen*, or "representations," in the same sense in which he used the word in his papers on electromagnetism, in which he discusses the different representations of electromagnetism that lead to Maxwell's equations. See the first part of the "Theoretical" section at the end of Paper No. 5.]

compare with one another with respect to permissibility, correctness, and appropriateness.

I

The customary representation of mechanics provides us with a first picture.[7] By this we mean the representation, varying in detail but identical in essence, contained in almost all textbooks that deal with the whole of mechanics, and in almost all courses of lectures that cover the whole of this science. This is the path along which the great army of students travel and by which they are inducted into the mysteries of mechanics. It closely follows the course of historical development and the sequence of discoveries. Its principal stages are distinguished by the names of Archimedes, Galileo, Newton, Lagrange. The conceptions on which this representation is based are the ideas of space, time, force and mass. In it force is introduced as the cause of motion, existing before motion and independently of it. Space and force first appear by themselves, and their relations are treated in statics. Kinematics, or the science of pure motion, confines itself to connecting the two ideas of space and time. Galileo's conception of inertia supplies a connection only between space, time, and mass. Not until Newton's Laws of Motion do the four fundamental ideas become connected with one another. These laws contain the seed of future developments; but they do not furnish any general expression for the influence of rigid spatial connections. Here d'Alembert's principle extends the general results of statics to the case of motion, and closes the series of independent fundamental statements that cannot be deduced from each other. From here on everything is deductive inference. In fact, the above-mentioned ideas and laws are not only necessary but sufficient for the development from them of the whole of mechanics as a necessary result of thought; and all other so-called principles can be regarded as propositions and corollaries deduced by special assumptions. Hence the above ideas and laws give us, in the sense in which we have used the words, a first system of principles of mechanics, and at the same

---

[7][The German here is "Ein erstes Bild liefert uns die gewöhnliche Darstellung der Mechanik." Throughout this Introduction we consistently translate *Darstellung* as "representation." Occasionally Hertz seems to use the word *Bild* also in the sense of "representation."]

time the first comprehensive picture of the natural motions of material bodies.

Now, at first sight, any doubt as to the logical permissibility of this picture may seem very far-fetched. It seems almost inconceivable that we should find logical imperfections in a system that has been thoroughly and repeatedly considered by many very able intellects. But before we abandon the investigation on this account, we would do well to inquire whether the system has always given satisfaction to these able intellects. It is really wonderful how easy it is to attach to the fundamental laws considerations that are quite in accordance with the usual modes of expression in mechanics, and which are still an undoubted hindrance to clear thinking. Let us endeavor to give an example of this. We swing in a circle a stone tied to a string, and in so doing we are conscious of exerting a force on the stone. This force constantly deflects the stone from its straight path. If we vary the force, the mass of the stone, and the length of the string, we find that the actual motion of the stone is always in accordance with Newton's second law. But now the third law requires an opposing force to the force exerted by the hand upon the stone. With regard to this opposing force, the usual explanation is that the stone reacts on the hand in consequence of centrifugal force, and that this centrifugal force is in fact exactly equal and opposite to that which we exert. Now is this mode of expression permissible? Is what we call centrifugal force anything else than the inertia of the stone? Can we, without destroying the clearness of our conceptions, take the effect of inertia twice into account, firstly as mass, and secondly as force? In our laws of motion, force was a cause of motion, and was present *before* the motion. Can we, without confusing our ideas, suddenly begin to speak of forces that arise through motion and are effects of the motion? Can we behave as if we had already asserted anything about forces of this new kind in our laws, as if by calling them forces we could invest them with the properties of forces? These questions must clearly be answered in the negative. The only possible explanation is that, properly speaking, centrifugal force is not a force at all. Its name, like the name *vis viva*, is accepted as a historic tradition; it is convenient to retain it, although we should rather apologize for its retention than endeavor to justify it. But what now becomes of the demands of the third law, which requires

Hertz's Introduction to His Principles of Mechanics   329

a force exerted by the inert stone on the hand, and which can only be satisfied by an actual force, not by a mere name?[8]

I do not regard these as artificial difficulties mischievously raised; they are objections that press for an answer. Is not their origin to be traced back to the fundamental laws? The force spoken of in the definition and in the first two laws acts upon a body in one definite direction. The meaning of the third law is that forces always connect two bodies and are directed from the first to the second as well as from the second to the first. It seems to me that the conception of force assumed and produced in us by the third law on the one hand, and the first two laws on the other hand, are slightly different. This slight difference may be enough to produce the logical obscurity of which the consequences are manifest in the above examples. It is not necessary to discuss further examples. We can appeal to general observations as evidence in support of the above-mentioned doubt.

As such, in the first place, I would mention the experience that it is exceedingly difficult to expound to thoughtful hearers the very introduction to mechanics without being occasionally embarrassed, without feeling tempted now and again to apologize, without wishing to get as quickly as possible over the rudiments, and on to examples which speak for themselves.[9] I fancy that Newton himself must have felt this embarrassment when he gave his rather forced definition of mass as being the product of volume and density. I fancy that Thomson and Tait[10] must also have felt it when they remarked that this is really more a definition of density than of mass, and nevertheless contented themselves with it as the only definition of mass. Lagrange, too, must have felt this embarrassment and the wish to get on at all costs; for he briefly introduces his Mechanics with the explanation that a force is a cause that imparts "or tends to impart" motion to a body; and he must certainly have felt the logical difficulty

---

[8][On this paragraph see the critical comments of G.F. FitzGerald in Paper No. 17.]

[9][Any physicist who has ever taught a course in introductory mechanics knows exactly what Hertz means here, and the situation does not seem that much better one hundred years after Hertz wrote these perceptive words.}

[10][William Thomson and P.G. Tait, *Treatise on Natural Philosophy* (Oxford, Clarendon Press, 1867), vol. 1, p. 162.]

of such a definition.[11] I find further evidence in the demonstrations of the elementary propositions of statics, such as the law of the parallelogram of forces, of virtual velocities, etc. Of such propositions we have numerous proofs given by eminent mathematicians. These claim to be rigid proofs; but, according to the opinion of other distinguished mathematicians, they in no way satisfy this claim. In a logically complete science, such as pure mathematics, such a difference of opinion is utterly inconceivable.

Weighty evidence seems to be furnished by the statements, which one hears with wearisome frequency, that the nature of force is still a mystery, that one of the chief problems of physics is the investigation of the nature of force, and so on. In the same way electricians [i.e., those doing research on electricity and magnetism] are continually attacked as to the nature of electricity. Now, why is it that people never in this way ask what is the nature of gold, or what is the nature of velocity? Is the nature of gold better known to us than that of electricity, or the nature of velocity better than that of force? Can we by our conceptions, by our words, completely represent the nature of any thing? Certainly not. I fancy the difference must lie in this. With the terms "velocity" and "gold" we connect a large number of relations to other terms; and between all these relations we find no contradictions that offend us. We are therefore satisfied and ask no further questions. But we have accumulated around the terms "force" and "electricity" more relations than can be completely reconciled among themselves. We have an obscure feeling of this and want to have things cleared up. Our confused wish finds expression in the confused question as to the nature of force and electricity. But the answer that we want is not really an answer to this question. It is not by finding out more and fresh relations and connections that it can be answered; but by removing the contradictions existing between those already known, and thus perhaps by reducing their number. When these painful contradictions are removed, the question as to the nature

---

[11][Gustav Kirchhoff, from whom Hertz took courses in Berlin, regarded the term "force" as a metaphysical concept fraught with ambiguities, which he wanted to avoid at all costs. He therefore built up his mechanics on the basis of only three fundamental quantities—space, time and mass—as Hertz does here. He then introduced force later in terms of Newton's second law: $F = ma$. On this see Kirchhoff (1877), pp. iii–iv.]

Hertz's Introduction to His Principles of Mechanics 331

of force will not have been answered; but our minds, no longer vexed, will cease to ask illegitimate questions.

I have thrown such strong doubts on the permissibility of this picture that it might appear to be my intention to contest, and finally to deny, its permissibility. But my intention and conviction do not go as far as this. Even if the logical uncertainties, which have made us worry about the reliability of our fundamental ideas, do actually exist, they certainly have not prevented a single one of the numerous triumphs that mechanics has won in its applications. Hence, they cannot consist of contradictions between the essential characteristics of our picture, nor, therefore, of contradictions between those relations of mechanics that correspond to the relations of things. They must rather lie in the unessential characteristics that we have ourselves arbitrarily worked into the essential content provided us by nature. If so, these dilemmas can be avoided. Perhaps our objections do not relate to the content of the picture devised, but only to the form of the representation of this content. It is not going too far to say that this representation has never attained scientific completeness; it still fails to distinguish thoroughly and precisely between those elements in the sketched picture that arise from the necessities of thought, from experience, and from arbitrary choice. This is also the opinion of distinguished physicists who have thought over and discussed[12] these questions, although it cannot be said that all of them are in agreement.[13] This opinion also finds confirmation in the increasing care with which the logical analysis of the elements is carried out in the more recent textbooks of mechanics.[14] We are convinced, as are the authors of these textbooks and the physicists referred to, that the existing defects are only defects in form; and that all indistinctness and uncertainty can be avoided by the suitable arrangement of definitions

---

[12] See E. Mach, *The Science of Mechanics*, p. 244. See also in *Nature* 48 (1893), pp. 62, 101, 117, 126, and 166; and *Proc. Phys. Soc.* 12 (1893), p. 289, a discussion on the foundations of dynamics introduced by Prof. Oliver Lodge and carried on in the Physical Society of London.

[13] See Thomson and Tait, *Treatise on Natural Philosophy*, No. 205 and following. [The full reference is Thomson and Tait (1867).]

[14] See E. Budde, *Allgemeine Mechanik der Punkte und starren Systeme*, pp. 111–138 (Berlin, 1890). The representation there given shows at the same time how great are the difficulties encountered in avoiding discrepancies in the use of the elementary concepts.

and notations, and by due care in the mode of expression. In this sense we admit, as everyone does, the permissibility of the content of mechanics. But the dignity and importance of the subject demand not simply that we should readily take for granted its logical clearness, but that we should endeavor to show it by a representation so perfect that there should no longer be any possibility of doubting it.[15]

Upon the correctness of the picture under consideration we can pronounce judgment more easily and with greater certainty of general assent. No one will deny that within the whole range of our experience up to the present the correctness is perfect; that all those characteristics of our picture, which claim to describe observable relations of things, do really and correctly correspond to them. Our assurance, of course, is restricted to the range of previous experience; as far as future experience is concerned, there will still be occasion to return to the question of correctness. To many this will seem to be excessive and absurd caution: to many physicists it appears simply inconceivable that any further experience whatsoever should find anything to alter in the firm foundations of mechanics. Nevertheless, that which is derived from experience can again be annulled by experience.[16] This overly optimistic opinion of the fundamental laws must obviously arise from the fact that the elements of experience are to a certain extent hidden in them and blended with the unalterable elements that are necessary consequences of our thought. Thus the logical indefiniteness of the representation, which we have just censured, has one advantage. It gives the foundations an appearance of immutability; and perhaps it was wise to introduce it in the beginnings of the science and to allow it to remain for a while. The correctness of the picture in all cases was carefully provided for by making the reservation that, if need be, facts derived from experience should determine definitions or *vice versa*. In a perfect science such groping, such an illusion of certainty, is inadmissible. Mature knowledge regards logical clearness as of prime importance; only

---

[15][This is Hertz's aim in his *Principles of Mechanics*, to show the logical clearness and consistency of mechanics "by a representation so perfect that there should no longer be any possibility of doubting it."]

[16][The development of the theory of relativity and of quantum theory in the decade after Hertz wrote this analysis shows the insight he had into what the real problems of physics were, and how they might develop in ways unthought of during his lifetime.]

logically clear pictures are tested as to correctness; only correct pictures are compared to decide on appropriateness. By pressure of circumstances the process is often reversed. Pictures are found to be suitable for a certain purpose; are next tested as to their correctness; and only in the last instance are they purged of implied contradictions.

If there is any truth in what we have just stated, it seems only natural that the system of mechanics under consideration should prove most appropriate in its applications to those simple phenomena for which it was first devised, i.e., especially to the action of gravity and the problems of practical mechanics. But we should not be content with this. We should remember that we are not here representing the needs of daily life or the standpoint of past times; we are considering the whole range of present physical knowledge, and are, moreover, speaking of appropriateness in the special sense defined in the beginning of this Introduction. Hence we are at once bound to ask: Is this outlined picture perfectly distinct? Does it contain all the characteristics that our present knowledge enables us to distinguish in natural motions? Our answer is a decided "No." All the motions of which the fundamental laws admit, and which are discussed in mechanics as mathematical exercises, do not occur in nature. Of natural motions, forces, and fixed connections, we can predicate more than the accepted fundamental laws do. Since the middle of this century we have been firmly convinced that no forces actually exist in nature that would involve a violation of the principle of the conservation of energy. The conviction is much older that only such forces exist as can be represented as a sum of mutual actions between infinitely small elements of matter. Again, these elementary forces are not free. We can assert as a property, which they are generally admitted to possess, that they are independent of absolute time and place. Other properties are disputed: whether the elementary forces can only consist of attractions and repulsions along the line connecting the acting masses; whether their magnitude is determined only by the distance or whether it is also affected by the absolute or relative velocity; whether the latter alone comes into consideration, or the acceleration or still higher differential coefficients as well—all these properties have been sometimes presumed, at other times questioned.

Although there is such difference of opinion as to the precise properties that are to be attributed to the elementary forces, there is a general agreement that more of such general properties can be assigned, and can from existing observations be deduced, than are contained in the fundamental laws. We are convinced that the elementary forces must, so to speak, be of a simple nature. And what here holds for the forces, can be equally asserted of the fixed connections of bodies that are represented mathematically by equations of condition between the coordinates, and whose effect is determined by d'Alembert's principle. It is mathematically possible to write down any finite or differential equation between coordinates and to require that it shall be satisfied; but it is not always possible to specify a natural, physical connection corresponding to such an equation. We often feel, indeed sometimes are convinced, that such a connection is by the nature of things excluded. And yet, how are we to restrict the permissible equations of condition? Where is the limiting line between them and the conceivable ones? To consider only finite equations of condition, as has often been done, is to go too far; for differential equations that are not integrable can actually occur as equations of condition in real-world problems.

In short, then, so far as the forces as well as the fixed relations are concerned, our system of principles embraces all the natural motions; but it also includes very many motions that are not natural. A system that excludes the latter, or even a part of them, would depict more of the actual relations of things to each other, and would therefore in this sense be more appropriate. We are next bound to inquire as to the appropriateness of our picture in another direction. Is our picture simple? Is it sparing in nonessential details—ones we ourselves have added, permissibly and yet arbitrarily, to the essential and natural ones? In answering this question our thoughts again turn to the idea of force. It cannot be denied that in very many cases the forces, which are introduced into our mechanics to handle physical problems, only serve as idling wheels[17] that completely suspend operation when they are required to represent actual facts. In the simple relations with which mechanics originally dealt, this is not the case. The weight of a stone and the force exerted by the arm seem to be

---

[17][We have translated *leergehende Nebenräder* as "idling wheels," which is the literal meaning of the phrase, whereas Jones and Walley prefer the metaphorical "sleeping partners." Ludwig Wittgenstein used Hertz's metaphor of "idling wheels" frequently in his *Tractatus*. On this see Barker (1980), p. 254, especially footnote 27.]

as real and as readily and directly perceptible as the motions that they produce. But it is otherwise when we turn to the motions of the stars. Here the forces have never been the objects of direct perception; all our previous experience relates only to the apparent positions of the stars. Nor do we expect to perceive the forces in the future. The future experiences that we anticipate again relate only to the positions of these luminous points in the heavens. It is only in the deduction of future experiences from the past that the forces of gravitation enter as transitory aids in the calculation, and then disappear from consideration. Precisely the same is true of the discussion of molecular forces, of chemical actions, and of many electric and magnetic actions.

And if after more mature experience we return to the simple forces, whose existence we never doubted, we learn that these forces, which we had perceived with convincing certainty, were after all not real. More sophisticated mechanics tells us that what we believed to be simply the tendency of a body toward the earth, is not really that at all; it is the result, imagined only as a single force, of an inconceivable number of actual forces that attract the atoms of the body toward all the other atoms in the universe. Here again the actual forces have never been the objects of previous experience; nor do we expect to come across them in future experiences. Only during the process of deducing future experiences from the past do they glide quietly in and out. But even if the forces have only been introduced into nature by ourselves, we should not on that account regard their introduction as inappropriate. We have felt sure from the beginning that nonessential relations could not be altogether avoided in our pictures. All that we can ask is that these relations should, as far as possible, be restricted, and that a wise discretion should be observed in their use.

But has physics always been sparing in the use of such relations? Has it not rather been compelled to fill the world to overflowing with forces of the most various kinds, with forces that have never been observed in the phenomena, even with forces that only came into action in exceptional cases? We see a piece of iron resting on a table and we accordingly imagine that no causes of motion—no forces—are there present. Physics, which is based on the mechanics considered here and necessarily determined by this basis, teaches us otherwise. Through the force of gravitation every atom of the iron is attracted by every other atom in the universe. But every atom of the iron is magnetic and is therefore connected by new forces with every other magnetic atom in the universe. Again, bodies in the universe contain electricity in motion, and this latter exerts further complicated forces that attract every atom of the iron. Insofar as the parts of the iron themselves contain electricity, we have additional forces to take into consideration; and on top of these there are, moreover, various kinds of molecular forces. Some of these forces are not small; if only a part of these forces were effective, this part would suffice to tear the iron into pieces. But, in fact, all the forces are so adjusted among themselves that the effect of the whole lot is zero; that, in spite of a thousand existing causes of motion, no motion takes place; and so the iron remains at rest. Now if we place these ideas before unprejudiced persons, who will believe us? Whom shall we convince that we are speaking of actual things, not fictions of a wild imagination? And it is for us to reflect whether we have really depicted the state of rest of the iron and its particles in a simple manner. Whether complications can be entirely avoided is questionable; but there can be no question that a system of mechanics that does avoid or exclude them is simpler, and in this sense more appropriate, than the one here considered; for this latter system not only allows such ideas, but imposes them directly upon us.

Let us now collect together as briefly as possible the doubts that have occurred to us in considering the customary mode of representing the principles of mechanics. As far as the form is concerned, we consider that the logical value of the separate statements is not defined with sufficient clearness. As far as the facts are concerned, it appears to us that the motions considered in mechanics do not exactly coincide with the natural motions under consideration. Many properties of the natural motions are not attended to in mechanics; many relations that are considered in mechanics are probably absent in nature. Even if these objections are acknowledged to be well founded, they should not lead us to imagine that the customary representation of mechanics is on that account either bound to, or likely to, lose its value and its privileged position. But they sufficiently justify us in looking out for other representations less liable to censure in these respects, and more closely conformable to the things that have to be represented.

II

There is a second picture of mechanical processes that is of much more recent origin than the first. Its development from, and in association with, the latter is closely connected with advances that physical science has made during the past few decades. Up to the middle of this century its ultimate aim was apparently to explain natural phenomena by tracing them back to innumerable actions-at-a-distance between the atoms of matter. This mode of conception corresponded completely to what we have spoken of as the first system of mechanical principles. Each one of the two was conditioned by the other. Now, toward the end of the century, physics has shown a preference for a different mode of thought. Influenced by the overpowering impression made by the discovery of the principle of the conservation of energy, it likes to treat the phenomena that occur in its domain as transformations of energy into new forms, and to regard as its ultimate aim the tracing back of the phenomena to the laws of the transformation of energy. This mode of treatment can also be applied from the beginning to the elementary phenomena of motion. There thus arises a new and different representation of mechanics, in which from the start the idea of force withdraws in favor of the idea of energy. It is this new picture of the elementary processes of motion that we shall denote as the second, and to it we shall now devote our attention.

In discussing the first picture we had the advantage of being able to assume that it stood out plainly before the eyes of all physicists. With the second picture this is not the case. It has never yet been portrayed in all its details. So far as I know, there is no textbook of mechanics that from the start teaches the subject from the standpoint of energy, and introduces the idea of energy before the idea of force. Perhaps there has never yet been a lecture on mechanics prepared according to this plan. But to the founders of the theory of energy it was evident that such a plan was possible. The remark has often been made that in this way the idea of force with its attendant difficulties could be avoided; and in special scientific applications chains of reasoning frequently occur that belong entirely to this mode of thought. Hence we can very well sketch the rough outlines of this picture; we can give the general plan according to which such a representation of mechanics must be arranged.

We here start, as in the case of the first picture, from four independent fundamental ideas; the relations of these to one another will form the content of mechanics. Two of them, space and time, have a mathematical character; the other two, mass and energy, are introduced as physical entities that are present in some given amount and cannot be destroyed or increased. In addition to explaining these matters, it will, of course, also be necessary to indicate clearly by what concrete experiences we ultimately establish the presence of mass and energy.[18] We here assume this to be possible and to be done. It is obvious that the amount of energy connected with given masses depends upon the state of these masses. But it is as a general experience that we must first lay down that the energy present can always be split up into two parts, of which one is determined solely by the relative positions of the masses, while the other depends upon their absolute velocities. The first part is defined as potential energy, the second as kinetic energy. The form of the dependence of the kinetic energy on the velocity of the moving bodies is in all cases the same, and is well known. The form of the dependence of the potential energy on the position of the bodies cannot be generally stated, since it is affected by the special nature and peculiar character of the masses involved. It is the problem of physics to ascertain from previous experience the proper form [of the potential energy] for the bodies that surround us in nature.

Up to this point there came into consideration essentially only three elements—space, mass, energy—considered in relation to one another. In order to settle the relations of all four fundamental ideas, and thereby the course in time of the phenomena, we make use of one of the integral principles of ordinary mechanics, which involve in their statement the idea of energy. It is not of much importance which of these we select; we can and shall choose Hamilton's principle. We thus lay down as the sole fundamental law of mechanics, in accordance with experience, the proposition that every system of natural bodies moves just as if it were assigned the task of attaining given positions in given times, and in such a manner that the average over the whole time of the difference between the kinetic and the potential energy

---

[18][There is a foreshadowing of Percy Bridgman's (1882–1961) idea of "operational definitions" here in Hertz's emphasis on "concrete experiences" to establish the presence of mass and energy.]

should be as small as possible.[19] Although this law may not be simple in form, it nevertheless represents without ambiguity the transformation of energy, and enables us to predetermine completely the course of actual phenomena for the future. In stating this new law we lay down the last of the indispensable foundations of mechanics.

All that we can further add are only mathematical deductions and certain simplifications of notation which, although expedient, are not necessary. Among these latter is the idea of force, which does not enter into the foundations. Its introduction becomes expedient when we consider not only masses that are connected with constant quantities of energy, but also masses that give up energy to other masses or receive energy from them. Still, it is not by any new experience that it [the idea of force] is introduced, but by a definition that can be formed in more than one way. And accordingly the properties of the force so defined are not to be obtained from experience, but are to be deduced from the definition and the fundamental laws. Even the confirmation of these properties by experience is superfluous, unless we doubt the correctness of the whole system. Hence the idea of force as such cannot in this system involve any logical difficulties, nor can it come into question in estimating the correctness of the system; it can only increase or diminish its appropriateness.

Somewhat after the manner indicated would the principles of mechanics have to be arranged in order to adapt them to the concept of energy. The question now is, whether this second picture is preferable to the first. Let us therefore consider its advantages and disadvantages.

It will be best for us here to consider first the question of appropriateness, since it is in this respect that the improvement is most obvious. For, to begin with, our second picture of natural motions is decidedly more distinct: it shows more of their peculiarities than does the first picture. When we wish to deduce Hamilton's

---

[19][The difference between the kinetic energy K and the potential energy V is called the Lagrangian function L, so that $K - V = L$. For a conservative mechanical system, Hamilton's principle is usually stated as follows: The time integral of the Lagrangian function has a stationary value for the actual path of the motion compared with all other possible paths having the same end-points and corresponding to the same time interval. In most practical cases the stationary value is a minimum.]

principle from the general foundations of mechanics we have to add to the latter certain assumptions as to the acting forces and the character of contingent fixed connections. These assumptions are of the most general nature, but they indicate a corresponding number of important limitations of the motions represented by the principle. And, conversely, we can deduce from the principle a whole series of relations, especially of mutual relations between every kind of possible force, which are wanting in the principles of the first picture; in the second picture they are present and likewise occur in nature—which is the important point. To prove this is the object of the papers published by von Helmholtz under the title, "*Über die physikalische Bedeutung des Prinzips der kleinsten Wirkung.*"[20] It would be more correct to say that the thing that has to be proved is precisely the discovery which is demonstrated and communicated in that paper. For it is truly a discovery to find that from such general assumptions, conclusions so distinct, so weighty, and so valid can be drawn. We may then appeal to that paper for confirmation of our statement; and, inasmuch as it represents the furthest advance of physics at the present time, we may spare ourselves the question whether it may be possible to conform yet more closely to nature, say by limiting the permissible forms of potential energy. We shall simply emphasize this, that in respect to simplicity as well, our present picture avoids the stumbling-blocks that endangered the appropriateness of the first [picture]. For if we ask ourselves the real reasons why physics at the present time prefers to express itself in terms of energy, our answer will be: because in this way it best avoids talking about things of which it knows very little, and which do not at all affect the essential statements under consideration. We have already had occasion to remark that in tracing back phenomena to forces we are compelled to turn our attention continually to atoms and molecules. It is true that we are now convinced that ponderable matter consists of atoms; and we have definite notions of the magnitude of these atoms and of their motions in certain cases. But the form of the atoms, their connection, their motion in most cases—all these are entirely hidden from us; their number is in all cases immeasurably great. So that, although our conception of atoms is in itself an important and interesting object for

---

[20][H. von Helmholtz, *Journal für die reine und angewandte Mathematik*, 100 (1887), pp. 137–166, 213–222. See also Helmholtz (1892).]

further investigation, it is in no way especially apt to serve as a known and secure foundation for mathematical theories. To an investigator like Gustav Kirchhoff, who was accustomed to rigid reasoning, it almost gave pain to see atoms and their vibrations willfully thrown into the middle of a theoretical deduction. The arbitrarily assumed properties of the atoms may not affect the final result; the result may still be correct. Nevertheless the details of the deduction are in great part presumably false; the deduction is a proof in appearance only. The earlier mode of thought in physics scarcely allowed any choice or any way of escape.

Herein lies the advantage of the conception of energy and of our second picture of mechanics: that in the assumptions of the problems there enter only characteristics that are directly accessible to experience, parameters or arbitrary coordinates of the bodies under consideration; that the consideration proceeds with the aid of these characteristics in a finite and complete form; and that the final result can again be directly translated into tangible experience. Beyond energy itself in its few forms, no auxiliary constructions enter into consideration. Our statements can be limited to the known peculiarities of the system of bodies under consideration, and we need not conceal our ignorance of the details by arbitrary and ineffectual hypotheses. All the steps in the deduction, as well as the final result, can be defended as correct and significant. These are the merits which have endeared this method to present-day physics. They are peculiar to our second picture of mechanics; in the sense in which we have used the words, they are to be regarded as advantages with respect to simplicity, and thus with respect to appropriateness.

Unfortunately we begin to be uncertain as to the value of our system when we test its correctness and its logical permissibility. The question of correctness at once gives rise to legitimate doubts. Hamilton's principle can be deduced from the accepted foundations of Newtonian mechanics, but this does not by any means guarantee its accordance with nature. We have to remember that this deduction only follows if certain assumptions hold good; and also that our system claims not only to describe certain natural motions correctly, but to embrace all natural motions. We must therefore investigate whether these special assumptions, which are made in addition to Newton's laws, are universally true; and a single example from nature to the contrary would invalidate the correctness of our system as such, although it would not disturb in the least the validity of Hamilton's principle as a general proposition. The doubt is not so much whether our system includes the whole multiplicity[21] of forces, as whether it embraces the whole multiplicity of rigid connections that may arise between the bodies of nature. The application of Hamilton's principle to a material system does not exclude the existence of fixed connections between the chosen coordinates. But at any rate it requires that these connections be mathematically expressible by finite equations between the coordinates; it does not permit the occurrence of connections that can only be represented by differential equations. But nature itself does not appear to exclude entirely connections of this kind. They arise, for example, when three-dimensional bodies roll on one another without slipping. By such a connection, examples of which frequently occur, the position of the two bodies with respect to each other is only limited by the condition that they must always have one point of their surfaces in common; the motion of the bodies is thus decreased by one degree of freedom. From the connection, then, there can be deduced more equations between the changes in the coordinates than between the coordinates themselves; hence there must be among these equations at least one non-integrable differential equation. Now Hamilton's principle cannot be applied to such a case; or, to speak more correctly, the application that is mathematically possible leads to results that are physically false.

Let us restrict our consideration to the case of a sphere rolling without slipping upon a horizontal plane under the influence of only its inertia. It is not difficult to see, without calculation, what motions the sphere can actually execute. We can also see what motions would correspond to Hamilton's principle: these would have to take place in such a way that with constant *vis viva* the sphere would attain given positions in the shortest possible time. We can thus convince ourselves, without calculation, that the two kinds of motions exhibit very different characteristics. If we choose any initial and final positions of the sphere, it is clear that there is always one definite motion from one position to the other for which the time of motion, i.e., the Hamilton's integral, is a minimum. But, as a matter of fact, a

---

[21][In the original English translation the German word *Mannigfaltigkeit* is translated as "manifold" throughout this section. For reasons of clarity we prefer to render it as "multiplicity," since in mathematics "manifold" has a very precise, technical sense, which is not what Hertz intends here.]

natural motion from every position to every other position is not possible without the cooperation of forces, even if the choice of the initial velocity is perfectly free. And even if we choose the initial and final positions so that a natural free motion between the two is possible, this will nevertheless not be the one which corresponds to a minimum of time. For certain initial and final positions the difference can be very striking. In this case a sphere moving in accordance with the [Hamilton] principle would decidedly have the appearance of a living thing, steering its course consciously toward a given goal, while a sphere following the law of nature would give the impression of an inanimate mass spinning steadily towards it. It would be of no use to replace Hamilton's principle by the principle of least action or by any other integral principle, for there is but a slight difference of meaning between all these principles, and in the aspect here being considered they are quite equivalent. Only in one way can we defend the system and preserve it from the charge of incorrectness. We must decline to admit that rigid connections of the kind referred to here do actually and strictly occur in nature. We must show that all so-called rolling without slipping is really rolling with a little slipping, and is therefore a case of friction. We have to rest our case upon this, that generally friction between surfaces is one of the processes that we have not yet been able to trace back to clearly understood causes; that the forces that come into play have only been ascertained quite empirically; and hence that the whole problem is one of those that we cannot at present handle without making use of force and the roundabout methods of ordinary mechanics. This defense is not quite convincing. For rolling without slipping does not contradict either the principle of energy or any other generally accepted law of physics. The process is one which is so nearly realized in the visible world that even integration machines are constructed on the assumption that it strictly takes place. We have scarcely any right, then, to exclude its occurrence as impossible, at any rate from the mechanics of still unknown systems, such as the atoms or the parts of the ether.[22] But, even if we admit that the connections in question are only approximately realized in nature, the failure of Hamilton's principle still creates difficulties in these cases.

---

[22][Note that Hertz's mind seems to return to the ether frequently in this Introduction. The "rigid connections" he discusses here would be necessary if the ether were to be the medium for the transmission of transverse electromagnetic waves.]

We are bound to require of every fundamental law of our mechanical system that, when applied to approximately correct relations, it should always lead to approximately correct results, not to results that are entirely false. For, otherwise, since all the rigid connections which we draw from nature and introduce into our calculations correspond only approximately to the actual relations, we should get into a state of hopeless uncertainty as to which admitted of the application of the law and which did not. And yet we do not wish to abandon entirely the defense we have proposed. We should prefer to admit that the doubt is one that affects the appropriateness of the system, not its correctness, so that the disadvantages that arise from it may be outweighed by other advantages.

The real difficulties first meet us when we try to arrange the elements of the system in strict accordance with the requirements of logical permissibility. In introducing the idea of energy we cannot proceed in the usual way, starting with force and proceeding from this to force-functions, to potential energy, and to energy in general. Such an arrangement would belong to the first representation of mechanics. Without assuming any previous consideration of mechanics, we have to specify by what simple, direct experiences we propose to define the presence of a store of energy, and to determine its amount. In what precedes we have only assumed, not shown, that such a determination is possible. At the present time many distinguished physicists tend to attribute to energy so many of the properties of a substance that they assume that every smallest portion of it is associated at every instant with a given place in space, and that through all the changes of place and all the transformations of the energy into new forms, it retains its identity. These physicists must have the conviction that definitions of the required kind can be found; and it is therefore permissible to assume that such definitions can be given. But when we try to throw them into a concrete form, satisfactory to ourselves and likely to command general acceptance, we become perplexed. This mode of conception as a whole does not yet seem to have arrived at a satisfactory and conclusive result. At the very beginning there arises a special difficulty, from the circumstance that energy, which is alleged to resemble a substance, occurs in two such totally dissimilar forms as kinetic and potential energy. Kinetic energy itself does not really require any new fundamental determination, for it can be deduced from the ideas of velocity and mass; on the other hand, potential energy, which does need to be determined independently, does not lend itself at all well to any definition that ascribes to it the properties of a substance. The amount of a substance is necessarily a positive quantity; but we never hesitate in assuming the potential energy contained in a system to be negative. When the amount of a substance is represented by an analytical expression, an additive constant in the expression has the same importance as the rest; but in the expression for the potential energy of a system an additive constant never has any meaning. Lastly, the amount of any substance contained in a physical system can only depend on the state of the system itself; but the amount of potential energy contained in a given system depends on the presence of distant masses, which perhaps have never had any direct influence on the system. If the universe, and therefore the number of such distant masses, is infinite, then the amount of many forms of potential energy contained in even finite quantities of matter is infinitely great. All these are difficulties that must be removed or avoided by the desired definition of energy. We do not assert that such a definition is impossible, but as yet we cannot say that it has been framed. The most prudent thing to do is to regard it for the present as an open question whether this system can be developed in a logically unexceptional form.

It may be worthwhile discussing here whether there is any justification for another objection that might be raised as to the permissibility of this second system. In order that a picture of certain external things may in our sense be permissible, not only must its characteristics be consistent among themselves, but they must not contradict the characteristics of other pictures already established in our knowledge. On the strength of this it seems inconceivable that Hamilton's principle, or any similar proposition, should really play the part of a fundamental law of mechanics and be a fundamental law of nature. For the first thing that is to be expected of a fundamental law is simplicity and plainness, whereas Hamilton's principle, when we come to look into it, proves to be an exceedingly complicated statement. Not only does it make the present motion dependent on consequences that can only exhibit themselves in the future, thereby attributing intentions to inanimate nature; but, what is much worse, it attributes to nature intentions that are void of meaning. For the integral, whose minimum is required by Hamilton's principle, has no simple physical meaning;[23] and for nature it is an unintelligible aim to make a mathematical expression a minimum, or to bring its variation to zero. The usual answer, which physics nowadays keeps ready for such attacks, is that these considerations are based upon metaphysical assumptions; that physics has renounced these, and no longer recognizes it as its duty to meet the demands of metaphysics. It no longer attaches weight to the reasons which used to be urged from the metaphysical side in favor of principles that indicate design in nature, and thus it cannot lend an ear to objections of a metaphysical character against these very same principles. If we had to decide on such a matter, we should not think it unfair to place ourselves rather on the side of the attack than of the defense. A doubt that makes an impression on our mind cannot be removed by calling it metaphysical; every thoughtful mind as such has needs that scientific men are accustomed to denote as metaphysical.[24] Moreover, in the case in question, as indeed in all others, it is possible to show what are the sound and just sources of our needs. It is true that we cannot *a priori* demand from nature simplicity, nor can we judge what in her opinion is simple. But with regard to pictures of our own creation we can lay down requirements. We are justified in deciding that if our pictures are well adapted to the things, the actual relations of the things must be represented by simple relations between the pictures. And if the actual relations between the things can only be represented by complicated relations between the pictures, which are not even intelligible to an unprepared mind, we decide that these pictures are not sufficiently well adapted to the things. Hence our requirement of simplicity does not apply to nature, but to the pictures thereof that we fashion; and our repugnance to a complicated statement as a fundamental law only expresses the conviction that, if the contents of the statement are correct and comprehensive, it can be stated in a simpler form by a more suitable choice of the fundamental pictures. The same conviction finds expression in the desire we feel to penetrate from the external acquaintance with such a law to the deeper and real meaning we are convinced it possesses. If this view is correct, the objection brought forward does really justify a doubt as to the system; but it does

---

[23][On this see footnote 19 above.]

[24][This section indicates the profound respect for philosophy that characterized Hertz's outlook.]

not apply so much to its permissibility as to its appropriateness, and comes under consideration in deciding as to the latter. However, we need not return to the consideration of this.

If we once more glance over the merits that we were able to claim for this second picture, we come to the conclusion that as a whole it is not quite satisfactory. Although the whole tendency of recent physics moves us to place the idea of energy in the foreground, and to use it as the cornerstone of our structure, it yet remains doubtful whether in so doing we can avoid the harshness and ruggedness that were so disagreeable in the first picture of mechanics. In fact, I have discussed this second mode of representation at some length, not in order to urge its adoption, but rather to show why, after due trial, I have felt obliged to abandon it.

III

A third arrangement of the principles of mechanics is that which will be explained at length in this book. Its principal characteristics will be at once stated, so that it may be criticized in the same way as the other two. It differs from them in this important respect, that it only starts with three independent fundamental concepts, namely, those of time, space, and mass.[25] The problem that it has to solve is to represent the natural relations between these three, and between these three only. Difficulties have hitherto been encountered in connection with a fourth idea, such as the idea of force or of energy; this, as an independent fundamental conception, is here avoided. G. Kirchhoff

---

[25][It is worth noting that Hertz never raises objections similar to those he has raised against "force" with respect to the concept of "mass." First Hertz defines mass independently of experience as follows: "The number of material particles in any space compared with the number of material particles in some chosen space at a fixed time, is called the *mass* contained in the first space." (*Mechanics*, p. 46, no. 4.) Later, in the text proper of his book he gives a more operational definition: "The mass of bodies . . . is determined by weighing. The unit of mass is the mass of some body settled by arbitrary convention." (*Mechanics*, p. 140, no. 300). Although this may seem to introduce the concept of force (since weight is a force), it seems unobjectionable, since Hertz is merely assuming that, if the weights of two objects at any point on the surface of the earth are equal, then their masses are also equal; if $m_1 g = m_2 g$, then $m_1 = m_2$.]

has already made the remark in his *Textbook of Mechanics*[26] that three independent quantities are necessary and sufficient for the development of mechanics. Of course, the resulting deficiency in the fundamental concepts necessarily requires some addition. In our representation we try to fill up the void that occurs by the use of an hypothesis, which is not stated here for the first time, but which is not customarily introduced in the very elements of mechanics. The nature of this hypothesis may be explained as follows.

If we try to understand the motions of bodies around us, and to refer them to simple and clear rules, paying attention only to what can be directly observed, our attempt will in general fail. We soon become aware that the totality of things visible and tangible does not form an universe conformable to law, in which the same results always follow from the same conditions. We become convinced that the complexity of the actual universe must be greater than that of the universe which is directly revealed to us by our senses. If we wish to obtain a picture of the universe (*Weltbild*) that is well-rounded, complete, and conformable to law, we have to presuppose, behind the things that we see, other invisible things—to imagine other things concealed beyond the limits of our senses. These deep-lying influences we recognized in the first two representations; we imagined them to be entities of a special and peculiar kind, and so, in order to include them in our representation, we created the ideas of force and energy. But another way lies open to us. We may admit that there is a hidden something at work, and yet deny that this something belongs to a special category. We are free to assume that this hidden something is nought else but motion and mass again, motion and mass that differ from the visible ones not in themselves, but in relation to us and to our usual means of perception. Now this mode of conception is just our hypothesis. We assume that it is possible to conjoin with the visible masses of the universe other masses obeying the same laws, and of such a kind that the whole thereby becomes intelligible and conformable to law. We assume this to be possible everywhere and in all cases, and that there are no causes whatever of the phenomena other than those hereby admitted. What we are accustomed to denote as force and as energy now become nothing more than an action of mass and motion, but not necessarily of mass and motion recognizable by our coarse senses.

---

[26][Kirchhoff (1877), p. iii–iv.]

Such explanations of force in terms of processes of motion are usually called dynamical; and we have every reason for saying that physics at the present time regards such explanations with great favor. The forces connected with heat have been traced back with certainty to the concealed motions of tangible masses. Through Maxwell's labors the supposition that electromagnetic forces are due to the motion of concealed masses has become almost a conviction.[27] Lord Kelvin gives a prominent place to dynamical explanations of force; in his theory of vortex atoms he has endeavored to present a picture of the universe in accordance with this viewpoint. In his investigation of cyclical systems von Helmholtz has treated the most important form of concealed motion fully, and in a manner that admits of general application; through him "hidden mass" [*verborgene Masse*] and "hidden motion" [*verborgene Bewegung*] have become current as technical expressions in German. But if this hypothesis is capable of gradually eliminating the mysterious forces from mechanics, it can also entirely prevent their entering into mechanics. And if its use for the former purpose is in accordance with present tendencies in physics, the same must hold good of its use for the latter purpose. This is the leading thought from which we start. By following it out we arrive at the third picture, the general outlines of which will now be sketched.

We first introduce the three independent fundamental concepts of time, space, and mass as objects of experience, and we specify the concrete sensible experiences by which time, mass, and space are to be determined. With regard to the masses we stipulate that, in addition to the masses recognizable by the senses, hidden masses can by hypothesis be introduced. We next bring together the relations that always obtain between these concrete experiences, and that we have to retain as the essential relations between the fundamental concepts. To begin with, we naturally connect the fundamental concepts in pairs. Relations between space and time alone form the subject of kinematics. There exists no connection between mass and time alone. Experience teaches us that between mass and space there exists a series of important relations. For we find certain purely spatial connections between the masses of nature: from the very beginning

---

[27][Hertz seems to mean that Maxwell's theory and Hertz's own experiments require an ether, and that the ether consists of concealed masses performing concealed motions in a way that enables electromagnetic forces to propagate through space.]

onwards through all time, and therefore independently of time, certain positions and certain changes of position are prescribed and associated with these masses as being possible, and all others as impossible. Respecting these connections we can also assert generally that they only apply to the relative position of the masses among themselves; and further that they satisfy certain conditions of continuity, which find their mathematical expression in the fact that the connections themselves can always be represented by homogeneous linear equations between the first differentials of the quantities by which the positions of the masses are denoted. To investigate in detail the connections of definite material systems is not the business of mechanics, but of experimental physics. The distinguishing characteristics that differentiate the various material systems of nature from one another are, according to our conception, simply and solely the connections of their masses.

Up to this point we have only considered the connections of the fundamental concepts in pairs; we now address ourselves to mechanics in the stricter sense, in which all three have to be considered together. We find that their general connection, in accordance with experience, can be epitomized in a single fundamental law, which exhibits a close analogy with the usual law of inertia. In accordance with the mode of expression that we shall use, it can be represented by the statement;

> Every natural motion of an independent material system consists herein, that the system follows with uniform velocity one of its straightest paths.[28]

Of course, this statement only becomes intelligible when we have given the necessary explanation of the mathematical mode of expression used; but the sense of the law can also be expressed in the usual language of mechanics. The law condenses into one single statement the usual law of inertia and Gauss's Principle of Least Constraint. It therefore asserts that if the connections of the system

---

[28][For Hertz a "straightest path" is one executed by a mechanical system for which the deviations from the path of minimum curvature are least. These deviations are produced by the rigid connections linking the observed masses with the concealed masses. Note that this is the path of a system of particles, not of any individual particle. Hertz defines carefully in the body of his book exactly what he means by this "system path."]

could be momentarily destroyed, its masses would become dispersed, moving in straight lines with uniform velocity; but since this is impossible, they tend as nearly as possible to such a motion. In our picture [the third], this fundamental law is the first proposition derived from experience in mechanics proper; it is also the last. From it, together with the admitted hypothesis of hidden masses and normal connections, we can derive all the rest of mechanics by purely deductive reasoning. Around it we group the remaining general principles, according to their relations to it and to each other, as corollaries or as partial statements. We endeavor to show that the contents of mechanics, when arranged in this way, do not become less rich or diverse than are its contents when it starts with four fundamental concepts [as in the first two pictures]; at any rate not less rich or diverse than is required for the representation of nature.

We soon find it convenient to introduce into our system the idea of force. However, it is not as something independent of us and apart from us that force now makes its appearance, but as a mathematical aid whose properties are entirely in our power. It cannot, therefore, in itself contain anything mysterious for us. Thus, according to our fundamental law, whenever two bodies belong to the same system, the motion of the one is determined by that of the other. The idea of force now comes in as follows. For assignable reasons we find it convenient to divide the determination of the one motion by the other into two steps. We thus say that the motion of the first body determines a force, and that this force then determines the motion of the second body.[29] In this way force can with equal justice be regarded as being always a cause of motion, and at the same time a consequence of motion. Strictly speaking, it is a middle term conceived only between two motions. According to this conception, the general properties of force must clearly follow from the fundamental law as a necessary consequence of reasoning; and if we see these properties confirmed in experiences that are possible, we can in no way feel surprised unless we are sceptical about our fundamental law. Precisely the same is true of the idea of energy and of any other aids that may be introduced.

What has hitherto been stated relates to the physical content of this picture, and nothing further need be said in this regard; but it will

---

[29][For example, the motion of a charged body produces a magnetic field, and this magnetic field then interacts with the motion of a second charged body to change its motion.]

be convenient to give here a brief explanation of the special mathematical form in which it will be represented. The physical content is quite independent of the mathematical form, and as the content differs from what is customary, it is perhaps not quite judicious to present it in a form that is itself unusual. But the form as well as the content differ only slightly from such as are familiar; and they are, moreover, so well suited that they mutually assist each other. The essential characteristic of the terminology used consists in this, that instead of always starting from single points, it from the beginning conceives and considers whole systems of points.[30] Everyone is familiar with the expressions "position of a system of points," and "motion of a system of points." There is nothing unnatural in continuing this mode of expression, and denoting by its path the aggregate of the positions traversed by a system in motion. Every smallest part of this path is then a path-element. Of two path-elements one can be a part of the other; they then differ in magnitude and only in magnitude. But two path-elements that start from the same position may belong to different paths. In this case neither of the two forms part of the other; they differ in other respects than that of magnitude, and thus we say that they have different directions. It is true that these statements do not suffice to determine without ambiguity the characteristics of "magnitude" and "direction" for the motion of a system. But we can complete our definitions geometrically or analytically so that their consequences shall neither contradict themselves nor the statements we have made; and so that the magnitudes thus defined in the geometry of the system shall exactly correspond to the magnitudes that are denoted by the same names in the geometry of a point, with which, indeed, they always coincide when the system is reduced to a point.

Having determined the characteristics of magnitude and direction, we next call the path of a system straight if all its elements have the same direction, and curved if the direction of the elements changes from position to position. As in the geometry of a point, we measure curvature by the rate of variation of the direction with position. From these definitions we at once get a whole series of relations; and the number of these increases as soon as the freedom of

---

[30][On the mathematical expression for the displacement of a system of points, see the last few pages of FitzGerald's review of Hertz's *Mechanics* (Paper No. 17).]

Hertz's Introduction to His Principles of Mechanics

motion of the system under consideration is limited by its connections. Certain classes of paths that are distinguished among the possible paths by peculiarly simple properties then claim special attention. Of these the most important are those paths which at each of their positions have the least possible curvature; these we shall denote as the straightest paths of the system. These are the paths that are referred to in the fundamental law, and that have already been mentioned in stating it. Another important type consists of those paths that form the shortest connection between any two of their positions; these we shall denote as the shortest paths of the system. Under certain conditions the concepts of straightest and shortest paths coincide. The relation is perfectly familiar in connection with the theory of curved surfaces; nevertheless it does not hold good in general and under all circumstances.

The compilation and arrangement of all the relations that arise here belong to the geometry of systems of points. The development of this geometry has a peculiar mathematical attraction, but we pursue it only as far as is required for the immediate purpose of applying it to physics. A system of n points presents a $3n$-manifold of motion, although this may be reduced to any arbitrary number by the connections of the system. Hence there arise many analogies with the geometry of a space of many dimensions; and these in part extend so far that the same propositions and notations can apply to both. But we must note that these analogies are only formal, and that, although they occasionally have an unusual appearance, our considerations refer without exception to concrete structures of space as perceived by our senses. Hence all our statements represent possible experiences; if necessary, they could be confirmed by direct experiments, namely, by measurements made with models. Thus we need not fear the objection that in building up a science dependent on experience, we have gone outside the world of experience.

On the other hand, we are bound to answer the question how a new, unusual, and comprehensive mode of expression justifies itself, and what advantages we expect from using it. In answering this question we specify as the first advantage that it enables us to render the most general and comprehensive statements with great simplicity and brevity. In fact, propositions relating to whole systems do not require more words or more ideas than are usually employed in referring to a single point; the latter, indeed, does not need independent investigation, as it only appears occasionally as a simplification and a special case. If it is urged that this simplicity is only artificial, we reply that in no other way can simple relations be secured than by artificial and well-considered adaptation of our ideas to the relations that have to be represented. But in this objection there may be involved the imputation that the mode of expression is not only artificial, but far-fetched and unnatural. To this we reply that there may be some justification for regarding the consideration of whole systems as being more natural and obvious than the consideration of single points. For, in reality, the material particle is simply an abstraction, whereas the material system is presented directly to us. All actual experience is obtained directly from systems; and it is only by processes of reasoning that we deduce conclusions as to possible experiences with single points.

As a second merit, although not a very important one, we specify the advantage of the form in which our mathematical mode of expression enables us to state the fundamental law. Without this we should have to split it up into Newton's first law and Gauss's principle of least constraint. Both of these together would represent accurately the same facts [as our fundamental law]; but in addition to these facts they would by implication contain something more, and this something more would be too much. In the first place they suggest the idea, which is foreign to our system of mechanics, that the connections of the material system might be destroyed, whereas we have denoted them as being permanent and indestructible throughout. In the second place we cannot, in using Gauss's principle, avoid suggesting the idea that we are not only stating a fact, but also the cause of this fact. We cannot assert that nature always keeps a certain quantity, which we call a constraint, as small as possible, without suggesting that this quantity signifies something that is for nature itself a constraint— an uncomfortable feeling. We cannot assert that nature acts like a judicious calculator reducing his observations, without suggesting that deliberate intention underlies the action. There is undoubtedly a special charm in such suggestions; and Gauss felt a natural delight in giving prominence to it in his beautiful discovery,[31] which is of fundamental importance in our mechanics. Still, it must be confessed that the charm is that of mystery; we do not really believe that we can

---

[31][i.e., Gauss's Principle of Least Constraint.]

Hertz's Introduction to *His* Principles of Mechanics

solve the enigma of the world by such half-suppressed allusions. Our own fundamental law entirely avoids any such suggestions. It exactly follows the form of the customary law of inertia, and like this it simply states a bare fact without any pretense of proving it. And as it thereby becomes plain and unvarnished, in the same degree does it become more honest and truthful.

Perhaps I am prejudiced in favor of the slight modification that I have made in Gauss's principle, and see in it advantages that will not be manifest to others. But I feel sure of general assent when I state as the third advantage of our method that it throws a bright light upon Hamilton's method of treating mechanical problems by the aid of characteristic functions. During the sixty years since its discovery this mode of treatment has been well appreciated and much praised; but it has been regarded and treated more as a new branch of mechanics, and as if its growth and development had to proceed in its own way and independently of the usual mechanics. In our form of the mathematical representation, Hamilton's method, instead of having the character of a side branch, appears as the direct, natural and, if one may say so, self-evident continuation of the elementary statements in all cases to which it is applicable. Further, our mode of representation gives prominence to the fact that Hamilton's mode of treatment is not based, as is usually assumed, on the special physical foundations of mechanics. Rather it is fundamentally a purely geometrical method, which can be established and developed quite independently of mechanics, and which has no closer connection with mechanics than any other of the geometrical methods employed in it. It has long since been remarked by mathematicians that Hamilton's method contains purely geometrical truths, and that a peculiar mode of expression, suitable to it, is required in order to express these [truths] clearly. But this fact has only come to light in a somewhat perplexing form, namely, in the analogies between ordinary mechanics and the geometry of many-dimensional spaces, which have been discovered by following out Hamilton's ideas. Our mode of expression gives a simple and intelligible explanation of these analogies. It allows us to take advantage of them, and at the same time it avoids the unnatural admixture of supra-sensible abstractions with a branch of physics.

We have now sketched the content and form of our third picture of mechanics insofar as this can be done without encroaching upon the contents of our book, and sufficiently to enable us to submit it to criticism with respect to its permissibility, its correctness, and its appropriateness. I think that as far as logical permissibility is concerned it will be found to satisfy the most rigid standards, and I trust that others will be of the same opinion. This virtue of the representation I consider to be of the greatest importance, indeed of unique importance. Whether this outlined representation is more appropriate than another, whether it is capable of including all future experience, even whether it only embraces all present experience—all this I regard almost as nothing compared with the question whether it is in itself complete, pure and free from contradictions. For I have not attempted this task because mechanics has shown signs of inappropriateness in its application, nor because it conflicts with experience in any way, but solely in order to rid myself of the oppressive feeling that to me its elements were not free from things obscure and unintelligible. What I have sought is not the only picture of mechanics, nor yet the best picture; I have only sought to find an intelligible picture and to show by an example how this is possible and what it must look like.

Perfection is, of course, unattainable for us along any route; and I must confess that, in spite of the pains I have taken with it, the picture I have developed is not so convincingly clear but that it may be exposed to doubt on some points or may require defense.[32] And yet it seems to me that, of objections of a general nature, there is only a single one that is so pertinent that it is worthwhile to anticipate it and exclude it. It relates to the nature of the rigid connections, which we assume to exist between the masses and which are absolutely indispensable in our system. Many physicists will at first be of the opinion that by means of these connections forces are introduced into the elements of mechanics, and are introduced in a way that is concealed and therefore not permissible. For, they will assert, rigid connections are not conceivable without forces; they cannot come into existence except by the action of forces. To this we reply: Your assertion is correct for the mode of thought of ordinary mechanics, but it is not correct independently of this mode of thought; it does not carry conviction to a mind that considers the facts without prejudice

---

[32][Hertz's *Mechanics* does contain inconsistencies that might have been avoided if he had been allowed the time to discuss and argue out his ideas with other physicists before the book went to press. On this see his letter of 19 Nov. 1893 to his parents, in *Erinnerungen*, p. 343.]

and as if for the first time. Suppose we find in any way that the distance between two material particles remains constant at all times and under all circumstances. We can express this fact without making use of any other conceptions than those of space; and the value of the fact stated, as a fact, for the purpose of predicting future experience and for all other purposes, will be independent of any explanation that we may or may not possess of it. In no case will the value of the fact be increased, or our understanding of it improved, by putting it in the form: "Between these masses there acts a force that keeps them at a constant distance from each other," or "Between them there acts a force that makes it impossible for their distance to alter from its fixed value." But it will be urged that this latter explanation, although apparently only a ludicrous circumlocution, is nevertheless correct. For all the connections of the actual world are only approximately rigid; and the appearance of rigidity is only produced by the action of the elastic forces that continually annul the small deviations from the position of rest. To this we reply as follows: With regard to rigid connections that are only approximately realized, our mechanics will naturally only state as a fact that they are approximately satisfied; and for the purpose of this statement the idea of force is not required. If we wish to proceed to a second approximation and take into consideration the deviations, and with them the elastic forces, we shall make use of a dynamical explanation for these as for all forces. In seeking the actual rigid connections we shall perhaps have to descend to the world of atoms. But such considerations are out of place here; they do not affect the question whether it is logically permissible to treat of fixed connections as independent of forces and preceding them. All that I wished to show was that this question must be answered in the affirmative, and this I believe I have done. This being so, we can deduce the properties and behavior of the forces from the nature of the fixed connections without being guilty of a *petitio principii*.[33] Other objections of a similar kind are possible, but I believe they can be removed in much the same way.

By way of giving expression to my desire to prove the logical purity of this system in all its details, I have thrown its representation into the older synthetic form. For this purpose the form used has the merit of compelling us to specify beforehand, definitely even if

---

[33][i.e., of begging the question.]

monotonously, the logical value that every important statement is intended to have. This makes it impossible to use the convenient reservations and ambiguities into which we are enticed by the wealth of combinations of ordinary speech. But the most important advantage of the form chosen is that it is always based on what has already been proved, never on what is to be proved later on; thus we are always sure of the whole chain if we sufficiently test each link as we proceed. In this respect I have endeavored to carry out fully the obligations imposed by this mode of representation. At the same time it is obvious that the form by itself is no guarantee against error or oversight; and I hope that any chance defects will not be the more harshly criticized on account of the somewhat presumptuous mode of presentation. I trust that any such defects will be capable of improvement and will not affect any important point. Now and again, in order to avoid excessive wordiness, I have consciously abandoned to some extent the rigid strictness which this mode of representation properly requires. Before proceeding to mechanics proper, as dependent on physical experience, I have naturally discussed those relations that follow simply and necessarily from the definitions adopted and from mathematics; the connection of these latter with experience, if any, is of a different nature from that of the former. Moreover, there is no reason why the reader should not begin with the second book.[34] The matter with which he is already familiar and the clear analogy with the dynamics of a particle will enable him easily to guess the content of the propositions in the first book. If he admits the appropriateness of the mode of expression used, he can at any time return to the first book to convince himself of its permissibility.

We next turn to the second essential requirement that our picture must satisfy. In the first place, there is no doubt that the system correctly represents a very large number of natural motions.[35] But this

---

[34][Book I of Hertz's *Principles of Mechanics* treats the "Geometry and Kinematics of Material Systems"; Book II, which is entitled "Mechanics of Material Systems," is devoted to what would conventionally be called dynamics. Hertz, however, by replacing forces by hidden masses tied together by rigid connections both among themselves and with the observable masses in the system, is really attempting to reduce dynamics to kinematics. On this see Klein (1972), p. 74.]

[35][Here by a "natural motion" Hertz seems to mean a motion that has been observed in nature. But in Book II, #312, of his *Mechanics* he defines a "natural motion" more technically: "Every motion of a free material system, or

does not go far enough: the system must include all natural motions without exception. I think that this too can be asserted of it, at any rate in the sense that no definite phenomena can at present be mentioned that would be inconsistent with the system. We must of course admit that we cannot rigorously examine all phenomena. Hence the system goes a little beyond the results of assured experience; it therefore has the character of a hypothesis that is accepted tentatively and awaits sudden refutation by a single example or gradual confirmation by a large number of examples. There are especially two places in which we go beyond assured experience: firstly, in our limitation of the possible connections; and secondly, in the dynamical explanation of force. What right have we to assert that all natural connections can be expressed by linear differential equations of the first order? With us this assumption is not a matter of secondary importance, which we might do without. Our system stands or falls with it, for it raises the question whether our fundamental law is applicable to connections of the most general kind. And yet, connections of a more general kind are not only conceivable, but they are permitted in ordinary mechanics without hesitation. There nothing prevents us from investigating the motion of a point, where its path is only limited by the supposition that it makes a given angle with a given plane, or that its radius of curvature is always proportional to another given length. These are conditions that are not permissible in our system. But why are we certain that they are excluded by the nature of things? We might reply that these and similar connections cannot be realized by any practical mechanism, and in this respect we might appeal to the great authority of Helmholtz's name. But in every example possibilities might be overlooked, and ever so many examples would not suffice to substantiate the general assertion.

It seems to me that the reason for our conviction should more properly be stated as follows. All connections of a system, which are not embraced within the limits of our mechanics, indicate in one sense or another a discontinuous succession of its possible motions. But, as a matter of fact, it is an experience of the most general kind that nature

---

of its parts, which is consistent with the fundamental law, we call a natural motion of the system in contradistinction to its conceivable and possible motions" (pp. 144–145 in the English edition). Therefore Hertz's assumption is that his fundamental law applies to all motions actually occurring in nature, and to no motions not observed in nature.]

exhibits continuity in infinitesimals everywhere and in every sense, an experience that has crystallized into firm conviction in the old proposition: *Natura non facit saltus.*[36] In the text I have therefore laid stress on this: that the permissible connections are defined solely by their continuity, and that their property of being represented by equations of a definite form is only deduced from this. We cannot attain to actual certainty in this way. For this old proposition is indefinite, and we cannot be sure how far it applies, how far it is the result of actual experience, and how far the result of arbitrary assumption. Thus the most conscientious plan is to admit that our assumption as to the permissible connections is of the nature of a tentatively accepted hypothesis.

The same may be said with respect to the dynamical explanation of force. We may indeed prove that certain classes of hidden motions produce forces that, like actions-at-a-distance in nature, can be represented to any desired degree of approximation as differential coefficients of force-functions. It can be shown that the form of these force-functions may be of a very general nature, and in fact we do not deduce any restrictions for them. But, on the other hand, it remains for us to prove that any and every form of the force-functions can be realized; and hence it remains an open question whether such a mode of explanation may not fail to account for some one of the forms occurring in nature. Here again we can only bide our time so as to see whether our assumption is refuted, or whether it acquires greater and greater probability by the absence of any such refutation. We may regard it as a good omen that many distinguished physicists tend more and more to favor the hypothesis. I may mention Lord Kelvin's theory of vortex-atoms; this presents us with an image of the material universe, which is in complete accord with the principles of mechanics. And yet our mechanics in no wise demands such great simplicity and limitation of assumptions as Lord Kelvin has imposed on himself. We need not abandon our fundamental propositions if we were to assume that the vortices revolved about rigid or flexible, but inextensible, nuclei; and instead of assuming simply incompressibility,

---

[36]["Nature does not make jumps." If this was an accepted proposition in mechanics when Hertz wrote this, it ceased to be such once Max Planck appeared on the scene.]

we might subject the all-pervading medium[37] to much more complicated conditions, the most general form of which would be a matter for further investigation. Thus there appears to be no reason why the hypothesis admitted in our mechanics should not suffice to explain the phenomena.

We must, however, make one reservation. In the text we take the natural precaution of expressly limiting the range of our mechanics to inanimate nature; how far its laws extend beyond this we leave as quite an open question. As a matter of fact, we cannot assert that the internal processes of life follow the same laws as the motions of inanimate bodies, nor can we assert that they follow different laws. According to appearances and general opinion there seems to be a fundamental difference. And the same feeling that impels us to exclude as foreign from the mechanics of the inanimate world every indication of an intention, of a sensation, of pleasure and pain, this same feeling makes us unwilling to deprive our picture of the animate world of these richer and more varied elements. Our fundamental law, although it may suffice for representing the motion of inanimate matter, appears (at any rate this is one's initial and normal impression) too simple and too narrow to account for even the lowest processes of life. It seems to me that this is not a disadvantage, but rather an advantage for our [fundamental] law. For while it allows us to survey the whole domain of mechanics, it shows us what are the limits to this domain. By giving us only bare facts, without attributing to them any appearance of necessity, it enables us to recognize that everything might be quite different. Perhaps such considerations will be regarded as out of place here. It is not usual to treat of them in the elements of the customary representation of mechanics. But there the complete vagueness of the forces introduced leaves room for free play. There is a tacit stipulation that, if need be, a contrast between the forces of animate and inanimate nature may be established later on. In our representation the outlines of the picture are from the first so sharply delineated that any subsequent perception of such an important division becomes almost impossible. We are therefore bound to refer to this matter at once, or to ignore it altogether.

---

[37][Here again Hertz seems to be referring to the ether, which was assumed to fill all space, and even to penetrate within atoms and molecules.]

As to the appropriateness of our third picture, we need not say much. With respect to its distinctness and simplicity, as the contents of this book will show, we may assign to it about the same position as to the second picture; and the merits to which we drew attention in the latter are also present here. But the permissible possibilities are somewhat more extensive than in the second picture. For we pointed out that in the latter certain rigid connections were wanting; by our fundamental assumptions, however, these are not excluded. And this extension is in accordance with nature, and is therefore a virtue; nor does it prevent us from deducing the general properties of natural forces, in which lay the significance of the second picture. The simplicity of this picture, as of the second, is very apparent when we consider their physical applications. Here, too, we can confine our consideration to any characteristics of the material system that are accessible to observation. From their past changes we can deduce future ones by applying our fundamental law, without any necessity for knowing the positions of all the separate masses of the system, or for concealing our ignorance by arbitrary, ineffectual, and probably false hypotheses. But as compared with the second picture, our third one exhibits simplicity in the sense that it adapts its concepts so closely to nature that the essential relations existing in nature are reproduced by simple relations between these concepts. This is seen not only in the fundamental law, but also in its numerous general corollaries, which correspond to the so-called principles of mechanics. Of course, it must be admitted that this simplicity only obtains when we are dealing with systems that are completely known, and that it disappears as soon as hidden masses come in. But even in these cases the reason for the complication is perfectly obvious. The loss of simplicity is not due to nature, but to our imperfect knowledge of nature. The complications that arise are not simply a possible, but a necessary, result of our special assumptions. It must also be admitted that the cooperation of hidden masses, which is the remote and special case from the standpoint of our mechanics, is the most common case in the problems that occur in daily life and in technology. Hence it will be well to point out again that we have only spoken of appropriateness in a special sense, in the sense of a mind that endeavors to embrace objectively the whole of our physical knowledge without considering the accidental position of man in nature, and to set forth this knowledge in a simple manner.

The appropriateness of which we have spoken does not refer to practical applications or the needs of mankind. In respect to these latter it is scarcely possible that the usual representation of mechanics, which has been devised expressly for them, can ever be replaced by a more appropriate system.[38] Our representation of mechanics bears toward the customary one somewhat the same relation that the systematic grammar of a language bears to a grammar devised for the purpose of enabling learners to become acquainted as quickly as possible with what they will need in daily life. The requirements of the two are very different, and they must differ widely in their arrangement if each is to be properly adapted to its purpose.

* * *

In conclusion, let us glance once more at the three pictures of mechanics [*drei Bilder der Mechanik*] that we have discussed, and let us try to make a final and conclusive comparison between them. After what we have already said, we may leave the second picture out of consideration. We shall place the first and third pictures on an equal footing with respect to permissibility [*Zulässigkeit*], by assuming that the first picture has been thrown into a form completely satisfactory from a logical point of view. This we have already assumed to be possible. We shall also place both pictures on an equal footing with respect to appropriateness [*Zweckmässigkeit*], by assuming that the first picture has been rendered complete by suitable additions, and that the advantages of the two are equal, although their aims are different. We shall then have as our sole criterion the correctness [*Richtigkeit*] of the pictures; this is determined by the things themselves and does not depend on our arbitrary choice. And here it is important to observe that only one or the other of the two pictures can be correct; they cannot both be correct at the same time. For if we try to express as briefly as possible the essential relations of the two representations [*Darstellungen*], we come to this: The first picture [*Bild*] assumes as the final constant elements in nature the relative accelerations of the masses with reference to one another; from these it incidentally deduces approximate, but only approximate, fixed relations between their positions. The third picture assumes as the

---

[38][This admission of the practical superiority of the conventional system of mechanics over his own system colored the reception Hertz's *Mechanics* received. Many physicists were led to consider his *Mechanics* as beautiful, but useless.]

strictly invariable elements of nature fixed relations between the positions; from these it deduces (when the phenomena require it) approximately, but only approximately, invariable relative accelerations between the masses. Now, if we could perceive natural motions with sufficient accuracy, we should at once know whether in them the relative accelerations, or the relative positions of the masses, or both, are only approximately invariable. We should then know which of our two assumptions is false, or whether both are false, for they cannot both be correct simultaneously. The greater simplicity is on the side of the third picture. What at first induces us to decide in favor of the first is that in actions-at-a-distance we can actually exhibit relative accelerations that, up to the limits of our observation, appear to be invariable; whereas all fixed connections between the positions of tangible bodies are quickly perceived by our senses as only approximately constant. But the situation changes in favor of the third picture as soon as a more refined knowledge shows us that the assumption of invariable distance-forces only yields a first approximation to the truth, a case that has already arisen in the sphere of electric and magnetic forces. And the balance of evidence will be entirely in favor of the third picture when a second approximation to the truth can be attained by tracing back the supposed actions-at-a-distance to motions in an all-pervading medium[39] whose smallest parts are subjected to rigid connections; a case that also seems to be nearly realized in the same sphere.

This is the field in which the decisive battle between these different fundamental assumptions of mechanics must be fought out. But in order to arrive at such a decision it is first necessary to consider thoroughly the existing possibilities in all directions. To develop them in one special direction is the object of this treatise. Therefore this treatise has been needed, even if we are still far from a possible decision, and even if the decision should finally prove unfavorable to the picture here developed.

---

[39][" *in einem raumerfüllenden Mittel*"—by which Hertz means the ether. This is a further indication of the direction in which he hoped the *Principles of Mechanics* would take his research, toward the construction of an acceptable mechanical theory of the ether.]

PAPER NO. 17

Review of Hertz's *Principles of Mechanics* by George Francis FitzGerald

(G.F. FitzGerald, "The Foundations of Dynamics," *Nature* 51 [Jan. 17, 1895], pp. 283–285. Reprinted with permission from *Nature* 51. Copyright © 1895 Macmillan Magazines Limited.)

This posthumous volume of Hertz's works,[1] edited by Prof. Lenard, with a preface by von Helmholtz, has a doubly melancholy interest. It is the last work of Hertz upon which he was engaged until a few days before his death, and it contains a preface which is almost the last work of von Helmholtz. The pupil died shortly before his master, and by the departure of such a pupil and of such a master, science, and with science mankind, have lost many prospects of advances in the near future.

In his preface,[2] von Helmholtz pays a touching tribute to the genius of his favorite pupil, from whom he hoped most, and who had drunk most deeply of his master's thoughts. In 1878 their intimacy began. At that time difficulties connected with various electrical theories of action-at-a-distance were occupying his thoughts, and he offered a prize for the best essay on induction in non-inductively wound coils. Weber's[3] theory would have involved an inertia of the electric current distinct from the magnetic inertia. The question is still interesting in connection with discharges between two charged conductors, one of which completely surrounds the other, when a dielectric between them is suddenly made conductive. There is then

---

[1][FitzGerald's review is of the 1894 German edition of Hertz's *Mechanics*; the text given here is the slightly revised version of the review contained in FitzGerald (1902), pp. 325–337. There is another review of the *Mechanics* in *Nature* on May 17, 1900, this time of the 1899 English edition and signed A.E.H.L. This is by the British physicist and authority on the theory of elasticity, Augustus Edward Hough Love (1863–1940).]

[2][Paper No. 14.]

[3][Wilhelm Weber (1804–1891). His theory of electromagnetism dominated German research in electromagnetism from 1846, when his first paper on the subject appeared, till Hertz's time. For Weber's theory, see footnote 36 in the Introductory Biography.]

no magnetic force. Is there *no* inertia? Can a medium become *suddenly* conducting? Is a conducting medium homogeneous? Is there inertia of ionic charges which represent the non-homogeneity of the medium? These questions still require answering; but in the seventies, in Germany, Maxwell's idea of magnetic force accompanying displacement currents was not generally received, and Helmholtz's question as to the induction in non-inductively wound coils really had reference to these displacements. Hertz won the prize by showing that at most only 1/20th or 1/30th of the extra-current could be due to electric inertia. By subsequent experiments on the possible effect of centrifugal force on the current in rapidly rotating plates, he reduced this estimate to a very much smaller value. Mr. Larmor[4] has suggested that any centrifugal force may be balanced by a tension in the length of the current, much in the same way that the tension of a running rope will balance centrifugal force in the curves round which it may be running. In every way the subject deserves further investigation, for it is intimately connected with the most fundamental questions as to the nature of electricity and its connection with matter.

The next thing to which Hertz devoted himself was a prize problem proposed, at von Helmholtz's suggestion, by the Berlin Academy. The problem was to investigate Maxwell's postulate that changing electric displacement was an electric current. This was the bud from which Hertz's great work sprang. Of it von Helmholtz says: "It is a pity we do not possess more such histories of the inner psychological development of knowledge.[5] Its author deserves our sincerest thanks for letting us see so deeply into the inmost working of his thoughts, and for recording even his temporary mistakes. By this work Hertz has settled forever the question as to electromagnetic actions being propagated by a medium, and the only outstanding question of the kind is as to gravitation, which we do not yet know how to logically explain as other than a pure action-at-a-distance."[6] It thus appears that von Helmholtz to the last was unconvinced as to the

---

[4][Joseph Larmor (1857–1942), the British physicist well-known for the "Larmor precession" and for an electron theory similar to that of Hendrik Lorentz. Larmor also edited FitzGerald's collected writings; see FitzGerald (1902).]

[5][This comment of Helmholtz is directed to Hertz's Introduction to his *Electric Waves*, Paper No. 5.]

[6][See footnote 13 to Paper No. 14.]

FitzGerald's Review of Principles of Mechanics

probability of any hypothesis like LeSage's[7] or Osborne Reynolds's.[8] He seems, on the other hand, to have been satisfied with the possibility of chemical actions being explained either by electromagnetic actions or by actions-not-at-a-distance. This latter term, of course, requires explanation as to what "at a distance" means. Any actions other than those of absolutely rigid bodies, such, for instance, as the fairly well-established forces of attraction of gaseous molecules for one another, and some of which can hardly be explained either by electricity, magnetism, or gravitation, seem to be actions-at-a-distance that require explanation just as much as gravitation.

Following this short history of the work of his pupil which, coming from such a master, must have a permanent interest to all, von Helmholtz gives a *résumé* of the last work of Hertz. In it there is attempted a continuously elaborated presentation of a complete and self-dependent system of mechanics, in which each particular application of this science is deduced from a single fundamental law, which can of course be itself only assumed as a plausible hypothesis. In order to explain how this is required, von Helmholtz gives a short history of the development of the science of mechanics. The first developments arose from the study of the equilibrium and motion of solid bodies in direct contact with one another, such as the simple machines: the lever, the inclined plane, the pulley. The law of virtual velocities gives the most fundamental general solution of all such problems. Galileo subsequently developed the knowledge of inertia and of moving force as an accelerating agent. It was, however, conceived by him as a succession of blows. Newton was the first who arrived at the notion of force acting at a distance, and its more accurate determination by the principle of action and reaction. It is well known how strenuously he and his contemporaries resisted this idea of pure action-at-a-distance. From this, men developed the methods of treating all problems of conservative forces with constant connections, whose most general solution is given by D'Alembert's principle. All the

---

[7][George Louis LeSage (1724–1803), a French-Swiss physicist who in 1782 proposed to account for gravitation by means of an ether of Descartes' original type, consisting of a cloud of extremely minute corpuscles.]

[8][The Irish physicist Osborne Reynolds (1842–1912) had proposed a mechanical model of a hydrodynamical and yet elastic-solid ether. He employed diagrams and demonstrations to indicate how discrete grains suitably packed could constitute an elastic medium.]

general principles of dynamics have been developed from Newton's hypothesis of permanent forces between material points and permanent connections between them. It was subsequently found that these laws held even when these foundations could not be proved, and it was thence deduced that all the laws of nature agreed with certain general characteristics of Newton's conservative forces of attraction, although it was not found possible to deduce all these generalizations from one common fundamental principle. Hertz has devoted himself to discovering such a fundamental principle for mechanics, from which all the laws of mechanics hitherto known as universally valid can be deduced; and he has carried this out with great acuteness, and by means of a very remarkable presentation of a peculiarly general kinematic conception. In working it out, he returns to the oldest mechanical theories, and supposes all actions to be by means of rigid connections. Of course he has to assume that there are innumerable imperceptible masses and invisible motions of these, in order to explain the apparent actions upon one another of bodies that are not in immediate contact. Though he has not given examples of how this may be the case, he evidently builds his expectation of being thus able to explain natural actions upon the existence of cyclical systems, rollers, etc., with invisible motions. The justification of such an assumption can only be obtained by its success. Von Helmholtz concludes this interesting preface by remarking how English physicists have so often based their work on dynamical and geometrical suppositions, as for example Lord Kelvin and his vortex atoms, Maxwell and his cells with rotating contents. These physicists, he says, "have clearly been more satisfactorily helped by such illustrations than by the mere most general representations of the facts and their laws as given by the system of physical differential equations. I must confess that I have restricted myself to this latter method of investigation, and have felt most confidence therein; and indeed I might not have arrived at any important results by the methods which eminent physicists such as the three mentioned have employed."[9]

Although so far it seems as if there were very little to choose between the old methods of supposing that natural actions can be

---

[9] [The three physicists to whom Helmholtz refers here are, of course, Kelvin, Maxwell, and Hertz. The original English version of Hertz's *Mechanics* provides a much better translation of Helmholtz's words than does FitzGerald here. For this translation see the next-to-last paragraph of Paper No. 14.]

explained by conservative forces between molecules, and such a system of rigid connections, Hertz in his introduction shows that he is dissatisfied with the hypotheses of these forces as entities, while von Helmholtz, by his silence, seems to hold the view that the old method was good enough for him. Hertz's method has, however, the advantage of turning our attention to something definite to be investigated and invented, namely, the structure of these rigid connections. It is apparently very closely related to Osborne Reynolds's[10] and "Waterdale's,"[11] suggestions as to the structure of the ether, namely that it consists of perfectly rigid particles in almost complete juxtaposition which, whether by their smoothness or by their rolling upon one another, waste no energy in internal heat motions.

In his own preface,[12] Hertz says that he has culled many things from many minds; nothing particular in his work is new; what he presents as new is the arrangement and collocation of the whole, and the logical, or rather philosophical, aspect thereby attained.

To these prefaces there follows a long introduction,[13] in which Hertz reviews and criticizes the present foundations of dynamics. The great road by which this domain is now entered is one that was laid by Archimedes, Galileo, Newton, and Lagrange. It is founded on our notions of space, time, force, and mass. Force is introduced prior to motion, as the independent cause thereof. Galileo's notion of inertia only involved space, time, and mass. Newton first introduced all four notions. To this D'Alembert's principle gave the analytical method of treating generally connected systems. Beyond it, all is deduction. Here Hertz introduces a discussion as to the so-called forces of inertia. From his discussing the case of a solid subject to centripetal acceleration by

---

[10][See footnote 8 above.]

[11][In the 27 April 1893 issue of *Nature* (vol. 47), there appeared a book review, entitled "Dynamics *in Nubibus*," of a book called *Waterdale Researches: Fresh Light on Dynamics* (London: Chapman and Hall, 1892). Both the book author and the reviewer are anonymous. "Waterdale" proposed that the ether consisted of absolutely hard spheres of two different sizes. The reviewer dismisses the book as the work of a crackpot, but then proceeds to devote five pages to analyzing it! The reviewer was FitzGerald himself, as he admitted in a letter to Oliver Heaviside on 16 Aug. 1893: "I mentioned this in a review of a cracked creature 'Waterdale' in *Nature* a few weeks ago." (I am grateful to Bruce J. Hunt for this information.)]

[12][Paper No. 15.]

[13][Paper No. 16.]

means of a string, the question is much more intricate than if he had taken the case of a body falling freely under gravity, where the force is applied directly by the Earth to each point of the body, and not, as in the case of the string, distributed to each part by stresses in the solid. Hertz seems to consider that there is some outstanding confusion in applying the principle of equality of action and reaction, and appears to hold that by this principle the action on the body requires some reaction *in the body* whose acceleration is the effect of the force. He does not seem fully to appreciate that action and reaction are always on *different* bodies. From his consideration of this, and from a general review of our conception of force, he concludes that there is something mysterious about it, that its nature is a problem in physics, like the nature of electricity. We have a quite distinct conception of velocity; why not of force? He concludes that the mystery is not due to our not having enough ideas to associate with the word, but to our trying to put too much into it. These mysteries, however, do not invalidate in any way the deductions that have been made; they only require us to seek out a new foundation for our dynamics. He goes on to criticize this method of filling nature with forces of which, being ultimately between molecules, we can have no direct experience. A piece of iron on a table is acted on by gravitation, cohesion, repulsion, magnetic, electromagnetic, electric, and chemical forces. Some of these would drag it to pieces if not balanced to a nicety by others. Is this a sound view of nature? Can we not get a more attractive one?

A second view may be elaborated by making our fundamental quantities space, time, mass, and energy. There is no book in which this view of nature is fully and consistently worked out, at least none that Hertz was acquainted with. He sketches how it might proceed. Besides the postulate of the conservation of energy we require some definition of potential energy and experimental relations connecting it with space; and in addition we have a choice of relations with kinetic energy, of which Hertz suggests the choice of the integral form of Hamilton's principle known as that of least action. This is, no doubt, a recondite idea to use as a fundamental postulate, but it only implicitly involves the idea of force, which then comes in merely as a definition. To this method, which certainly has several great advantages, Hertz makes a number of objections. In the first place he objects that it requires the equations of connection to be integral equations, and we know such actions as pure rolling of one hard body on another cannot be so expressed. We must, in order to specify the subsequent motion, know the rate of rotation round the normal axis through the point of contact, and this cannot be specified except in terms of differentials. To such motions we cannot apply the proposed principle of least action, and yet we can hardly dispute that such rolling is possible in nature. If we treat it as the limit of frictional sliding, we introduce the whole of the difficulties of force, or of the irregular heat actions which have not yet been fully made amenable to accurate dynamical treatment. Again, difficulties arise as to the foundation of this method. There is great difficulty in specifying energy itself. How can it be satisfactorily measured without returning to the first method, and introducing the idea of force? Some have conceived of energy as a sort of substance; but when we try to form concrete conceptions of what is occurring, we get involved in perplexities. The very existence of two forms of energy [i.e., kinetic and potential] is a very serious difficulty. Again, it is doubtful whether it can be sound to consider the integral of least action as a *fundamental* principle. It makes the present depend on the future. It sets the problem to nature to make a certain integral the minimum.

A good many of these objections could be got over by making all energy kinetic, which is what Hertz himself practically assumes in his own method.

This third method begins by assuming only three fundamental quantities, time, space, and mass, and puts aside as non-fundamental, force and energy. In order to explain how nature works, we already do postulate invisible underlying structures in nature. We postulate these in the atoms and molecules of matter. Hertz sees in all actions the working of an underlying structure whose masses and motions are producing the effects on matter that we perceive, and what we call force and energy are due to the actions of these invisible structures, which he implicitly identifies with the ether.[14]

We must, however, assume certain connections between the three quantities, time, space, and mass. Between time and mass there is no direct connection. Space and mass, Hertz considers, are connected by the existence of a given mass at each point of space. He cannot mean here to assume a complete plenum, which would make

---

[14][In relating Hertz's "hidden masses" and "hidden motions" to the ether, FitzGerald here seems to put his finger on Hertz's chief motivation in writing his *Principles of Mechanics*.]

serious difficulties in the way of the working of what he subsequently assumes to be a structure of rigid bodies; he must include a vanishingly small density at some points, though perhaps he may have had in view the filling of the interstices between his rigid bodies with a fluid. Anyway, he goes on to say that some connection is required between all three quantities, and for this purpose he postulates his great fundamental single law of motion, which he considers is an extension to systems of Newton's first law of motion for a single body; it is that a system, which is unconnected with any others, moves with constant swiftness along one of its straightest paths. "*Systema omne liberum perseverare in statu [suo] quiescendi vel movendi uniformiter in directissimam.*"[15] In order to understand what Hertz here means by the path of a system, and by its being straight or curved, requires further explanation; but from this principle, which is capable of analytical representation, and from the assumption that the connections of a system are all rigid, he deduces all the fundamental principles, conservation of areas, momentum, energy, least action, etc. In considering the motion of any part of a system, we find that we may conveniently introduce certain actions of the other parts of the system upon it which are measured by forces, which thus come in as mere definitions. He does not seem to investigate anywhere the question as to the danger of his rigid connections becoming tangled. Analytically a postulate that the points of two different bodies that act on one another are in contact is easily expressed, but it does not follow that when we come to invent actual rigid connections to produce the observed effects, they will do so for any length of time without jamming. It is a seductive theory that gravitation or electrical actions may be due to vortex filaments ending on atoms; but the tangling of the filaments is a very serious difficulty that has not been satisfactorily got over.

---

[15][This Latin sentence is taken directly from Hertz's *Mechanics* (German edition, No. 309 on p. 162; English edition, p.144), but there seems to be a problem with Hertz's Latin: the word *perseverare* should really be *perseverat* if the sentence is to make sense. With this change, the Latin agrees perfectly with Hertz's German and with the English translation: "Every free system persists in its state of rest or of uniform motion along its straightest path." FitzGerald also incorrectly writes "quo" instead of the "suo" in the original text of the *Mechanics.*]

FitzGerald's Review of Principles of Mechanics

Hertz does not seem to feel this as a serious difficulty, but he does notice an obvious objection that is sure to be raised, namely, that rigidity in itself postulates forces. To this he replies that rigidity in itself is merely a matter of definition and of fact. How is our view of the fact that two points are at a constant distance apart, improved by saying that there is a force between them? As, however, real bodies are only imperfectly rigid, Hertz concedes that it may be that when we learn more about these invisible connections, they may turn out not to be absolutely rigid. It is a matter for further investigation. This very same view might have been urged, and has been urged already with reference to actions like gravity. The *law* of gravity can be perfectly well described without any reference to the notion of force. We may say, every element of matter moves towards every other element in the universe with an acceleration inversely proportional to the square of their distances apart. We can describe the law kinetically, just as Hertz proposes to describe the law of motion of parts of a rigid body. There is no *necessity*, however convenient it may be, to introduce the notion of force; the other bodies in the universe are a sufficient cause for motion of each, without postulating an entity, force. The principal reason for introducing this notion was to account for a body acting where it was not; force was invented to get over this; the body produced force, and this force existed where the body did not, and there acted on other bodies. This whole difficulty seems, however, to be partly due to want of distinct ideas connected with the question of where a body *is*. We are so accustomed to consider a body as having a definite boundary, that we think there is a definite boundary in reality. All we know of the atoms and molecules, however, would lead us to conclude that round the center of each there is a very complexly structured region which may or may not change abruptly in structure, but which often extends to considerable distances from the atom, so that it is practically impossible to state absolutely where the atom ends and where the empty space begins. With this view of matter there is no serious reason why we may not rightly consider each atom as existing everywhere that it acts, that is, throughout the whole of space, for its action in causing gravitational accelerations exists, so far as we know, throughout space. A view of this kind entirely gets over any difficulty of a body acting where it is not; for all bodies *are* everywhere, and if we consider matter to be the cause of motion of other matter, there seems no very imperious necessity for imagining another cause which we call force.

There are two assumptions that Hertz makes which he considers can only be proved by their success. One is that all the connections in nature can be represented by linear differential equations. There are plenty of cases imaginable in which this would not be true, as, for example, connections depending on the curvature of the path. The other assumption is that forces can be represented by force functions. This, again, may not be a complete representation of nature.

Following this introduction comes the book itself, which is divided into two parts. The first part is purely kinematical; the second deals with the deductions from Hertz's fundamental postulate of motion in the straightest possible path.

The first part begins by explaining what is meant by the path of a system of points. To get at this we calculate the mean square of the displacements of a system of points when they are displaced; the square root of this, Hertz calls the displacement of the system of points. If there is a mass at each point, then the displacement of the system is the square root of the mean squares of the displacements of the points, each multiplied by the mass at it. Thus, if s be the displacement of the system, and s_1, s_2, etc. the displacements of each point of masses m_1, m_2, etc., then

$$(m_1 + m_2 + \ldots)s^2 = m_1 s_1^2 + m_2 s_2^2 + \ldots$$

By taking s_1, s_2, etc., as the displacements in the element of time, we evidently get a similar expression for the velocity of the system, and for its acceleration. The mean square of the velocity of the parts of a system is well known in connection with the principle of least action. Further than this, however, Hertz defines the angle between two displacements. This is defined by the equation

$$(m_1 + m_2 + \ldots)ss' \cos \varepsilon = (m_1 s_1 s_1' \cos \alpha_1 + m_2 s_2 s_2' \cos \alpha_2 + \ldots)$$

s and s' being the two displacements of the system as calculated above, and s_1, s_1' etc., the two displacements of each point, and α_1, α_2, the angles between these latter; then ε is the angle between s and s'. Hertz remarks that these can all be very interestingly expressed in terms of a space of multiple dimensions, in which analytical diagrams are supposed to be drawn. This, however, represents the real by the unattainable. There follow, then, several chapters expressing these

FitzGerald's Review of Principles of Mechanics

displacements in terms of various systems of coordinates, and discussions as to the conditions that the connections of a system should fulfil in order that they may be represented by equations not involving differentials. The curvature of the path is here studied. It is defined as

$$c = \frac{d\varepsilon}{ds},$$

and from this it follows that, representing $\frac{d^2x}{ds^2}$ by x'', etc.,

$$(m_1 + m_2 + \ldots) c^2 = \Sigma_1^n m_1 \left(x_1''^2 + y_1''^2 + z_1''^2 \right).$$

The problem then of making the path of the system straightest, is to make c a minimum consistently with the connections of the system. Now, in accordance with his assumption that the connections of the system are linear differential equations of the form

$$\Sigma_1^n (P_1 x_1' + Q_1 y_1' + R_1 z_1') = 0,$$

whose differentiation gives

$$\Sigma_1^n P_1 x_1'' + \Sigma_1^n \Sigma_1^n \frac{dP_1}{dx_2} x_1' x_2' + \ldots = 0,$$

we are to determine the minimum value of

$$c^2 = \Sigma_1^n \frac{m_1}{m} \left(x_1''^2 + y_1''^2 + z_1''^2 \right),$$

where $m = m_1 + m_2 + \ldots$.

In determining the variations of these, we must recollect that the positions and directions of the displacement, i.e., the first differentials of the system, are supposed given, and that it is *only* the second differentials that can be varied in order to make c a minimum. Calling, then, a system of indeterminate coefficients λ, μ, etc., corresponding to the equations of condition, we evidently get a system of equations of the form

$$\frac{m_1}{m}x_1'' + \sum_1^n P_1 \lambda = 0,$$

which are sufficient to determine the second differentials required.

From this form of result one can see how the ordinary equations of motion are derivable from the conception of the straightest path, and how, when dealing with part of a system, these indeterminate coefficients introduce what are equivalent to forces. This method of deducing the equations of motion lends itself particularly well to the deduction of the principles of least action, and the other general methods in dynamics. So far, he deals with free systems subject only to internal constraints. It is where he investigates how to deal with parts of systems that he requires to consider the nature of the constraints joining one part to another. For this purpose he defines two systems as coupled when coordinates can be so chosen that one or more of them are the same for both systems. Force is then defined as the action one system has on another. Now, when a coordinate is the same for two systems, one of the equations of condition is $p = p'$, p and p' being coordinates of the coupled systems, and for this equation the coefficient P becomes the same in the two systems, except as to sign, so that the equations of motion involve the indeterminate coefficient λ corresponding to this equation equally with reference to each system. It is thus that the equality of action and reaction appears, being thus bound up with the constant equality of the common coordinate. This seems to be where the assumption that the connections are rigid is introduced. When rigid bodies act upon one another by non-slipping contact, certainly the coordinates of the point of contact are common to the two systems. It is also quite evident that if we assume rigid bodies acting upon one another by contact only, we can have no potential energy, and all *necessity* for talking about the forces disappears. In Hertz's system there are no forces like Newton's acting between bodies which have no common coordinate, like the Earth and the Sun. We would have to invent connections to explain the motion before we could be certain that action and reaction are equal in this case.

The proof of the principle of virtual velocities by substituting for the forces between parts of a system a number of pulleys which produce the same effects, is quite analogous to Hertz's supposition that the actual connections are by rigid bodies. It is not, however, liable to the objection that the connections may become tangled, for it is only applied to the case of infinitesimal virtual displacements, while Hertz postulates the possibility of his connections existing as the real ones for all time, and throughout all finite displacements of the system.

The work considers many other matters, and shows how all the general methods in dynamics are deducible from his fundamental postulate of the straightest path. It includes discussions on how best to deal with systems whose connections do not involve differentials, how to treat cyclical coordinates, and many other matters. It is most philosophical and condensed, and gives one of the most—if not the most—philosophical presentations of dynamics that has been published. It is worthy of its author; what more can be said?

PART VI

Heinrich Hertz's Importance in the History of Physics

> Hertz's life work is now complete; not a single sentence will he be able to add to it in the future. From now on science will go forward without him.... But no one who works in his field will be able to escape his influence; a thousand times more important than the results of his work are the seeds which he scattered abroad in his writings, and which, in the proper soil, can develop into new shoots.
>
> *Max Planck (1894)*

Bust of Heinrich Hertz in the entrance courtyard at the Technical University in Karlsruhe. The inscription reads, "In this place Heinrich Hertz discovered electromagnetic waves in the years 1885–1889." Courtesy of IBM Corporation.

Introduction

The last paper in this collection is Max Planck's Memorial Address of 16 February 1894. It summarizes Hertz's life and work and provides perceptive insights into his character; evident throughout is Planck's profound admiration for Hertz both as a physicist and as a man. Planck's address climaxed in a ringing tribute from one great physicist to another:

> Spoken or unspoken, the name of Hertz will be among the first of this generation, as long as men pay attention to electromagnetic waves. But we, the Physical Society, we will sun ourselves in the glow of his name, yes, we will share in its glory, for he was truly one of us.

At the conclusion of this address, Helmholtz thanked Planck profusely for his moving tribute to Hertz. Planck later called this "one of those thrilling moments" that he would treasure all the days of his life.

The death of Hertz at age thirty-six must have affected Helmholtz almost as deeply as that of his own son, Robert, just five years earlier. Helmholtz—the complete scientist, devoted father, and wise mentor—might have taken some comfort from the beautiful words of the American playwright Zoë Akins (1886–1958): "Nothing seems so tragic to one who is old as the death of one who is young, and this alone proves that life is a good thing."

PAPER NO. 18

Heinrich Rudolf Hertz: A Memorial Address by Professor Max Planck at the Meeting of the Physical Society of Berlin on 16 Feb. 1894[1]

Translated by Joseph F. Mulligan

This new year has begun with a death, a deeply-moving tragedy that has struck the physics community with overwhelming force. Heinrich Hertz, after completing highly-regarded research of outstanding quality, while still in the prime of life, actively engaged in demanding work and with great plans for the future, has fallen victim to an insidious disease. In him one of the leaders of our discipline, the pride and the hope of our country, has been committed to the grave.

The Physical Society does not merely mourn the death of a scientist; it has shared with him a more intimate, personal relationship. Often has he spoken in this place; often in the course of our debates has he contributed clarifying and stimulating remarks; hardly a single paper was written by him here that he did not present to this smaller audience beforehand; and, even after he left Berlin, he retained ties to our society as a nonresident associate member.

His whole career conformed to the simplicity and rectitude of his personal appearance. Heinrich Rudolph Hertz was born on 22 February 1857 in Hamburg, the eldest son of a lawyer who is at present Senator and head of that city's justice administration. He received his first instruction in a private school, as was customary in Hamburg, and then transferred into the *Prima* grade of the Hamburg classical *Gymnasium*, the *Johanneum*. Even as a youth he demonstrated a multitude of aptitudes, especially for scientific and technical matters, along with a remarkable memory. One of his favorite activities was work at the carpenter's bench or the lathe, where he constructed all sorts of instruments for his own use, for example, a complete spectroscope. He also showed talent in drawing and painting, and preserved an interest in botany.

---

[1] [Planck (1894); reprinted in Planck (1958), vol. 3, pp. 268–288.]

Very early on he directed his attention—as indeed every independent and ambitious young man in the exact sciences does—to the deepest problems of astronomy, physics and mathematics, in which he naturally left his fellow students far behind. But in other fields of learning, especially in languages, he also revealed both interest and remarkable ability. He pursued his classical studies with great enthusiasm, and could in later years recite from memory lengthy passages from Homer or the Greek tragedians. Yes, even in Sanskrit and Arabic, which in his thirst for knowledge he had begun to study, he moved along so rapidly that his private tutor strongly advised his father to allow him to study philology, certain that one day he would accomplish something truly outstanding in this field.

Still, these facts reveal only half of his total personality. The other side was a very unusual, often expressed, sense of duty that Hertz displayed from his childhood. This, combined with extraordinary intellectual ability and a happy sense of humor, provided a solid foundation for his later life's unfoldment.

When he left the *Gymnasium* at Easter 1875, with his certificate of matriculation,[2] he went to Frankfurt am Main with the intention of devoting himself to engineering. In Frankfurt he worked as an unpaid assistant in the Building Department of the city on the construction of a new bridge over the Main river. He studied a semester at the Polytechnic in Dresden, and then served his year's tour of military duty in Berlin with the Railroad Guards Regiment. In autumn 1877, he resumed his studies, this time in Munich. It was here, during his adjustment to university life, that he finally decided in favor of pure science; not that he had lacked a great attraction to science in the past—he had never made a secret of it—but because earlier, lacking sufficient experience, he had underestimated his abilities along these lines.

Hertz spent the last, most fruitful years of his studies in Berlin. Here first of all Hermann von Helmholtz, and secondly Gustav Kirchhoff, by their example and teaching had a lasting influence on his scientific thinking, even in small details. For this reason throughout his life he demonstrated toward these two professors a particularly warm affection. His first important research, carried out in this Institute (at

---

[2][The *Reifezeugnis*, or proof of maturity, required for admission to any German university.]

that time newly built), was stimulated by one of the prize problems posed by the Philosophical Faculty for the year 1879. After Hertz completed his research on the problem, he was awarded the prize.

This research dealt with the experimental investigation of a possible kinetic energy associated with a current of moving electric charges.[3] If one accepts the view of Wilhelm Weber that in an electric current two light, but still massy, fluids, the positive and negative electricities, flow in opposite directions with equally great speeds and the same density, so the kinetic energy of this current produces an expression that appears as an additional term in the electrokinetic energy. This energy is determined through the self-potential and has the same effect as if the self-induction coefficient calculated from this potential were increased by a constant term.

Hertz succeeded in proving, by measuring the induced current which flowed in two thick parallel wires, first in the direction of the primary current, and then in the opposite direction, that the kinetic energy of the induced electricity, if it is at all different from zero, must in any case be smaller than a specified quantity of very small magnitude. Of course, he still had to propose this result provisionally, with the reservation that the density of the electric fluid not be directly proportional to the specific conductivity of the metal conductors used. Then he chose an experimental procedure (employing a Wheatstone bridge) such that, if this special condition were indeed valid, any possible inertia of the induced electricity would make no difference in the measurement. He kept this subject in mind a while longer, and by the use of a clever technique was actually able not only to get rid of the above limitation, but to reduce still further the upper limits on the result.[4]

He set a horizontal metal plate in rotation about a vertical axis and had a constant current enter and leave the plate at two fixed points. If the flowing electricity possessed the least inertia, it would have to appear in a sideways displacement of the lines of current flow. This, however, could not be detected. One can get an idea of the smallness of the effect by taking the behavior of an electrolyte for comparison. Here an electric current is linked to the transport of

---

[3][*Misc.Pprs.*, pp. 1–34. The papers of Hertz that are included in his *Miscellaneous Papers* are also discussed by Lenard in Paper No. 1, and by FitzGerald in Paper No. 4, of the present collection.]

[4][*Misc.Pprs.*, pp. 137–145.]

matter; the speed of the ions is, of course, very small, and their kinetic energy is also small, since it is proportional to the moving mass and the square of the velocity; and yet the easily calculated kinetic energy is still very great compared to that of the electricity itself, so that the inertia of the electricity compared to that of the ions can be neglected.

In March 1889 Hertz *promoviert*[5] with a theoretical dissertation on induction in solid conducting spheres, when they rotate between two magnets.[6] He solved the differential equations derived for the problem from Neumann's theory by expanding the induced potential in spherical harmonics. For small speeds of rotation, the self-induction can be completely ignored; for high rotation speeds, however, the self-induction is so large that the current in the inner layers [of the sphere] completely disappears and pulls back to the surface layer of the conductor.

Hertz, who by this time had been made an assistant in the Physical Institute, promptly published as an appendage to this research an essentially theoretical investigation of the distribution of electricity on the surface of a moving conductor, in this case a rotating sphere.[7]

This he followed with a treatment of the rotation of a sphere not in a magnetic, but in an electrostatic field. The resulting phenomena are essentially determined by the ratio of the specific electrical resistance of the conductor (measured electrostatically) to the time of rotation (a pure number). Two limiting cases were realized in nature: for metals the resistance is extremely small, but for insulators it is extremely great, compared to the time of rotation. In these two special cases no damping of the rotation occurs: in the first case because the potential inside the metal is everywhere constant and so no potential difference exists to drive the current; in the second case because no current exists at all. In general, however, damping does occur, because the electricity in the conductor constantly tends toward new configurations, and the current therefore experiences Joule heating. To demonstrate this, Hertz set a horizontal needle, which carried at its ends two tiny horizontal measuring plates, into horizontal torsional vibration above and very close to a badly-conducting horizontal glass

---

[5][That is, Hertz received his D.Phil. degree.]

[6][*Misc. Pprs.*, pp. 35–126.]

[7][*Misc. Pprs.*, pp. 127–136.]

plate. He in fact obtained a very remarkable increase of the damping at the instant that he charged the tiny plates electrostatically.

Fewer positive results were obtained from the next research project that Hertz carried out on the evaporation of mercury in a vacuum.[8] He distilled heated mercury vapor through a vacuum over into a condensing flask at a constant low temperature of about zero degrees centigrade. It occurred to him to write the speed of the evaporation as a fixed function of the temperature of the fluid surface and the vapor pressure. The complete carrying out of his plan, however, was undermined by the complicated conditions of the process and the difficulty in measuring the temperature and the pressure, which depended on these conditions. To indicate just one problem: he had at the beginning surmised that the pressure of the distilled mercury vapor in the whole room would be equal to that of the saturated vapor at the coldest spot in the room, i.e., at zero degrees centigrade—a result that would have to be valid if the pressure were everywhere the same. But closer observations revealed that this assumption was not even approximately valid. The temperature in the vicinity of the evaporation surfaces showed great differences from place to place, corresponding to the heat absorption occurring at these places. Finally he contented himself with setting fixed limits between which the observed phenomena occurred. Despite these problems, he was able to carry out, for a mercury-vapor vacuum pump, an involved calculation of the elasticity of the saturated mercury vapor at low temperatures, using the general principles of thermodynamics.[9]

At about this time he worked on problems in the theory of elasticity, beginning with some research on the contact of solid elastic bodies.[10] He demonstrated that the theory agreed completely with the conditions of equilibrium for two elastic bodies that had been pressed together, with respect to both the deformation and the strain; in particular, he showed that the surface of contact, a small surface of the second degree (which he called the pressure surface), was bounded by an ellipse whose dimensions increased as the cube root of the force with which the bodies were pressed together. He completed a number of similar pieces of research, of which only one may be mentioned

---

[8][*Misc. Pprs.*, pp. 186–199.]
[9][*Misc. Pprs.*, pp. 200–206.]
[10][*Misc. Pprs.*, pp. 146–162.]

here, since it contained a definition of the hardness of an object.[11] He proposed as the measure of hardness the normal pressure which existed at the center of the circular pressure surface of a body just exactly when the elastic limit is reached. Of course, this definition suffers from the same uncertainty that usually attends the determination of the elastic limit.

Not long afterwards he devoted himself again to experiments in his favorite field, that of discharge processes in gases, for which he appropriately suggested some very fruitful areas of research. A characteristic phenomenon, which occurred when a spark was passed through a dry gas at reasonably low pressure, surprised him; with further investigation this proved to be only a mechanical vortex of incandescent gas.[12] He then investigated more thoroughly the processes of the glow discharge, using a constant battery source.[13] The first question was this: Is the glow discharge always intermittent, as has been established in many cases as undoubtedly true when constant-voltage sources are used, or is it in many cases also rigorously continuous? By the application of ever more refined methods he was able finally to reach the conclusion that there are glow discharges which, if they are intermittent, occur at least two billion times per second, so that the laws of probability seem to indicate that they are truly continuous. A second question concerns the light emitted by the cathode. Do the cathode rays mark the path of the current? Indeed, has the current anything directly to do with the cathode rays? And, if not, what path do the current lines, i.e., the lines of the actual charges, take? The first part of the question, which appears related in some degree to the known influence of a magnet on the cathode rays, had to be answered in the negative after many different experiments; after he had further established that a magnet does affect the cathode rays, but the cathode rays do not affect a magnet, he was able to determine the location of the current lines in special cases by the exploration of the discharge field with a small magnetic needle. These current lines certainly differ from the cathode rays, since in some places they are directed at right angles to them.[14] According to this observation the

---

[11][*Misc.Pprs.*, pp. 163–183.]

[12][*Misc. Pprs.*, pp. 216–223.]

[13][*Misc.Pprs.*, pp. 224–254.]

[14][Some of Hertz's observations in this paper were adversely affected by residual gas in the cathode-ray tube due to the inability of his vacuum system to

current lines and the cathode rays are entirely independent one from the other; the working of a magnet on the cathode rays is not the Hall Effect, but something to be compared with the magnetic rotation of the polarization of light [the Faraday Effect].

Finally from the Berlin period we must mention a series of smaller experimental projects, with which Hertz busied himself from time to time. These included: the construction of a hygrometer,[15] whose principle rested on the increase in weight that calcium chloride experiences when it absorbs water vapor; an electrodynamometer,[16] which measured the intensity of an alternating current by the thermal expansion of a silver wire through which the current flowed (he used a torsion device for these measurements, which he also used for his research on the glow discharge); and finally some research on the behavior of benzene as an insulator and as a producer of residual charge.[17]

With regard to the newer viewpoint that the residual charge of a substance always can be traced back to defects in homogeneity in its interior, the idea came to Hertz to find an undoubtedly homogeneous body that nevertheless displayed residual charge. Pure commercial benzene, which he considered suitable for this purpose, did not satisfy this requirement; for as long as it showed a noticeable residual charge, it proved to be impure. Later in the Strassburg laboratory of Leo Arons further research has made it highly probable that the charge residue on the inside of a substance in fact arises in all cases from a defect in homogeneity, namely from the variation through space of the ratio between electrical conductivity and dielectric constant.

---

reach a low enough pressure. Later experiments by others achieved better vacuums and led to a revision of these observations. On this see the last few pages of FitzGerald's review of Hertz's *Miscellaneous Papers* (Paper No. 4).]

[15][*Misc.Pprs.*, pp. 184–185.]

[16][*Misc.Pprs.*, pp. 211–215.]

[17][*Misc.Pprs.*, pp. 255–260. In this paper Hertz shows that the so-called "residual charge" is an "after-effect of polarization." There is no net charge on the dielectric used, but there is a residual polarization of the dielectric after it is subjected to an electric field, making one end of the dielectric positive and the other negative. After the field is removed, in almost all cases the molecules of the dielectric gradually "relax" and the residual polarization disappears. The exceptions, called "electrets," are materials that can be permanently polarized in an electric field. They are the electric equivalents of permanent magnets.]

Each of the works just described, all of which were completed in a space of three or four years, reveal clearly the remarkable experimental technique, the richness of ideas and the basic training [of Hertz]—even allowing for the fact that the results obtained did not always correspond to the insight and hard work he dedicated to the project. Above all, they reveal the unusual degree of self-criticism of the experimenter, so that even today their study will furnish stimulation and instruction to every budding physicist. In this regard it is noteworthy how many of his propositions he clothed in negative terms: "There is no kinetic energy of induced electricity"; "The glow discharge is not always discontinuous"; "The cathode rays do not mark the path of the current," etc. It appears from this that he was not primarily interested in winning external fame by the discovery of new, astonishing facts—in many cases the results could have been expected to some degree on the basis of previous knowledge—but much more by employing versatile, unassailable and quite general methods to achieve the necessary clarity, and along with it, a correct grasp of the nature of the processes involved.

Linked to this approach is his reluctance to be satisfied with the statement of simple propositions. Rather, by the setting of limiting values, he established the extent to which the proposition would be valid when the conditions of the problem were slowly changed through a wide range of variables. This, indeed, is the mark of the real scientific experimenter. It is certainly more convenient (and it does sound better) to set forth a proposition simply as universally valid with no statement of limits on its validity, while keeping at one's disposal for later eventualities the limits that have been passed over in silence. This Hertz never did; he never avoided making such calculations for limiting values, calculations which are difficult to carry through both from a physical and a mathematical point of view because of the many different influences requiring consideration. Such calculations demand the broad outlook of a versatile, well-trained physicist.

In 1883 Hertz *habilitiert* at the University of Kiel and at the same time was appointed professor of theoretical physics.[18] Because of the

---

[18][*Habilitation* indicated that a scholar had submitted an additional piece of research (in Hertz's case his research on cathode rays) beyond his work for the doctorate, and that this work had been defended and approved. This made him a *Privatdozent*, with the right to teach at a university. Planck is wrong about Hertz being appointed professor of theoretical physics. His appointment

nature of this appointment and the inadequacy of the facilities he had for experimental research during this period, his efforts were directed more along theoretical lines. The ever-changing views of the beautiful Kiel harbor presented a continuous array of problems to a physicist, and Hertz often availed himself of the opportunity to take a trip around the harbor in a steam or sail boat with a group of colleagues of about his own age. His inner drive toward unifying his knowledge of nature led him, therefore, to carry out research projects derived from the natural world outside, and he began about this time to busy himself more earnestly with meteorological studies.

Earlier in his life Hertz had now and then studied the action of the stars on the tides;[19] now he worked out a graphical method for the determination of the adiabatic expansion of damp air.[20] One clearly obtains here two kinds of graphs from whose shapes a good overview of the desired variation can be obtained; for as long as the air is not saturated with water vapor, its adiabatic expansion is that of a perfect gas, with a graph of the first kind, up to the instant when saturation occurs. After that, on further expansion the air remains saturated, but a corresponding quantity of vapor precipitates out; as a result heat is generated and a change of state occurs, which is represented by a graph of the second kind.

While in Kiel Hertz published some studies which had essentially been completed in Berlin. These involved the equilibrium of floating elastic plates, for example, a flat sheet of ice with a weight at its center and floating on water—a problem which has interest in many different connections.[21] The complete solution of this problem on the basis of the general equations of the theory of elasticity teaches us the following: If the plate is of infinite extension, the weight in the middle produces an elastic indentation, and round about it an elevation. The final result is not a gradual return to the sheet's undisturbed level, but a remarkable periodic array of raised and lowered regions, whose

---

was as a *Privatdozent* to teach courses in theoretical physics, with a promise that this appointment might be changed to that of an *Extraordinarius*, or associate professor, in a few years. On this see Jungnickel and McCormmach (1986), vol. 2, pp. 46–47.]

[19][This work is in *Misc.Pprs.*, pp. 207–210.]

[20][*Misc.Pprs.*, pp. 296–312. On the importance of this work to meteorologists, see Garber (1976).]

[21][*Misc.Pprs.*, pp. 266–272.]

heights of course fall off quickly as they go out from the center. Even more noteworthy, the buoyancy of the water, produced by the depression created, is always exactly equal to the added weight, independent of the thickness and the specific gravity of the plate. Also, a large ice-sheet, even if it is very thin, can support that weight situated at its center, as long as the elastic limit is not exceeded; in other words, the limits of its bearing strength are determined not by its lightness but by the strength of the ice.

If one assumes that the plates are of limited area, the following paradoxical conclusions immediately arise. A limited-size plate, which has a specific gravity greater than water, will obviously sink if it is laid horizontally on a water surface. If one loads the plate sufficiently in the middle, however, then by virtue of the depression produced there the plate will float, and the greater the load, the more easily it floats, assuming that the plate does not break apart. If one gradually removes the load, the ability to float is continuously reduced and at a certain definite limit the plate sinks, together with the remainder of the load.

In this manner Hertz followed every problem through to its complete solution. Mathematical difficulties never frightened him off; he refused to admit such a possibility even in principle. From the beginning a physics problem must be reduced to its purest, simplest form by ignoring all extraneous complications that only impede the mathematical treatment. "Everything works out mathematically," he used to say, "if only the problem is properly posed." To be able to do this, one must indeed not only be a mathematician but also a physicist. If his analysis brought him to a result that was unexpected or contradicted accepted ideas, he did not hesitate for an instant to accept the logical consequences [of his analysis] as alone correct; and therefore accepted ideas had to be revised. Again, if the result of the calculation disagreed with an observed fact, his conviction of the necessary agreement of the laws of nature with those of human logic so ruled Hertz's life that such a situation placed him in a condition of the greatest disquiet. Then it often happened that he shut himself off totally from the outside world for hours at a time, entirely lost in his own thoughts (perhaps whistling a melody that grew and fell away again) until finally the error was discovered and he was again at peace with himself.

In this same period [at Kiel] Hertz began again to study electrodynamics—a field that henceforth he would never again abandon. He began with a paper on the relationship between Maxwell's equations and the fundamental equations of the opposing school of electrodynamics (that of Weber and Neumann).[22] This is a theoretical work of the first magnitude, which is fully deserving of being placed beside his later works in this field. Up to this time the conflict between the above-mentioned theories had been sought only in the differences in their predictions about the behavior of open currents. Hertz showed that it was possible to find, even in the case of uniform closed currents, a crucial point on which a decision can be made between the two theories. This point is the unity of the electric force, and also of the magnetic force.[23]

If only an electric force exists, and if this force, with which a rubbed ebony rod attracts or repels an electrically-charged pith ball, is the same as the force with which a moving or otherwise changed magnet induces an electric current in a conductor, then this same magnet must be able to set a charged pith ball in motion; then also the reverse must happen according to the mechanical principle of action and reaction: an electrostatically-charged body must work on a varying magnet to move it, and so finally a varying magnet must act on another varying magnet (in addition to the usual magnetic interaction) to move it, with an electric force that depends on the relative motion of the magnets, i.e., on the changes in their magnetism.

Now, the electrodynamics built up [by Weber and Neumann] on the basis of action-at-a-distance recognizes only such ponderomotive interactions between magnets that depend on their magnetism, but not on the variations of that magnetism in time. Thus it appears that this theory, as developed from the standpoint assumed here, is incomplete. The addition of the relevant missing term yields a fixed correction that is in general very small, since it contains the square of the speed of light in the denominator. But one cannot rest with this result. Besides the correction for the ponderomotive interaction, the principle of conservation of energy requires a correction for the inductive interaction. But since the induced forces should be identical

---

[22][Paper No. 3. The two German physicists referred to here are Wilhelm Weber (1804–1891) and Franz Neumann (1798–1895). For the contributions of Weber and Neumann to electrodynamics, see footnote 36 in the Introductory Biography.]

[23][For Hertz's ideas about "the unity of the electric force," and "the unity of the magnetic force," see the first few pages of Paper No. 3.]

in character to the ponderomotive forces, there results a new correction for the ponderomotive interaction, and so there is no end to the process. Indeed, every time the proper correction is introduced, one obtains not only for the ponderomotive, but also for the inductive, interaction of both the electric and the magnetic kind an infinite series, which contains terms with [increasing even powers of the reciprocal][24] of the light velocity and therefore, in general, converges. The remarkable thing is that this series satisfies perfectly the differential equations obtained by Maxwell for electromagnetic effects, according to which these effects travel through space with the speed of light. This peculiar derivation of the Maxwell theory from the assumption of a direct action-at-a-distance theory is naturally not to be taken as a proof of this theory, because no exact result can be obtained from an inexact assumption.[25] But [Hertz's] discussion was only intended to show that the assumption of an instantaneous action-at-a-distance, even if it yields a strong convergence for slowly varying effects, is basically irreconcilable with the proposition of the unity of the electric force (and also of the magnetic force), a proposition that is plausible because of its basic simplicity.

Maxwell's theory, however, did not display these defects. There can scarcely be any doubt that these considerations of Hertz weighed heavily in support of Maxwell's theory, even if it could not have been easy for him or for any German physicist to be content with the furthering of man's understanding by the abandonment of long-accepted theories [of German physicists]. To be as certain as possible about this result, Hertz for a time formulated his theoretical work from the standpoint of a theory developed by Helmholtz, which included all other theories as special cases.[26]

---

[24][Planck's text has been modified here (p. 279 in the German text), since it does not correspond to the equations of Hertz that he is discussing. Planck writes "... unendliche Reihen, die nach absteigenden geraden Potenzen der Lichtgeschwindigkeit fortschreiten...." He really means not "decreasing even powers of c," but "increasing even powers of $1/c$." This is clear from Hertz's equations in Paper No. 3. See, for example, the equation following the one numbered (10) in that paper, where Hertz has previously defined A as $1/c$ in footnote 4.]

[25][Hertz makes the same point himself in his paper.]

[26][See Helmholtz (1882–1895), vol.1, pp. 545–628; and Jungnickel and McCormmach (1986), vol. 2, pp. 22–26.]

A paper which appeared shortly after the one just discussed contained a minor contribution to the much-discussed question at that time of the dimensions of the electric and magnetic quantities.[27]

Despite the successes he had achieved during this period of theoretical research in Kiel, the longer Hertz remained there, the more strongly he was drawn to his favorite kind of work, experimental physics. He had outfitted a previously unused room in his living quarters as a kind of laboratory, containing some rather primitive instruments; he intended to carry out some researches on thermoelectricity there with the permission and support of the director of Kiel's Physical Institute, Professor Gustav Karsten. But just then he received a call to a prestigious position as director of his own Physical Institute at the *Technische Hochschule* in Karlsruhe. As soon as he could, in spring 1885, Hertz moved to Karlsruhe.

In Karlsruhe Hertz became acquainted with Elisabeth, the charming daughter of his colleague Doll, the Professor of Geodesy; some months later he brought her home as his wife. (His wife and two dependent daughters now mourn his death.) Then began the great epoch of Hertz's life, marked by work on electric waves that resulted in a constant stream of research publications. In this work he forced nature to reveal phenomena that no man before him had ever observed. An obvious question can of course be raised as to how much the success of this work was due to the design and subsequent completion of a definite plan, and how much to a fortunate coincidence of unforeseen external circumstances. Prescinding from the fact that the attempt to penetrate more deeply into nature's secrets seldom works out badly for the curious investigator who has that wonderful combination of ability and luck, the question is in this case altogether pointless. For, as far as any sort of definite answer can be given, it has already been provided by Hertz in the Introduction to his book on *Electric Waves*.[28] The situation is pretty much the same here as in all long-lasting undertakings: the external circumstances affect the progress of the work often in helpful, but just as often in unhelpful, ways. As an example of the latter we have the perturbations that the electric waves under observation experienced from the nearby bodies in the laboratory room. In the beginning these perturbations proved a

---

[27][*Misc.Pprs.*, pp. 291–295.]
[28][Paper No. 5.]

great obstacle to the correct interpretation of the observations; very probably they were the cause of the only error of critical importance that Hertz made in this newly embraced field of research, i.e., his conclusion that the electric waves travelled more slowly on wires than in free space. Later the same perturbations would be seen as a stroke of luck for Hertz because they brought to him the idea of producing standing waves by reflection from a wall—an idea that perhaps might not have occurred to him so quickly in a larger, less crowded laboratory. Therefore the external circumstances by themselves produced erratic results, sometimes hindering, sometimes fostering, the research effort, so that the sum total of their influence was neither negative nor positive, but zero. It was much more [Hertz's] intellect, however, which made the terms of this sum individually useful, which first gave to each term, as it were, the proper sign, and in this way led to an observable positive result.

For the exploitation of this new field of phenomena, two conditions had to be satisfied: first, the production of waves that varied so rapidly in time that their wavelength in air could be easily measured—for the most rapidly varying [high-frequency] waves up to that time had been observed by Feddersen, and these yielded wavelengths of the order of magnitude of a kilometer—and secondly, the devising of an instrument that could serve for their detection. Both problems Hertz solved in his publications on very fast [i.e., high-frequency] electric waves,[29] the first through the discovery that a spark discharge between two spheres under the proper conditions could produce very rapid free vibrations of the system of conductors consisting of the spheres and any nearby conductors. (In this he had been anticipated 17 years earlier by Wilhelm von Bezold, but without Hertz's knowing it.[30]) The second problem was solved through the discovery that the principle of resonance was also applicable to these electric vibrations. For Hertz the existence of resonance vibrations in a tuned secondary circuit would be the tool with which he would explore the [electromagnetic] field in the neighborhood of the original vibrating system. The determination of the properties of this field then made smooth the road to all the results that followed.

---

[29][The first of these was our Paper No. 6.]

[30][When this was brought to Hertz's attention, he was very apologetic and insisted on including Bezold's paper in his *Electric Waves* (pp. 54–62).]

In the beginning Hertz sought to obtain clarity about the complexities of the observed properties [of the electromagnetic field] by assuming a separate electrodynamic and a separate electrostatic force, which propagated with different speeds; later he recognized that this separation was unnecessary and in general even impossible, and that a complete clarification of all observed phenomena could be achieved if, in accord with Maxwell's ideas, one spoke no longer of electrostatic and electrodynamic forces, but simply of the electric field.

A striking but peripheral phenomenon led Hertz for a short time down a side alley. This was the effect that the primary spark had on the occurrence of the secondary spark. He succeeded in making giant strides toward clarifying the nature of the effect. This work,[31] reported in the true spirit of Faraday, can, when looked at by itself, be considered the model for the experimental handling of a new discovery. After he had found that it was only the ultraviolet radiation from the primary spark that acted on the secondary spark gap, he handed over the further investigation of this phenomenon to other researchers and pressed forward again with his main line of research.[32]

To be sure we have fresh in our memories how at that time paper followed paper in quick succession, piling up new observations and expanding our knowledge. We learned that electrical processes in insulators can also produce electrodynamic effects; that electromagnetic waves, which travel through air, interfere with those that travel through wires in different ways and in different places; also that the waves in air do not have infinitely-long wavelengths, i.e., do not propagate at infinite speed. We learned that, through reflection of electric waves by a conducting wall, standing waves can be produced and so the wavelengths can be directly measured;[33] and finally that electric waves propagate entirely in the same way as light waves and

---

[31][This work, the first observation of the photoelectric effect, is described in Paper No. 7.]

[32][Here Planck does not seem to give Hertz sufficient credit for recognizing the importance of the photoelectric effect, and for the six months of intense research activity he devoted to it. The discovery was indeed accidental, but Hertz recognized immediately its importance as a possible road to the better understanding of light and electricity. That is why he took six months out of his electromagnetic wave research to investigate it.]

[33][Paper No. 8.]

satisfy the same laws of reflection, polarization and refraction.[34] In brief, light waves with all their characteristics increased a million times over, have been observed in nature.[35] And the proof for all this was obtained through tiny little sparks, which must be looked at in partial darkness with a magnifying glass in order to observe them at all. What research scientist does not remember even today the feelings of wonder and amazement that came over him at the first news of these discoveries, the laws of which make no distinction between the great and the small; but we were also startled by the amazing abstractive power of the human intellect, which can combine the most cogent logical thought with the imagination of a true artist.

German scientists expressed their gratitude to Hertz at the convention at Heidelberg in the autumn of 1889, when Hertz delivered a popular address on the relationship between light and electricity.[36] In this talk he compared Maxwell's theory to a bridge that spanned with a bold arc the wide chasm between the regions of optical and electromagnetic phenomena, of molecular and cosmic wavelengths. Hertz at that time continued to urge that it should be through rapidly-varying electric waves that a new solid ground should be obtained where a chasm existed before, and that on this ground a solidly-based pillar should be erected for further support of the bridge. Since that time this pillar has been raised and broadened by outstanding research of many kinds; the bridge stands today stronger and prouder than ever; it serves no longer, as it once did, to carry only isolated clever speculations on occasional excursions; no, today the bridge can bear the heavy traffic of exact research, and carry its secrets unceasingly from one research field to another, and by so doing enrich both.

But not only the scientists, but the entire civilized world on both sides of the ocean turned their attention to this research. Hertz's name was soon on everyone's lips: talks were delivered and articles written about him; learned societies elected him a member or bestowed other honors on him; princes invited him to be their guest—but he always remained his old self, simple, scholarly, a true friend of his friends, a humble and grateful pupil of his former teachers; not acting out of

---

[34][Paper No. 9.]

[35][Planck means primarily that the wavelength has been increased a million times over, from the 500 nm of visible light to the 50 cm of electromagnetic waves of the kind Hertz produced.]

[36][Paper No. 10.]

cunning calculation, but from a natural disposition in which the highest intellectual cultivation was joined to the deepest feeling. His modesty was the expression of his inner being; he considered his accomplishments as simply the natural outcome of an inner drive, and as a matter of course he desired to avoid making any fuss.[37] It is no wonder that mistrust and detraction always remained far from such a disposition as his. When the successes of these years of his public life came to an end, there resulted a decline in a certain reserve, which to him as a basically aristocratic person was proper in his dealing with men, and which now made way for a more approachable kindliness as he grew in maturity.

Heartfelt and truly touching was his childlike respect for his teacher Helmholtz, a respect he displayed again and again at every opportunity, with an evident satisfaction that he had received the great blessing of being able to see clothed in personal form his own enthusiasm for the value and the truth of science. During his last sojourn in Berlin, at the Helmholtz celebration in November 1891, when he spoke at the banquet in the imperial palace as the representative of Helmholtz's students, he was granted the

---

[37][A striking example of Hertz's modesty and simplicity despite all his honors was the letter he sent to the editor of *The Electrician* on 22 June 1890 in response to a request for permission to publish a portrait of Hertz in that journal:

Dear Sir:

You will oblige me very much by postponing for a year, or better, for two years, the kind intention you have in respect to my portrait. I feel as if presenting my portrait now in so prominent a place follows rather too quickly the little work I have done. I should like to wait a little, and see if the general approbation which my work meets with is of a lasting kind. Too much honor certainly does me harm in the eyes of reasonable men, as I have sometimes occasion to observe. If your kind intention is the same in two years, or even in one year, I shall readily consent and help you in every respect.

With many kind thanks, and the assurance that I feel only too much honored. Yours truly.

H. Hertz

After Hertz's death this letter was published in *The Electrician* 32 (Friday, 12 Jan. 1894), p. 261.]

opportunity to give eloquent expression to his feelings.[38] In scientific questions he paid no regard to personalities; for him only facts and principles mattered, and they could originate from any conceivable source. He always showed the same objective kindness with respect to both the most learned remark, and the most naive, as long as it was sincerely intended. He could only be caustic and intolerant toward an approach to research that unfortunately is all too common today. This is research based on personal convenience and hurried work habits, which leads only to confusion.

In the year 1889 Hertz was called to fill the Professorship vacated by Clausius in Bonn.[39] This new position gave him for the first time duties of a different kind. In the Bonn Physical Institute a total reorganization was imperative: additional space was required, new equipment had to be purchased, and the work of the assistants more carefully supervised. Hertz's practical approach soon achieved some success, as demonstrated by a series of worthwhile pieces of research performed in the institute under his direction.[40] He even found time in the midst of his teaching duties to do some research himself. Previously he had, in an unusual experiment, obtained direct proof that his electrical waves, if they passed through wires, travelled not through the inner portion of the metal, but entirely in the outer layer of the wire; now he was able to bring under investigation not merely the electromotive but also the mechanical effects of these waves.[41]

Hertz more recently carried out a further investigation of cathode rays, to which thin metal sheets unexpectedly proved to be transparent, even if light rays were completely absorbed; on the other hand, substances transparent [to light] were entirely impermeable to the cathode rays.[42]

At this time [1890] Hertz devoted his attention chiefly to the theoretical elaboration of Maxwell's theory, in which project a

---

[38][The content of our Paper No. 2 contains a similar "eloquent expression" of Hertz's feelings for Helmholtz.]

[39][Rudolf Clausius died in 1888. He was an outstanding theoretical physicist, but weak on the practical side of running a physics institute. He also had little interest in experimental physics.]

[40][For example, by Philipp Lenard and Vilhelm Bjerknes. On Lenard and Bjerknes, see Part V of the Introductory Biography.]

[41][*El.Waves*, pp. 186–194.]

[42][Paper No. 13.]

colleague, who was at the same time lecturing on this topic, provided welcome stimulation.[43] He attempted to remove from the theory all those concepts that were not necessary for the complete description of the physical phenomena, but only served for the facilitation of the presentation or for purposes of calculation. Thus he retained, in general, only two variables: the electric force and the magnetic force[44], which at any instant of time fully determined the electromagnetic state of a region of space. The variations in these forces [fields] are tied together by certain differential equations, in which (aside from a few constants) only derivatives with respect to time and space occur. For most important phenomena in bodies at rest, a simple system of equations is derived in this way;[45] for moving bodies, on the other hand, there arises the question whether the velocity state of the substance at each point is completely determined by the velocity component of the ponderable matter, or whether at the beginning a special velocity must also be assigned to the light-ether.[46] Many accepted facts appeared to suggest such independent movement of the ether, but this subject had not been looked into enough to guarantee a sound basis for such an assumption. Hertz therefore decided to prescind entirely from this complication and assume that the ether participated fully in the motion of the ponderable matter. Then the ether velocity falls out of the theory, and there is no further need to consider the ether at all. Even if these equations [of Hertz] are not completely general, they provide a self-consistent formulation of electromagnetism and optical phenomena; many a theoretical physicist will still find it rewarding to pursue further the riches of their content in specialized fields of science.[47]

---

[43][The identity of this colleague is not certain, but Planck seems to be referring to Hermann Lorberg, who came to Bonn in 1890 as *Extraordinarius* in theoretical physics. Lorberg had published a number of papers on electrodynamics and lectured on that subject at Bonn. On this see Jungnickel and McCormmach (1986), vol. 2, p.36.]

[44][Today we would say more correctly: "the electric field and the magnetic field."]

[45][*El.Waves*, pp. 195–240.]

[46][*El.Waves*, pp. 241–268.]

[47][In addition to Planck, famous physicists like Einstein, Boltzmann and Sommerfeld found great substance and stimulation in these two theoretical papers published by Hertz in Bonn in 1890.]

Only for the class of electrochemical phenomena does Maxwell's theory not seem to surpass significantly the other theories of electricity. Hertz therefore concerned himself with this field only briefly, for, as he once wrote to me, he had not yet come to a satisfactory overall view of the subject. Here again we see at work the often-noted experience that the complete description of a phenomenon is only possible if it takes place over distances for which the bodies can be considered as continuous, while for events in the molecular world still unresolved questions always remain.

These investigations and his need to obtain a still deeper foundation for his understanding of nature drove his restless spirit toward an involvement with the general principles of mechanics. In this work,[48] which he completed at the end of his life with the expenditure of his last bit of bodily strength, he addressed once again the experts in the field. But even here he did not wish to rest. He proposed more new experiments, this time with currents driven by unusually high voltages, so high that those close to him worried about the danger inherent in the use of such voltages.[49]

But other events intervened. For the first time, in the summer of 1892 strange symptoms of disease began to appear, although up till that time he had enjoyed good health.[50] These symptoms included swelling of the nose and earaches, perhaps related to an infected tooth. While at first his malady was thought to be harmless, there was no notable improvement with treatment; rather, his difficulties increased as time went on. Finally an operation behind the ear was judged necessary, which led to the removal of a mass of pus collected in the

---

[48][Heinrich Hertz, *The Principles of Mechanics Presented in a New Form.* It is disappointing that Planck has so little to say about a book to which Hertz had devoted the last three years of his life, especially after Planck has devoted pages to far less important papers from Hertz's early days in Berlin.]

[49][The reference here is uncertain, but may be to Hertz's desire to use higher voltages for cathode-ray research. Once, in Berlin, he had constructed with his own hands one thousand batteries, to be combined in series to produce a voltage of close to 2,000 volts for use in studying cathode rays.]

[50][Planck was wrong about this. Hertz had never been a strong and healthy man. According to his mother, he was born "half-dead" and needed special care and nourishing food to survive. Later both at Kiel and in Karlsruhe he had a variety of problems with his eyes, ears, teeth and throat. On this see the section on the "Final Illness and Death" of Hertz in Part V of the Introductory Biography.]

ear-bone. With this done, it was hoped that the source of the malady had been removed. But the poison remained in his system and again produced a multitude of abscesses; additional operations on his upper jawbone brought about general improvement for a while, but were not able to bring the illness to an end. A vacation on the Riviera in the spring, and another one in Reichenfall in the fall of the preceding year, helped restore his bodily strength and mental outlook for a short while.

All this time his friends and colleagues were sincerely concerned for his condition, and greeted with pleasure any news of an improvement in his health. Still, by the beginning of the winter, unsettling rumors began to circulate; only reluctantly and in whispers would they be discussed in the presence of his friends; one would not, could not, face that most dreadful possible outcome. And still the same forces of nature, which had once revealed themselves to him because he had penetrated their invariable laws, now in accordance with the same unavoidable laws demanded his life and, together with his body, destroyed so many thoughts still dormant in his mind. On 7 Dec. [1893] he was forced to cancel his lectures, on which he had worked with the greatest possible energy up to that time. The last weeks he suffered increasing, and at the end unspeakable, pain, the whole time being fully conscious, until at last on the first day of the New Year he was released from his misery. An autopsy was not performed; the doctors gave blood-poisoning as the cause of death.

Hertz's life work is now complete; not a single sentence will he be able to add to it in the future. From now on science will go forward without him; whatever he would have perhaps been permitted to discover, that will—of this we have no doubt—be found sooner or later by others. But no one who works in his field will be able to escape his influence; a thousand times more important than the results of his work were the seeds which he scattered abroad in his writings and which, in the proper soil, can develop into new shoots.

Spoken or unspoken, Hertz's name will be among the first of this generation, as long as men pay attention to electric waves. But we, the Physical Society, we will sun ourselves in the glow of his name, yes, we will share in its glory, for he was truly one of us.

Bibliography

Part A: Books, Articles, and Addresses by Heinrich Hertz

Hertz, Heinrich, *Erinnerungen • Briefe • Tagebücher*. First (German) edition; edited by Johanna Hertz. Leipzig: Akademische Verlagsgesellschaft, 1927.

———, *Memoirs • Letters • Diaries*. Second (German-English) edition; arranged by Johanna Hertz; edited by Mathilde Hertz and Charles Susskind; with a biographical introduction by Max von Laue. Weinheim: Physik Verlag, and San Francisco: San Francisco Press, 1977.

Except for the personal letters and diaries in the above volumes, which were edited by Hertz's two daughters, almost all other published works of Hertz are contained in the three volumes of his collected works:

Hertz, Heinrich, *Gesammelte Werke*

Band I: *Schriften vermischten Inhalts* (edited, and with an introduction, by Philipp Lenard). Leipzig: J.A. Barth, 1895.

Band II: *Untersuchungen über die Ausbreitung der elektrischen Kraft*. Leipzig: J.A. Barth, 1892.

Band III: *Die Prinzipien der Mechanik in neuem Zusammenhange dargestellt* (edited by Philipp Lenard; with a Preface by Hermann von Helmholtz). Leipzig, J.A. Barth, 1894.

There are unabridged and unaltered reprints of these three volumes in German, published by Sändig Reprints Verlag, Vaduz, Liechtenstein (1987–1988).

Hertz, Heinrich, *Collected Works* (English translation of Hertz's *Gesammelte Werke*)

Vol. 1: *Miscellaneous Papers* (translated by D.E. Jones and G.A. Schott). London: Macmillan and Co., 1896.

Vol. 2: *Electric Waves, Being Researches on the Propagation of Electric Action with Finite Velocity through Space* (translated by D.E. Jones; with a Preface by Lord Kelvin). London: Macmillan and Co., 1893. *Reprint:* Dover Publications, New York, 1962.

Vol. 3: *The Principles of Mechanics Presented in a New Form* (translated by D.E. Jones and J.T. Walley). London: Macmillan and Co., 1899. *Reprint:* Dover Publications, Inc., New York, 1956 (with a new introduction by Robert S. Cohen).

For completeness, the titles, dates and journals in which the articles in the *Collected Works* were originally published are given below. In this listing the following abbreviations will be used:

Annalen: Both Poggendorff's and Wiedemann's *Annalen der Physik und Chemie*
Verhandlungen: *Verhandlungen der physikalischen Gesellschaft zu Berlin*
Sitzungsberichte: *Sitzungsberichte der Berliner Akademie der Wissenschaft*

Papers included in the present collection are indicated by an asterisk before the number of the paper.

Vol. 1: Miscellaneous Papers

1. "Experiments to Determine an Upper Limit to the Kinetic Energy of an Electric Current," *Annalen* 10 (1880), 414–448.

2. "On Induction in Rotating Spheres," *Inaugural Dissertation*, Berlin, March 15, 1880.

3. "On the Distribution of Electricity over the Surface of Moving Conductors," *Annalen* 13 (1881), 266–275.

4. "Upper Limit for the Kinetic Energy of Electricity in Motion," *Annalen* 14 (1881), 581–590.

5. "On the Contact of Elastic Solids," *Journal für die reine und angewandte Mathematik* 92 (1881), 156–171.

6. "On the Contact of Rigid Elastic Solids and on Hardness," *Verhandlungen des Vereins zur Beförderung des Gewerbefleisses* 61 (1882), 449–463.

7. "On a New Hygrometer," *Verhandlungen*, conference of January 20, 1882.

8. "On the Evaporation of Liquids, and Especially of Mercury, *in vacuo*," *Annalen* 17 (1882), 177–193.

9. "On the Pressure of Saturated Mercury Vapor," *Annalen* 17 (1882), 193–200.

10. "On the Continuous Currents Which the Tidal Action of Heavenly Bodies Must Produce in the Oceans," *Verhandlungen*, conference of January 5, 1883.

Bibliography

11. "Hot-Wire Ammeter of Small Resistance and Negligible Inductance," *Zeitschrift für Instrumentenkunde* 3 (1883), 17–19.

12. "On a Phenomenon Which Accompanies the Electric Discharge," *Annalen* 19 (1883), 78–86.

13. "Experiments on the Cathode Discharge," *Annalen* 19 (1883), 782–816.

14. "On the Behavior of Benzene with Respect to Insulation and Residual Charge," *Annalen* 20 (1883), 279–284.

15. "On the Distribution of Stress in an Elastic Right Circular Cylinder," *Zeitschrift für Mathematik und Physik* 28 (1883), 125–128.

16. "On the Equilibrium of Floating Elastic Plates," *Annalen* 22 (1884), 449–455.

*17. "On the Relations between Maxwell's Fundamental Electromagnetic Equations and the Fundamental Equations of the Opposing Electromagnetics," *Annalen* 23 (1884), 84–103.

18. "On the Dimensions of Magnetic Pole in Different Systems of Units," *Annalen* 24 (1885), 114–118.

19. "A Graphical Method of Determining the Adiabatic Changes of Moist Air," *Meteorologische Zeitschrift* 1 (1884), 421–431.

*20. "On the Relations between Light and Electricity"—A Lecture delivered at the Sixty-Second Conference of German Scientists and Physicians in Heidelberg on 20 Sept. 1889 (first published by Emil Strauss in Bonn in 1889).

*21. "On the Passage of Cathode Rays through Thin Metallic Layers," *Annalen* 45 (1892), 28–32.

*22. "Hermann von Helmholtz: On the 31st of August, 1891," Supplement to the *Münchener Allgemeine Zeitung* for August 31, 1891.

Vol. 2: Electric Waves

*1. "Introduction"—written by Heinrich Hertz in 1892 to accompany the first German edition of his *Electric Waves*.

*2. "On Very Rapid Electrical Oscillations," *Annalen* 31 (1887), 421–448; 543–544.

3. (W. von Bezold: "Researches on the Electric Discharge.")

*4. "On an Effect of Ultraviolet Light on the Electric Discharge," *Sitzungsberichte*, 9 Juni 1887; *Annalen* 31 (1887), 983–1000.

5. "On the Action of a Rectilinear Electric Oscillation on a Neighboring Circuit," *Annalen* 34 (1888), 155–170.

6. "On Electromagnetic Effects produced by Electrical Disturbances in Insulators," *Sitzungsberichte*, 10 Nov. 1887; *Annalen* 34 (1888), 273–285.

7. "On the Finite Velocity of Propagation of Electromagnetic Actions," *Sitzungsberichte*, 2 Feb. 1888; *Annalen* 34 (1888), 551–569.

*8. "On Electromagnetic Waves in Air and Their Reflection," *Annalen* 34 (1888), 610–623.

9. "The Forces of Electric Oscillations, Treated According to Maxwell's Theory," *Annalen* 36 (1889), 1–22.

10. "On the Propagation of Electric Waves by Means of Wires," *Annalen* 37 (1889), 395–408.

*11. "On Electric Radiation," *Sitzungsberichte*, 13 Dec. 1888; *Annalen* 36 (1889), 769–783.

12. "On the Mechanical Action of Electric Waves on Wires," *Annalen* 42 (1891), 407–415.

13. "On the Fundamental Equations of Electromagnetics for Bodies at Rest," *Annalen* 40 (1890), 577–624.

14. "On the Fundamental Equations of Electromagnetics for Bodies in Motion," *Annalen* 41 (1890), 369–399.

Vol. 3: Principles of Mechanics

This is the only one of the three volumes of Hertz's *Collected Works* that is not a collection of previously published articles. It was written during the last three years of Hertz's life and consists of two parts:

Book I. The Geometry and Kinematics of Material Systems
Book II. The Mechanics of Material Systems

The only parts of Hertz's *Principles of Mechanics* included in the present collection are Helmholtz's Preface, the Author's Preface, and Hertz's long and important Introduction.

Some Other Publications by Heinrich Hertz

Abstract by Hertz of his Inaugural Dissertation: "On Induction in Rotating Spheres"; translated from Wiedemann's *Beiblätter*, No. 8 (1880), pp. 622–624, and appearing in *Phil.Mag.* 10 (1880), 451–452.

"Recherches sur les ondulations électriques," *Archives des Sciences physiques et naturelles* 21 (April 1889), 281–308. A summary of Hertz's researches on electromagnetic waves, especially prepared for the *Archives* at the invitation of Professor Sarasin in Geneva.

Starting in 1880 and continuing until 1892, the *Philosophical Magazine*, the *Journal de Physique*, and the *American Journal of Science* contain abstracts of many of Hertz's papers from the *Annalen* during that period.

Bibliography

Reprints, Translations and Collections of Hertz's Papers

English Translations

"On the Propagation of Electric Waves by Means of Wires," *Phil.Mag.* 28 (1889), 117–127; (Paper No. 10 in *Electric Waves*).

"On Electric Radiation," *Phil.Mag.* 27 (1889), 289–298; (Paper No. 11 in *Electric Waves*).

"On the Relation between Light and Electricity," *American Journal of Physics* 25 (1957), 335–343. A new translation of Hertz's Heidelberg address with an introduction by E.C. Watson (Paper No. 20 in *Miscellaneous Papers*).

German Books and Articles

Gustav Hertz (editor), *Über sehr schnelle elektrische Schwingungen.* Leipzig: Geest und Portig, 1971 (Vol. 251 of *Ostwald's Klassiker der exakten Wissenschaften*). This contains the original German versions of Papers No. 2, 8 and 11 in *Electric Waves*, plus Hertz's 1889 Heidelberg address.

Josef Kuczera (editor), *Die Prinzipien der Mechanik.* Leipzig: Geest und Portig, 1984 (Vol. 263 of *Ostwald's Klassiker der exakten Wissenschaften*). This volume contains the German text of all the preliminary materials to Hertz's *Mechanics* (but not the two parts of the main text), with an introduction and notes by Josef Kuczera, plus Hertz's 31 Aug. 1891 tribute to Helmholtz.

Robert Rompe und Hans-Jürgen Treder (editors): *Zur Grundlegung der theoretischen Physik; Beiträge von H. von Helmholtz und H. Hertz.* Berlin: Akademie Verlag, 1984. Included in this volume are Hertz's Preface and Introduction to his *Principles of Mechanics*, with notes by the editors.

E. Wildhagen (editor): *Raum und Kraft.* Berlin: Deutsche Buch-Gemeinschaft, 1932. A selection of the more accessible papers of Helmholtz and Hertz; contains Hertz's introductions to his *Electric Waves* and to his *Mechanics*, and his Heidelberg address.

Drei Bilder der Mechanik: Einleitung zu *Die Prinzipien der Mechanik* von Heinrich Hertz. Mit Beiträgen von Franz Wolf und Helmut Hönl (*Physikalische Schriften*, Heft 7). Mosbach/Baden: Physik Verlag, 1959. Contains Hertz's Introduction to his *Principles of Mechanics*.

Part B: Books, Articles, and Addresses Relating to Hertz, His Life, and His Work

Aitken, Hugh G.J. 1985. *Syntony and Spark—The Origins of Radio.* Princeton: Princeton University Press.

Andrade, E.N. da C. 1947. "Obituary of Professor P. Lenard," *Nature* 160: 895–896.

Appleyard, Rollo. 1930. *Pioneers of Electrical Communication.* London: Macmillan and Co. *Reprint:* Books for Libraries Press, Freeport N.Y., 1968.

Barker, Peter. 1980. "Hertz and Wittgenstein," *Stud. Hist. Phil. Sci.* 11: 243–256.

Bevilacqua, Fabio. 1984. "H. Hertz's Experiments and the Shift Towards Contiguous Propagation in the Early Nineties," *Rivista di Storia della Scienza* 1: 239–256.

Beyerchen, Alan. 1977. *Scientists under Hitler.* New Haven: Yale University Press.

Bjerknes, Vilhelm. 1923. *Untersuchungen über elektrische Resonanz.* Leipzig: J.A. Barth.

Blackmore, John T. 1972. *Ernst Mach. His Work, Life and Influence.* Berkeley: University of California Press.

Blanchard, J. 1938. "Hertz, the Discoverer of Electric Waves," *Proceedings of the I.R.E.* 26: 505–515.

Boerger, G. ed. 1988. *Heinrich Hertz. Commemorative Essays on the 100th Anniversary of His Pioneering Experiments on Electromagnetic Waves.* Berlin: Heinrich-Hertz-Institut.

Boltzmann, Ludwig. 1890. "On the Experiments of Hertz," *Phil. Mag.* (5) 30, 126–127.

———. 1974. *Theoretical Physics and Philosophical Problems: Selected Writings* (edited by Brian McGuinness). Boston: D. Reidel.

Bonfort, Helene. 1894. "Sketch of Heinrich Hertz," *Report of the Smithsonian Institution for 1894,* pp. 719–726. Washington, D.C.: Government Printing Office. Reprinted from *Popular Science Monthly* 45, No. 3 (July, 1894).

Bonn University. 1895. *Chronik der Rheinischen Friedrich-Wilhelms Universität zu Bonn,* 20 (1894–95).

———. 1958. *In Memoriam Heinrich Hertz* (Addresses to the Bonn Faculty on 22 Jan. 1958 by Maximilian Steiner, Walther Gerlach, and Wolfgang Paul; also included are letters from Hertz to Aloys Schulte for the years 1889–1892, edited by Max Braubach). Bonn: Peter Hanstein Verlag.

———. 1970. *150 Jahre Rheinischen Friedrich-Wilhelms Universität zu Bonn, 1818–1968*. Bonn: H. Bouvier and Ludwig Röhrscheid Verlag.

Boorse, Henry A. and Lloyd Motz. eds. 1966. *The World of the Atom*. Two vols. New York: Basic Books.

Bopp, Fritz and Walther Gerlach. 1957. "Heinrich Hertz zum hundersten Geburtstag am 22 Feb. 1957," *Die Naturwissenschaften* 44: 49–52.

Bordeau, Sanford P. 1982. *Volts to Hertz... The Rise of Electricity*. Minneapolis: Burgess Publishing Co.

Born, Max. 1962. *Einstein's Theory of Relativity*. New York: Dover Publications. (This is a revised and enlarged version of the first English edition published by the Methuen Co. in 1924.)

Braithwaite, Robert Bevan. 1953. *Scientific Explanation*. Cambridge: Cambridge University Press.

Brocke, Bernhard vom. 1991. "Friedrich Althoff: A Great Figure in Higher Education Policy in Germany," *Minerva* 29 (No. 3): 269–293.

Brown, Sanford C. 1979. *Benjamin Thompson, Count Rumford*. Cambridge, Mass.: MIT Press.

Bryant, John H. 1988. *Heinrich Hertz: The Beginning of Microwaves*. New York: Institute of Electrical and Electronics Engineers.

Buchwald, Jed Z. 1985a. *From Maxwell to Microphysics: Aspects of Electromagnetic Theory in the Last Quarter of the Nineteenth Century*. Chicago: University of Chicago Press.

———. 1985b. "Modifying the Continuum: Methods of Maxwellian Electrodynamics," in Harman (1985), pp. 225–241.

———. 1990. "The Background to Heinrich Hertz's Experiments in Electrodynamics," in T.H. Levere and W.R. Shea. eds. *Nature, Experiment and the Sciences* (Vol. 120 of Boston Studies in the Philosophy of Science), pp. 275–306. Boston: Kluwer Academic Publishers.

———. 1992. "The Training of German Research Physicist Heinrich Hertz," in M.J. Nye *et al.* eds. *The Invention of Physical Science*. Boston: Kluwer Academic Publishers.

Cahan, David. 1989. *An Institute for an Empire: The Physikalisch-Technische Reichsanstalt, 1871–1918*. New York: Cambridge University Press.

Cassirer, Ernst. 1950. *The Problem of Knowledge* (translated by William H. Woglom and Charles W. Hendel). New Haven: Yale University Press.

Cath, P.G. 1957. "Heinrich Hertz (1857–1894)," *Janus* 46: 141–150.

Cazenobe, Jean. 1980. "Comment Hertz a-t-il eu l'idée des ondes hertziennes?" *Revue de Synthèse* 101: 345–382.

———. 1982. "Les incertitudes d'une découverte: L'onde de Hertz de 1888 a 1900," *Archives Internationale d'Histoire des Sciences* 32: 236–265.

———. 1983. "La viseé et l'obstacle: Étude et documents sur la 'préhistoire' de l'onde hertzienne." *Cahiers d'Histoire et de Philosophie des Sciences* 5 (248 pp.). Paris: Centre de Documentation Sciences Humaines.

———. 1984. "Maxwell, Précurseur de Hertz?", *Recherche* 15, 972–986.

Cohen, Robert S. 1956. "Hertz's Philosophy of Science: An Introductory Essay," in Dover reprint edition of Hertz's *Principles of Mechanics* (no page numbers).

Cooke, Roger Marvin. 1974. *Non-Objectivity in Classical Mechanics: An Essay in the Foundations of Mechanics from the Viewpoint of Heinrich Hertz.* Unpublished Ph.D. Dissertation, Yale University.

D'Agostino, Salvo. 1971. "Hertz and Helmholtz on Electromagnetic Waves," *Scientia* 106: 622–648 (Italian, 622–636; English 637–648).

———. 1975. "Hertz's Researches on Electromagnetic Waves," *Hist. Stud. Phys. Sci.* 6: 261–323.

———. 1989. "Pourquoi Hertz et non pas Maxwell a-t-il découvert les ondes électriques?", *Centaurus* 32: 66–76 (1989)

———. 1990. "Boltzmann and Hertz on the *Bild*-Conception of Physical Theory," *History of Science* 28: 380–398.

Darrigol, Olivier. 1993. "The Electrodynamic Revolution in Germany as Documented by Early German Expositions of 'Maxwell's Theory'." *Archive for History of Exact Sciences*, vol. 45, no. 3, pp. 189–280.

Dilthey, Wilhelm. 1961. "Anna von Helmholtz," in Pechel, Rudolf. ed.: *Deutsche Rundschau: Acht Jahrzehnte deutschen Geisteslebens*, pp. 261–272. Hamburg: Rütten and Loening.

Doll, Max. 1894. "Mittheilungen über Heinrich Hertz," *Verhandlungen des Naturwissenschaftlichen Vereins zu Karlsruhe* 11 (1888 bis 1895), pp. 355–359.

Doncel, Manuel G. 1991. "On the Process of Hertz's Conversion to Hertzian Waves," *Arch.Hist.Exact.Sci.* 43, 1–27.

Doran, Barbara Guisti. 1975. "Origin and Consolidation of Field Theory in Nineteenth-Century Britain," *Hist.Stud.Phys.Sci.* 6, 133–260.

Dugas, René. 1955. *A History of Mechanics* (translated by J.R. Maddox). Neuchatel, Switzerland: Editions du Griffon.

Ebert, Hermann. 1894. "Gedächtnis für Heinrich Hertz," *Sitzungsberichte der physik-med. Societät zu Erlangen*, 7 März 1894. An English translation by James L. Howard appeared in *The Electrician* 33 (1894), pp. 272–274; 299; 332–335.

Einstein, Albert. 1979. *Autobiographical Notes* (translated and edited by Paul A. Schilpp). Centennial Edition. La Salle, Illinois: Open Court Publishing Co.

———. 1987. *The Collected Papers of Albert Einstein.* Vol. 1. *The Early Years (1879-1902)*, (John Stachel *et al.*, eds.). Princeton: Princeton University Press.

Fahie, J.J. 1899. *A History of Wireless Telegraphy, 1838-1899.* London: William Blackwood and Sons.

FitzGerald, George Francis. 1891. "Hertz's Experiments," *Nature* 43, 536-538; 44, 12-14, 31-35. Reprinted in *Annual Report of the Smithsonian Institution for 1892*, pp. 293-327. Washington, D.C.: Government Printing Office, 1893.

———. 1893. Review of H. Hertz's *Electric Waves*, in *Nature* 48 (Oct. 5, 1893): 538-539.

———. 1895. Review of H. Hertz's *Principles of Mechanics*, in *Nature* 51 (Jan. 17, 1895): 283-285.

———. 1896. Review of H. Hertz's *Miscellaneous Papers*, in *Nature* 55 (Nov. 5, 1896): 6-9.

———. 1902. *The Scientific Writings of George Francis FitzGerald* (edited by Joseph Larmor). London: Longmans, Green and Co.

Friedburg, Helmut. 1988. "Die Karlsruher Experimente von Heinrich Hertz," *Fridericiana* (Zeitschrift der Universität Karlsruhe) 41: 39-57.

Friedman, Robert M. 1982. "Constituting the Polar Front," *Isis* 73: 343-362.

———. 1989. *Appropriating the Weather: Vilhelm Bjerknes and the Construction of a Modern Meteorology*. Ithaca, N.Y.: Cornell University Press.

Fuchs, Walter Robert. 1969. *Knaurs Buch der Modernen Physik* (with an introduction by Max Born). Munich: Droemer.

Garber, Elizabeth. 1976. "Thermodynamics and Meteorology (1850- 1900)," *Annals of Science* 33: 51-65.

Gerhard-Multhaupt, Reimund. 1988. "Hertz's Experimental Confirmation of Maxwell's Theory between the Years 1886 and 1889," in Boerger (1988), pp. 41-67.

Gerlach, Walther. 1958. "In Memoriam Heinrich Hertz," in Bonn University (1958), pp. 13-38.

———. 1968. "Heinrich Rudolf Hertz, 1857-1894," Sonderdruck aus Bonn University (1970).

Gillispie, Charles Coulston. ed. 1970-1978. *Dictionary of Scientific Biography*. 15 vols. New York: Scribner's.

Goldstein, Eugen. 1921. "Erinnerungen eines Laboratoriumspraktikanten," *Die Naturwissenschaften.* 9: 708–711.

———. 1925. "Aus vergangenen Tagen der Berliner Physikalischen Gesellschaft," *Die Naturwissenschaften* 13: 39–45.

Great Soviet Encyclopedia, article "Heinrich Hertz," vol. 6, pp. 596–597.

Grigor'ian, Ashot Tigranovich and A.N. Wjalzen. 1968. *Genrikh Gerts* [Heinrich Hertz Omnibus Volume]. Moscow: Verlag der Akademie der Wissenschaften der USSR.

Hamburg University. 1957. *Gedenkfeier der Freien u. Hansestadt Hamburg am 24 Feb. 1957* (contains addresses on H. Hertz by H. Raether, H. Wenke, and Gustav Hertz). Hamburg: Kayser and Konrad.

Harman, P.M. 1982. *Energy, Force and Matter: The Conceptual Development of Nineteenth-Century Physics.* Cambridge: Cambridge University Press.

———. 1985. ed. *Wranglers and Physicists. Studies on Cambridge Mathematical Physics in the Nineteenth Century.* Manchester: Manchester University Press.

Havas, P. 1966. "A Note on Hertz's 'Derivation' of Maxwell's Equations," *Am.J.Phys.* 34: 667–669.

Heilbron, J.L. 1986. *The Dilemmas of an Upright Man: Max Planck as Spokesman for German Science.* Berkeley, California: University of California Press.

Heimann, Peter M. 1971. "Maxwell, Hertz, and the Nature of Electricity," *Isis* 62: 149–157.

Helmholtz, Hermann von. 1847. "The Conservation of Force: A Physical Memoir," in Kahl (1971), pp. 3–55.

———. 1882–1895. *Wissenschaftliche Abhandlungen.* Three vols. Leipzig: J.A. Barth.

———. 1892. "The Principle of Least Action in Electrodynamics," *Sitzungsberichte der Berliner Akademie der Wissenschaft* (12 Mai 1892).

———. 1899. "Preface by H. von Helmholtz" to H. Hertz's *Principles of Mechanics.* London: Macmillan and Co.

Hermann, Armin. 1969. "Heinrich Rudolf Hertz, Physiker," *Neue Deutsche Biographie,* 8: 713–714. Berlin: Duncker and Hemblot.

———. 1977. "Wissenschaftspolitik und Entwicklung der Physik im Deutschen Kaiserreich," in Gunter Mann and Rolf Winau. eds. *Medizin, Naturwissenschaft, Technik und das Zweite Kaiserreich.* Göttingen: Vandenhoeck und Ruprecht, pp. 52–63.

Hertz, Gerhard. 1988. "Heinrich Hertz. Persönliche und historische Hintergründe der Entdeckung," *Fridericiana* (Zeitschrift der Universität Karlsruhe), 41: 3–37.

Hertz, Gustav. 1957. "Die Entdeckungen von Heinrich Hertz und ihre Auswirkungen," in Hamburg University (1957), pp. 21–34.

———. 1971a. ed. *Über sehr schnelle elektrische Schwingungen* (Vol. 251 of Ostwald's Klassiker der exakten Wissenschaften). Leipzig: Geest und Portig.

———. 1971b. "Heinrich Hertz, sein Leben und sein Werk," in 1971a, pp. 9–31.

Hesse, Mary B. 1955. "Action at a Distance in Classical Physics," *Isis* 46, 337–353.

———. 1962. *Forces and Fields: The Concept of Action at a Distance in the History of Physics*. New York: Philosophical Library.

Hirosige, Tetu. 1966. "Electrodynamics before the Theory of Relativity," *Japanese Stud.Hist.Sci.* 5: 1–49.

———. 1969. "Origins of Lorentz's Theory of Electrons and the Concept of the Electromagnetic Field," *Hist.Stud.Phys.Sci.* 1: 151–209.

Hoepke, Klaus-Peter. 1988. "Die Universität Fridericiana Karlsruhe und Hertz," *Fridericiana* (Zeitschrift der Universität Karlsruhe) 41: 59–79.

Hon, Giora. 1987. "'The Electrostatic and Electromagnetic Properties of the Cathode Rays Are Either *Nil* or Very Feeble' (1883): A Case Study of an Experimental Error," *Stud. Hist. Phil. Sci.* 18: 367–382.

Hunt, Bruce J. 1983. "Practice vs. Theory: The British Electrical Debate, 1888–1891," *Isis* 74: 341–355.

———. 1991. *The Maxwellians*. Ithaca, N.Y.: Cornell University Press.

Janik, Allan and Stephen Toulmin. 1973. *Wittgenstein's Vienna*. New York: Simon and Schuster.

Jones, D.E. 1894. "Heinrich Hertz," *Nature* 49: 265–266.

Jungnickel, Christa and Russell McCormmach. 1986. *Intellectual Mastery of Nature: Theoretical Physics from Ohm to Einstein.* 2 vols. Chicago: University of Chicago Press.

Kahl, Russell. ed. 1971. *Selected Writings of Hermann von Helmholtz*. Middletown, Conn.: Wesleyan University Press.

Karlsruhe Technical University. 1988. *Heinrich Hertz Symposium*: "100 Jahre Elektromagnetische Wellen." See Wiesbeck, W.

Kayser, Heinrich. 1895. "Das physikalische Institut," in Bonn University, pp. 39–40.

———. 1936. *Erinnerungen aus meinem Leben* (a 344-page unpublished typescript. Copies are in the libraries of the American Philosophical Society and of Harvard University.)

Kirchhoff, Gustav. 1877. *Vorlesungen über mathematische Physik.* Vol.1, *Mechanik.* 2nd edition. Leipzig: B.G. Teubner.

Kirsten, Christa and Hans-Günther Körber. eds. 1975. *Physiker über Physiker* (Wahlvorschläge zur Aufnahme von Physikern in die Berliner Akademie, 1870 bis 1929). Berlin: Akademie Verlag.

———. 1979. *Physiker über Physiker II* (Antrittsreden, Erwiderungen bei der Aufnahme von Physikern in die Berliner Akademie, Gedächtnisreden, 1870 bis 1929). Berlin: Akademie Verlag.

Klein, Martin J. 1970. *Paul Ehrenfest. Vol.1, The Making of a Theoretical Physicist.* Amsterdam: North-Holland Publishing Co.

———. 1972. "Mechanical Explanation at the End of the Nineteenth Century," *Centaurus* 17: 58–82.

Kleinpeter, Hans. 1905. *Die Erkenntnistheorie der Naturforschung der Gegenwart.* Unter Zugrundelegung der Anschauungen von Mach, Stallo, Clifford, Kirchhoff, Hertz, Pearson, und Ostwald. Leipzig: J.A. Barth.

Koenigsberger, Leo. 1902–1903. *Hermann von Helmholtz.* 3 vols. Braunschweig: F. Vieweg.

———. 1906. *Hermann von Helmholtz* (An abridged one-volume English translation by Frances A. Welby of the original German edition). Oxford: Clarendon Press. *Reprint:* Dover Publishing Co., New York, 1965.

Konen, Heinrich. 1933. "Das Physikalische Institut," in *Geschichte der Rheinischen Friedrich-Wilhelms-Universität zu Bonn am Rhein.* Zweiter Band: *Institute und Seminare, 1818–1933,* pp. 345–355. Bonn: Friedrich Cohen Verlag.

Kuczera, Josef. 1975. *Heinrich Hertz, Entdecker der Radiowellen.* Leipzig: B.G. Teubner Verlagsgesellschaft.

———. 1984. "Einführung," in H. Hertz, *Die Prinzipien der Mechanik in neuem Zusammenhange dargestellt.* Leipzig: Geest and Portig, 1984, pp. 11–36 (This is vol. 263 of *Ostwald's Klassiker der exakten Wissenschaften*).

Kurylo, F. and C. Susskind. 1981. *Ferdinand Braun. A Life of the Nobel Prizewinner and Inventor of the Cathode-Ray Oscilloscope.* Cambridge, Mass.: MIT Press.

Lampariello, C. 1955. "Das Leben und das Werk von Heinrich Hertz," *Arbeitsgemeinschaft für Forschung des Landes Nordrhein-Westfalen,* 43: 7–27. Köln: Westdeutscher Verlag.

Bibliography

Lanczos, Cornelius. 1949. *The Variational Principles of Mechanics.* Toronto: University of Toronto Press.

Laue, Max von. 1961. "Heinrich Hertz, 1857-1894," in *Gesammelte Schriften und Vorträge* 3, 247-256. Braunschweig: F. Vieweg.

Lehmann, Otto. 1892. "Geschichte des physikalischen Instituts der Technischen Hochschule Karlsruhe," in *Festgabe zum Jubiläum der vierzigjährigen Regierung Seiner Königlichen Hoheit des Grossherzogs Friedrich von Baden,* pp. 207-265. Karlsruhe: G. Braun Hofbuchdruckerei.

Lemmerich, J. 1988. "Die Hertzsche Entdeckung im Briefwechsel zwischen Hermann von Helmholtz, Emil DuBois-Reymond und Karl Runge," *Phys.Blätter* 44: 218-220.

Lenard, Philipp. 1894. "Heinrich Hertz," *The Electrician* 33, 415-417.

———. 1896. Introduction to Hertz's *Miscellaneous Papers,* pp. ix-xxvi. London: Macmillan and Co.

———. 1943. *Erinnerungen eines Naturforschers, der Kaiserreich, Judenherrschaft und Hitler erlebt hat* (Unpublished 218-page typescript, of which there is a copy at the University of Stuttgart's Lehrstuhl für Geschichte der Naturwissenschaften und Technik).

———. 1954. "Heinrich Hertz," in Lenard's *Great Men of Science* (second edition; translated by H.S. Hatfield), pp. 358-371. London: G. Bell and Sons.

———. 1957. "Hertz and Lenard," [Brief von Lenard an Max Wolf am 27 Jan. 1894]. *Phys.Blätter* 13, 567-569.

Lindsay, Robert Bruce and Henry Margenau. 1957. *Foundations of Physics.* New York: Dover Publications.

Lodge, Oliver J. 1894a. "Heinrich Hertz," *The Electrician* 32, 273.

———. 1894b. "Hertz's Experiments on Cathode Rays," *The Electrician* 32: 722-723.

———. 1894c. "The Work of Hertz," *Nature* 50 (June 7, 1894), 133-139; 160-161.

———. 1900. *Signalling through Space without Wires: The Work of Hertz and His Successors.* Third edition. London: *The Electrician* Printing and Publishing Co. Reprint: Arno Press, New York, 1974.

Lorentz, Hendrik A. 1937. *Collected Papers.* The Hague: Martinus Nijhoff.

Ludwig, Hubert. 1894. *Worte am Sarge von Heinrich Rudolf Hertz am 4 Jan. 1894 im Auftrage der Universität gesprochen.* Bonn: University of Bonn.

McClelland, Charles E. 1980. *State, Society and University in Germany, 1700-1914.* New York: Cambridge University Press.

McCormmach, Russell. 1970. "H.A. Lorentz and the Electromagnetic View of Nature," *Isis* 61: 459–497.

———. 1975. "Heinrich Rudolf Hertz," *Dictionary of Scientific Biography*, vol. 5, pp. 340–350.

———. 1982. *Night Thoughts of a Classical Physicist*. Cambridge, Mass.: Harvard University Press.

———. 1986. See Jungnickel, Christa.

Mach, Ernst. 1960. *The Science of Mechanics* (translated by Thomas J. McCormack; sixth English edition based on the ninth German edition). Chicago: Open Court Publishing Co.

Maxwell, James Clerk. 1873. *A Treatise on Electricity and Magnetism*. Two volumes. Oxford: Clarendon Press. Reprint of third edition: Dover Publications, New York, 1952.

Merz, John T. 1904–1912. *A History of European Thought in the Nineteenth Century*. Four vols. London: William Blackwood and Sons. Reprint: Dover Publications, New York, 1965.

Morrison, Philip and Emily Morrison. 1957. "Heinrich Hertz," *Scientific American* 197 (Dec. 1957): 98–106.

Mulligan, Joseph F. 1987. "The Influence of Hermann von Helmholtz on Heinrich Hertz's Contributions to Physics," *Am.J.Phys.* 55: 711–719.

———. 1989a. "Heinrich Hertz and the Development of Physics," *Physics Today* 42 (March 1989): 50–57.

———. 1989b. "Hermann von Helmholtz and His Students," *Am. J. Phys.* 57: 68–74.

———. 1992. "Doctoral Oral Examination of Heinrich Kayser, Berlin, 1879," *Am.J.Phys.* 60: 38–43.

Nahin, Paul J. 1988. *Oliver Heaviside, Sage in Solitude*. New York: IEEE Press.

———. 1990. "Oliver Heaviside," *Scientific American* 261 (June, 1990): 122–129.

Nichols, E.L. 1894. "Heinrich Hertz," *Phys.Rev.* 1: 383–386.

Nobel Lectures: Physics. Vol.1, 1901–1921. New York: Elsevier, 1967.

O'Hara, James G. 1988. "The Career of Hertz and the State of the Exact Sciences in Imperial Germany," in W. Wiesbeck (ed.), pp. 44–63.

O'Hara, James G. and W. Pricha. 1987. *Hertz and the Maxwellians*. London: Peter Peregrinus Ltd.

Olesko, Kathryn M. 1991. *Physics as a Calling. Discipline and Practice in the Königsberg Seminar for Physics*. Ithaca, N.Y.: Cornell University Press.

Bibliography

Planck, Max. 1894. "Heinrich Rudolf Hertz. Rede zu seinem Gedächtnis am 16 Feb. 1894," *Verhandlungen der physikalischen Gesellschaft zu Berlin* 13: 9–29. Reprinted in Planck (1958), vol. 3, pp. 268–288.

———. 1906. "Paul Drude. Gedächtnisrede am 30 Nov. 1906," in Planck (1958), vol. 3, pp. 289–320.

———. 1931. "Maxwell's Influence on Theoretical Physics in Germany," in J.J. Thomson (1931), pp. 45–65.

———. 1950. *A Scientific Autobiography and Other Papers* (translated by Frank Gaynor). London: William and Norgate Ltd.

———. 1958. *Physikalische Abhandlungen und Vorträge*. Three vols. Braunschweig: F. Vieweg.

Poggendorff, Johann Christian. 1971. *Poggendorff's biographisch-literarisches Handwörterbuch zur Geschichte der exakten Wissenschaften*. Band VIIa—Supplement, pp. 283–286 (on Hertz). Berlin: Akademie Verlag.

Poincaré, Henri. 1891. *Électricité et Optique*. Vol.2, *Les Théories de Helmholtz et les Expériences de Hertz*. Paris: Georges Carré.

———. 1894a. "Poincaré on Maxwell and Hertz," *Nature* 50 (May 3, 1894): 8–11.

———. 1894b. "Light and Electricity According to Maxwell and Hertz," *Annual Report of the Smithsonian Institution for 1894*, pp. 129–139. Washington, D.C.:Government Printing Office.

———. 1897. "Hertz on Classical Mechanics," English translation in A. Danto and S. Morgenbesser: *Philosophy of Science*. New York: World Book Co., 1960, pp. 366–373.

———. 1900. *La Théorie de Maxwell et les Oscillations Hertziennes*. Chartres: Imp. Durand.

Pupin, Michael. 1926. *From Immigrant to Inventor*. New York: Charles Scribner's Sons.

Reden, Wolf von. 1988. *Heinrich Hertz and the "Principles of Mechanics,"* in Boerger (1988), pp. 69–95.

Rosenfeld, Leon. 1957. "The Velocity of Light and the Evolution of Electrodynamics," *Nuovo cimento*, vol. 4, supplement 5, pp. 1630–1669.

Sachse, A. 1928. *Friedrich Althoff und sein Werk*. Berlin: Mittler und Sohn.

Saunders, S. and H.R. Brown. eds. 1991. *The Philosophy of Vacuum*. New York: Oxford University Press.

Schleiermacher, A. 1902. "Heinrich Hertz: Biographische Skizze," *Verhandlungen des Naturwissenschaftlichen Vereins in Karlsruhe* 15: 21–32.

Schuster, Arthur. 1911. *The Progress of Physics during 33 Years (1875–1908)*. Cambridge: Cambridge University Press.

———. 1925. "Biographical Byways," *Nature* 115: 342–343.

Simpson, Thomas K. 1966. "Maxwell and the Direct Experimental Test of His Electromagnetic Theory," *Isis* 57: 411–432.

Sommerfeld, Arnold. 1952. *Lectures on Theoretical Physics*. Vol.1: *Mechanics*. New York: Academic Press.

———. 1954. *Vorlesungen über Theoretische Physik*. Band III: *Elektrodynamik*. Leipzig: Akademische Verlagsgesellschaft Geest und Portig.

Stachel, John. 1987. "Einstein and Ether-Drift Experiments," *Physics Today* 40 (May 1987), 45–47.

Stuewer, Roger H. 1971. "Hertz's Discovery of the Photoelectric Effect," *Proceedings of the Thirteenth International Congress on the History of Science, Moscow, 1971*, vol. 6, pp. 35–43.

Susskind, Charles. 1962. *Popov and the Beginnings of Radiotelegraphy*. San Francisco: San Francisco Press.

———. 1964. "Observations of Electromagnetic-Wave Radiation before Hertz," *Isis* 55: 32–42.

———. 1965. "Hertz and the Technological Significance of Electromagnetic Waves," *Isis* 56: 342–345.

———. 1981. See Kurylo, F.

———. 1988. "Heinrich Hertz and the Discovery of Electromagnetic-Wave Propagation," *Endeavour* 12: 84–85.

Taton, René. 1965. *Science in the Nineteenth Century* (translated by A.J. Pomerans). London: Thames and Hudson.

Thiele, Joachim. ed. 1968. "Ernst Mach und Heinrich Hertz: Zwei unveröffentlichte Briefe aus dem Jahre 1890," *Zeitschrift für Geschichte der Naturwissenschaften, Technik und Medizin* 5: 132–134.

Thomson, George. 1970. "An Unfortunate Experiment: Hertz and the Nature of Cathode Rays," *Notes and Records of the Royal Society of London* 25: 237–242.

[Thomson, Joseph John. ed.]. 1931. *James Clerk Maxwell: A Commemoration Volume*. Cambridge: Cambridge University Press.

Thomson, William (Lord Kelvin). 1893. "Preface to the English Edition" of Hertz's *Electric Waves*. London: Macmillan.

Thomson, William (Lord Kelvin) and P.G. Tait. 1867. *Treatise on Natural Philosophy*. Oxford: Clarendon Press. Reprint: Dover Publications, New York, 1962, under new title: *Principles of Mechanics and Dynamics*.

Bibliography

Trouton, Frederick T. 1889. "Repetition of Hertz's Experiments," *Annual Report of the Smithsonian Institution for 1889,* pp. 191–203. Washington, D.C.: Government Printing Office.

———. 1891. "The Influence the Size of the Reflector Exerts in Hertz's Experiments," *Phil.Mag.* (5) 32: 80–90.

Tunzelmann, G.W. de. 1889. "Hertz's Researches on Electrical Oscillations," *Annual Report of the Smithsonian Institution for 1889,* pp. 145–190. Washington, D.C.: Government Printing Office.

Unsöld, A. 1970. "H. Hertz's *Prinzipien der Mechanik,*" *Phys.Blätter* 26: 337–342.

Wallner, Friedrich. 1981. "Boltzmann, Hertz and Wittgenstein," *Proceedings of International Conference on Boltzmann* (Vol. 8 of Works of Boltzmann, edited by Sexl and Blackmore).

Wenig, Otto. ed. 1970. *Verzeichnis der Professoren und Dozenten der Rheinischen Friedrich-Wilhelms Universität zu Bonn, 1818–1968.* Bonn: H. Bouvier Verlag und Ludwig Röhrscheid Verlag.

Wheaton, Bruce R. 1978. "Philipp Lenard and the Photoelectric Effect, 1889–1911," *Hist.Stud.Phys.Sci.* 9: 299–322.

Whittaker, Edmund T. 1951. *A History of the Theories of Aether and Electricity.* Two vols. London: Thomas Nelson and Sons. Reprint: Dover Publications, New York, 1989.

Wiesbeck, Werner. ed. 1988. *Heinrich Hertz Symposium*: "100 Jahre Electromagnetische Wellen" (Karlsruhe, March 14–15, 1988). Berlin: VDE Verlag.

Wilson, Andrew D. 1989. "Hertz, Boltzmann, and Wittgenstein Reconsidered," *Stud.Hist.Phil.Sci.* 20: 245–263.

Wolf, Franz. 1950. "Heinrich Hertz," in *Die Technische Hochschule Fridericiana Karlsruhe: Festschrift zur 125-Jahrfeier, 1950,* pp. 67–73.

———. 1967. "Heinrich Hertz," *Fridericiana: Zeitschrift der Universität Karlsruhe,* Heft 1 (November, 1967), pp. 5–18.

———. 1968. "Aus der Geschichte der Physik in Karlsruhe," *Phys.Blätter.* 24: 388–400.

Woodruff, A.E. 1962. "Action at a Distance in Nineteenth-Century Electrodynamics," *Isis* 53: 439–459.

———. 1968. "The Contributions of Hermann von Helmholtz to Electrodynamics," *Isis* 59: 300–311.

Zatzkis, H. 1965. "Hertz's Derivation of Maxwell's Equations," *Am.J.Phys.* 33: 898–904.

Zenneck, Jonathan. 1929. "Heinrich Hertz: Rede bei der Heinrich Hertz Feier der Universität Bonn am 19 Nov. 1927," *Deutsches Museum Abhandlungen und Berichte* 1 (Heft 2): 1–36. Berlin: VDI Verlag.

———. 1946. "Zum 90. Geburtstag von Heinrich Hertz," *Die Naturwissenschaften* 33: 225–230.

Index

Abitur, 6–7
Action-at-a-distance theories, 128
  and Faraday, 310–12
  and FitzGerald, 366
  and gravitational forces, 286, 314n
  and Helmholtz, 311–12, 314
  and Hertz, 80, 181–82, 286, 314n
  and Maxwell, 311
  and unity of electric force, 394
Adiabatic expansion of damp air, 107, 391
Aether. *See* Ether.
Akins, Zoë, 381
Althoff, Friedrich (1839–1908)
  and Hertz's call to Bonn, 47
  and Hertz's research in Bonn, 52
  interactions of, with Hertz, 25n, 25–26, 47–48
  and physics in German universities, 25
  recipient of honorary degree from Harvard, 25n
Ampère, André Marie (1775–1836)
  force law of, 127–28
  and unity of magnetic force, 128
Annalen der Physik und Chemie
  changing editors of, 14n
  changing titles of, 14n
Archimedes, 327
Arons, Leo (1860–1919), 389
Asphalt prism, 36, 159
Atoms
  Hertz on existence of, 340–41
  Kirchhoff on, 341

Baden-Baden, 74
Berlin, 12–13
  Hertz's military service in, 10
  Hertz's studies and research in, 13–18
  population of, 13
  status in Germany, 12–13
Berlin Academy of Sciences, 44
  1879 prize-problem of, 14–15, 366
  Hertz elected to membership in, 49, 293–95
Berlin Physical Society
  M. Planck's address to, 383–403
Berlin University
  foundation of, 13
  Hertz's decision to leave, 18
  Hertz's offer of professorship at, 47
  Hertz's studies at, 13–18
  history of, 13
  Physics Institute of, 13
Bezold, Wilhelm von (1837–1907), 91n, 295n
  and electromagnetic waves, 164–65, 396
Bilder
  meaning of, 67, 323
  and models, 324n
  necessary characteristics of, 324–26
  and representations, 326–27
Bjerknes, Vilhelm (1862–1951), 400n
  and application of Hertz's *Mechanics*, 73n
  and damping of electromagnetic oscillations, 179–80, 215n
  and electric resonance, 40, 62, 210n
  as "father of modern meteorology," 62
  and nature of cathode rays, 61–62
  opinion of Hertz, 62–65
  as research student of Hertz, 62–65

use of his research by M. Planck, 81
and weather fronts, 62
"Black Year in German Physics (1894)," 83
Blondlot, M. (1849–1930), 175n
Boltzmann, Ludwig (1844–1906), 191n
and Hertz's *Mechanics*, 69
on importance of Hertz's contributions, 83
Bonn, 50
Hertz in, 47–48, 299–300
Bonn University
cathode-ray research at, 57n
Hertz as professor at, 47–78, 299–300
Borchardt, Carl Wilhelm (1817–1880)
as teacher of Hertz, 322
Borchardt's *Journal*, 100n
Born, Max
and Hertz's electrodynamics of moving bodies, 80–81
Braithwaite, Robert Bevan
on Hertz's philosophical outlook, 70
Braun, Ferdinand (1850–1918), 24–25
and electrotechnology in Karlsruhe, 27
and wireless telegraphy, 27, 80
Bridgman, Percy (1882–1961)
and operational definitions, 338n
Buchwald, Jed Z., 22n, 127n, 145n
Bunsen burner, 253n
Bunsen cells, 196

Cambridge University
Hertz's visit to, 56
Capacitance, 34, 200
Carnot, Nicholas Sadi (1796–1832), 292
Cathode rays, 102–5
and V. Bjerknes, 61–62
and charged particles, 57, 304n
Crookes' experiments on, 155n
difference of, from light, 301
diffusion of, 304
effect of electrostatic fields on, 57, 155
effect of magnetic fields on, 57, 304
and the ether, 57
FitzGerald on, 62n, 153–56
and flow of matter, 148
and Eugen Goldstein, 56, 304
Hertz and Helmholtz on, 154n
Hertz's research on, 56–59, 111, 153–56, 299, 301–4
and Philipp Lenard, 57–59, 61–62
metals used to produce, 302
and multiple metallic layers, 302–3
nature of, 57, 61, 155–56, 301
passage of, through thin films, 156, 299, 301–4
properties of, 303–4
and J.J. Thomson, 62, 82, 155n
use of aluminum leaf with, 302
and x-rays, 156
Centrifugal force
and inertia, 328–29
Cherusker student corporation, 9
Circuits. *See* Currents.
Clark University, 48–49
Clausius, Rudolf (1822–1888), 290, 292, 308, 400n
death of, 47
professor of physics at Bonn, 12n
Cohen, Eduard, 5
Cohen, Robert
on Hertz as role model for scientists, 84
Cohn, Emil (1854–1944), 74
Concealed masses. *See* Hidden masses.
Concealed motions. *See* Hidden motions.
Condensers. *See* Capacitance, Leyden jars.
Connections (mechanical)

Index

and forces, 333–34, 356–57, 371–72
permissible, 358–60
rigid, 327, 349–50, 356–57
tangling of rigid, 372
Conservation of energy, 291, 337
FitzGerald's comments on, 152–53
Helmholtz's formulation of, 132–33
Cornu, M.A. (1841–1902), 180n
Coulomb's law, 185
Crookes, William (1832–1919)
and particle nature of cathode rays, 155n
Currents, 189n
closed and open, 128n, 187, 190, 308–9, 311n
in Hertz's 1884 paper, 128–29
importance of open, 128n, 309
magnetic, and changing polarization, 130–31
See also Electric current.
Cyclical systems, 318

D'Agostino, Salvo, 72n, 127n, 133n, 163n
d'Alembert's Principle, 317, 327
Damping
effect on oscillation frequency of, 179–80
of electrical oscillations, 215, 291
research of V. Bjerknes on, 179–80, 215n
Damping ratio
calculation of, 218–19
Darboux, Jean-Gaston (1842–1917), 321
Darstellungen. *See* Representations.
Darwin, Charles (1809–1882), 292
Death of Hertz. *See* Hertz, Heinrich, death of.
Dielectrics, 143
and empty space, 143n
Diffraction effects

in Hertz's experiments, 31, 34n, 176n, 261n, 262
Poincaré on, 34n
Displacement current, 125, 190–91, 311n, 366
Doll, Elisabeth
marriage of, to Heinrich Hertz, 28–29
personal qualities of, 29
Doll, Max (Hertz's father-in-law), 28
Dresden Polytechnic, 8–10
Hertz's courses at, 9–10
Drude, Paul (1863–1906), 14n
DuBois-Reymond, Emil (1818–1896), 13

Ehrenfest, Paul (1880–1933)
and application of Hertz's *Mechanics*, 73n
Einstein, Albert (1879–1955)
and Hertz's work, 80–81, 82
Elastic spheres,
Hertz's research on, 99–100, 150–51, 387–88
Electric current, 133
homogeneity of, in conductors, 149–50
inertia of, 13, 14, 92–97, 148–50, 293, 309–10, 365–66, 385–86
particle nature of, 148–50
skin-effect of, 40, 400
See also Currents.
Electric discharges. *See* Cathode-rays; Sparks.
Electric double layer, 129–30
Electric field
in electromagnetic radiation, 263–64
Electric field-intensity, 191n
Electric force, 134–36
and electric field, 191n, 245n
meaning of, for Hertz, 191n, 245n
Electric oscillations. *See* Oscillations, electric.

Electric radiation. *See*
    Electromagnetic waves.
Electric Waves (H. Hertz). *See*
    Electromagnetic waves in
    air; FitzGerald; Helmholtz;
    Hertz; Planck.
Electricity
    and action-at-a-distance, 274–75
    and the ether, 273, 286
    Faraday and, 275–77
    Maxwell and, 190–91, 277–79
    nature of, 274–75, 286–87, 330–31
    and ponderable matter, 274–75
Electrochemical phenomena
    Hertz's interest in, 402
Electrodynamometer
    construction of, by Hertz, 389
Electrolysis, 148
Electromagnetic field
    and electromagnetic radiation, 263–64
    equivalence of Hertz's and Maxwell's equations for, 142–44
    Hertz-Heaviside equations for, 54, 142
    incompleteness of other theories for, 144–45
    properties of, 397
    time for propagation of, 276–77, 280–81
Electromagnetic radiation. *See*
    Radiation, electromagnetic.
Electromagnetic waves in air
    and action-at-a-distance, 286
    acoustic and optical analogies with, 253–54, 282–83, 284–85
    apparatus used in experiments on, 257–59, 268–71
    application to wireless telegraphy of, 40–43
    beams of, 257, 259–61
    comparison of experiment with theory for, 250
    corrected velocity of, 241n
    detection of, 258–59, 261, 270–71, 271n
    diagram of, 244
    diffraction of, 262
    electric and magnetic fields in, 247–48, 263–64
    experiments on, 242–55, 257–71
    finite velocity of, 241
    generation of, 257–58
    at higher frequencies, 257, 259
    interference of, 241–42, 283–84
    and Maxwell's theory, 254–55
    measurement of electric field in, 248n
    measurement of magnetic field in, 241–42, 247–48, 250, 283–84
    period of, 257
    polarization of, 262–64, 265–66, 285
    rectilinear propagation of, 261–62, 284
    reflection of, 241–44, 264–66
    refraction of, 266–68
    similarity to light of, 268, 276, 284–85, 313–14
    with source between detector and reflector, 251–53
    and standing waves, 212–16, 245–50, 263–64
    transverse nature of, 283
    velocity measurement of, 283–84
    wavelength measurement of, 241–42, 250, 283–84
    See also Hertz, Heinrich.
Electromagnetic waves on wires, 172–73, 178, 283–84
    See also Lodge, Oliver J.
Electromagnetism, theories of, 183–92, 289–90, 299, 307–12
    action-at-a-distance, 184–85
    and the ether, 289–90
    and Faraday, 275–77, 310–12

Index

and Helmholtz, 177–78, 186–87, 308–10
and Hertz, 190–92
and Maxwell, 190–92, 254, 277–79, 294, 310–12
and F. Neumann, 308
and other German physicists, 308
and W. Weber, 275, 307–8
Elster and Geitel
and the photoelectric effect, 166
Energy
infinite value of potential, 345
kinetic, 338, 344
kinetic, of an electric current, 92–97, 148–50
negative potential, 345
potential, 338, 344–45
See also Conservation of energy.
Engineering
Hertz's change to physics from, 11, 89–91, 306–7
Ether, luminiferous, 194, 286
and cathode-rays, 57, 58, 304n
contradictions in properties of, 274
and electric polarization, 186–87, 189–90
and the electrodynamics of moving bodies, 54–55, 400–1
FitzGerald on, 290–91, 292, 371
Helmholtz on, 313, 317n
and Hertz's *Mechanics*, 65–67, 69, 319, 343n, 349n, 361n, 364
and hidden masses and hidden motions, 69
importance of, to Hertz, 273–74, 286, 289–90
Maxwell and, 143, 278
as medium for electromagnetic waves, 273, 276
and nature of electricity and matter, 286–87
structure of, 289–91
and theories of electromagnetism, 290–91
and theories of gravitation, 366–67
Evaporation
Hertz's research on, 101–2, 151–52, 295, 387
Expansion, adiabatic. *See* Adiabatic expansion.

Faraday, Michael (1791–1867), 58n, 79, 170, 310–11
and action-at-a-distance, 311–12
and electric and magnetic fields, 276
and electromagnetic waves, 276–77
and idea of electric field, 128
and lines of force, 276
and nature of electricity, 275–77, 310–12
on relations between light, electricity, and magnetism, 275–77
and time for propagation of electromagnetic fields, 276
Faraday Effect, 276n
Feddersen, Berend Wilhelm (1832–1918), 193, 194n
and electromagnetic waves, 396
Fermi, Enrico (1901–1954)
and order-of-magnitude calculations, 220n
Field theory. *See* Electromagnetic field; Faraday; Maxwell's electromagnetic theory.
First Railway Guards Regiment
Hertz's service in, 10
First (force) representation of mechanics, 327–36, 369–70
characteristics of, 328–34
concept of force in, 328–31, 333–36
criticisms of, 328–31
exposition of, 327–28

Hertz's conclusions about, 336
multiplicity of complicated forces in, 335–36
FitzGerald, George Francis (1851–1901)
account in *Nature* of Hertz's experiments, 291n
and action-at-a-distance theories, 366–67
career of, 126
and Hertz's *Mechanics*, 66–67, 365–77
and Hertz's visit to London, 56
as leader of the Maxwellians, 126
and Lorentz-FitzGerald contraction, 165n
and Newton's third law, 369–70, 376
and prediction of electromagnetic waves, 165
as professor at Trinity College, Dublin, 126
review of *Electric Waves*, 162, 289–92
on Hertz's experiments, 291–92
on importance of Hertz's experiments, 289, 291–92, 291n
on Maxwell's theory and Maxwell's equations, 290n
review of *Miscellaneous Papers*, 125–26, 147–57
review of *Principles of Mechanics*, 300, 365–77
assumptions implicit in, 374
on deaths of Hertz and Helmholtz, 365–69
on displacement of a system of points, 374
on Helmholtz's Preface, 365
on Hertz's forceless mechanics, 368–69, 371–74, 376–77
on Hertz's fundamental law, 367, 372, 377
on Hertz's Introduction, 369–74
on Hertz's Preface, 369
on main text of the *Mechanics*, 374–77
on Newton's third law, 369–70, 376
on philosophical nature of the *Mechanics*, 369, 377
Floating elastic plates
Hertz's research on, 105–6, 152, 295, 391–92
Force
concept of, in mechanics, 320n
confusion about nature of, 334–35
conservative, 317
definition of, by Lagrange, 329–30
FitzGerald's view of, 373
Helmholtz's view of, 320–21
Hertz's view of, 320n, 347–53
Kirchhoff's view of, 330n
Mach's view of, 321–22
nature of, 68–69, 328–31
Newton's view of, 317
properties of, 333–36
in second representation of Hertz's *Mechanics*, 339
See also Electric Force; Magnetic Force.
Force-functions, 360–61
Frankfurt am Main
Hertz's engineering apprenticeship in, 7–8
Hertz's homesickness in, 7–8
Frequencies, electrical
calculation of, from circuit parameters, 216–21
dependence of, on capacitance and inductance, 200n
use of Hz as unit for, 42–43
Fresnel mirror, 253
Friedrich-Wilhelm University (Berlin). *See* Berlin University.
Friedrich-Wilhelm University (Bonn). *See* Bonn University.

Index

Fundamental law of mechanics, 68, 350, 362
  and law of inertia, 350–51, 354
  and Gauss's Principle of Least Constraint, 350–51, 354–55
  use of, in deriving all of mechanics, 350–51

Galileo
  on inertia and force, 316, 327
Gauss, Carl Friedrich (1777–1855), 113
  and Principle of Least Constraint, 68, 350, 354
Geissler, Heinrich (1814–1879), 57n
Geissler tubes, 57, 103, 155n
Geitler, Josef von (Hertz's cousin), 63n
Germany
  Helmholtz's stature in, 114–15
  Hertz's status in, 83–84
  science in nineteenth-century, 113–14
Gewerbeakademie, 100, 306
Giessen
  call of Hertz to, 47
Goldstein, Eugen (1850–1930), 18n
  and cathode rays, 56, 153, 304
  and Hertz, 56
Grassmann, Hermann (1809–1877), 308
Gravitation, 185
  and action-at-a-distance, 286, 314n
  ether theories of, 366–67
  force of, 373
  Newton's law of universal, 275
Gutta-percha, 258

Habilitation, 390n
  of Hertz, 18, 20, 104–5
Hagen, Carl Ernst (1851–1923), 17–18
Hagenbach and Zehnder
  criticism of Hertz's research by, 180, 262n
Hall, G. Stanley (1846–1924)
  meeting of, with Hertz, 48–49
  and position for Hertz at Clark University, 49
Hallwachs, Wilhelm (1859–1922), 39n
  and photoelectric effect, 38–39, 166, 240n
Hallwachs effect, 39n
Hamburg, 3
  life of Hertz in, 3–7
Hamilton, William Rowan (1805–1865)
  and characteristic functions, 355
  and Hertz's *Mechanics*, 355
Hamilton's Principle, 320–21
  and fundamental laws of Newtonian mechanics, 341–43, 345–47
  as integral principle of mechanics, 338–39
  and Lagrangian functions, 339n
Hardness
  and cracking of materials, 150–51
  Hertz's research on, 99–100, 150–51, 387–88
Heaviside, Oliver, 53n, 56, 268n
  correspondence with Hertz, 53–54
  and Hertz-Heaviside form of Maxwell's equations, 54
  research on electromagnetic theory, 53–54
Heidelberg address, 156n, 273–87, 398
  Hertz's difficulty in preparing, 52–53, 273
  qualities of, 46, 53, 162
  reception of, 52–53, 108, 398
  references to, 33, 36
Helmholtz, Anna von (wife), 17n
Helmholtz, Hermann von (1821–1894), 161, 359
  and action-at-a-distance theories, 311–12, 314
  approach to mechanics of, 320–21
  career of, 115

and cathode rays, 59, 105n
and conservation of energy,
   118–20, 152–53
contributions to science, 114–20
and cyclical systems, 320
and derivation of Maxwell's
   equations, 169
appointed director of the
   *Physikalisch-Technische
   Reichsanstalt (PTR)*, 43–44, 47
on discovery in physics, 107
and electrical oscillations 193n,
   216, 218
on electromagnetic theories,
   307–12
and the ether, 313, 317n
and Hamilton's Principle, 320–
   21
and Heinrich Hertz, 83–84, 113–
   21, 147, 156–57, 161, 293–
   95, 306–7, 314, 381, 384–85,
   399–400
on Hertz's ability and character,
   295, 305, 312–13, 314
and Hertz's *Electric Waves*, 293–
   95, 312–14
and Hertz's *Principles of
   Mechanics*, 70–73, 299–300,
   305–18, 320–21
and Hertz's research on
   electromagnetism, 43–46,
   125, 163–64, 168–69, 183–
   84, 313, 394
and human physiology, 116–18
and Leyden-jar discharges, 281n
on Maxwell's and Lord Kelvin's
   models for electro-
   magnetism, 318
as mentor to Hertz, 13–16, 17–
   18, 24–25, 43–46, 47–48,
   70–73, 91–94, 97, 98–99,
   105, 111–12, 120–21
nomination of Hertz to Berlin
   Academy by, 293–95
and the ophthalmoscope, 115–
   16
personal qualities of, 120–21

philosophical views of, 117–18
Preface to Hertz's *Principles of
   Mechanics*, 300, 305–18
and Principle of Least Action,
   70, 72, 119–20, 339–40
as professor of physics in Berlin,
   13
as research mentor, 120–21,
   156–57
and search for a unifying
   principle of physics, 70, 72,
   119
seventieth birthday celebration
   of, 72–73
as a teacher, 120–21, 156–57
Helmholtz, Robert (son)
   death of, 45, 381
Henry, Joseph (1797–1878)
   and Leyden-jar oscillatory
   discharge, 281n
Hertz, Anna Elisabeth Pfefferkorn
   (1835–1910) (mother), 4–5,
   6
   and Hertz's early life, 87
Hertz, Elisabeth (Doll), 14n
   marriage of Heinrich and, 28–
   29, 315
Hertz, Gustav (brother), 85
Hertz, Gustav Ferdinand (1827–1914)
   (father), 3
Hertz, Heinrich (1857–1894)
   Abitur of, 6–7
   and action-at-a-distance, 80,
      181, 286, 314
   and Friedrich Althoff, 25–26, 47–
      48
   as assistant to Helmholtz, 99
   in Berlin, 12–18, 91–106, 300–7,
      384–90
   and Berlin Academy of Sciences,
      293–95
   and von Bezold, 164–65
   biography of, 3–84
   and V. Bjerknes, 62–65
   in Bonn, 47–48, 50–78, 108, 299–
      300, 400–3

Index

and cathode-rays, 56–59, 61, 82, 102–5, 108, 111–12, 153–56, 299, 301–4
character of, 55, 63, 76, 305, 312–15, 383–84, 398–400
childhood and early education of, 3–7, 306, 383–84
chronology of life of, xix-xxi
confirmation of Maxwell's electromagnetic theory by, 182, 254, 284
contributions to physics of, xiv, 36–40, 41n, 78–84, 312–14, 385–99
death of, 75–78, 315, 383, 402–3
doctoral degree of, 15–16, 98
doctoral dissertation of, 15, 97–98, 386–87
in Dresden, 8–10
1884 paper on electromagnetism, xv, 21–23, 44, 127–45
and Einstein, 80–81, 82
Electric Waves, 108, 289–92, 312–14, 395
Hertz's Introduction to, 163–92
See also Electromagnetic waves in air.
and electromagnetic theory, 53–55, 79–80, 106–7, 125, 183–92, 275–79, 293–94, 299, 392–94, 400–2
emotional problems of, 20, 23
and engineering, 7–11, 89–91
Erinnerungen, xiv, 4n
and the ether, 54–55, 58, 80–81, 273n, 276, 286, 317n, 319, 343n, 349n, 361n
and experiments on electromagnetic waves, 29–37, 163–82, 193–221, 241–55, 257–71, 293–94, 312–14, 395–98, 400
and G.F. FitzGerald, 56, 165–66
See also FitzGerald, George Francis.
in Frankfurt, 7–8

Habilitation of, 18–20, 104–5
in Hamburg, 3–7
and Heaviside, 53–54, 56
Heidelberg address, 33, 36, 46, 52–53, 108, 156n, 162, 273–87, 398
and Helmholtz, 15, 16–18, 19, 24–25, 44–46, 47–48, 72–73, 83–84, 91–94, 97, 99, 105, 111–12, 113–21, 147, 156–57, 161, 183, 293–95, 306–7, 314, 381, 384–85, 399–400
on Helmholtz's ability and character, 113–21, 397–99
and Helmholtz's influence on his research on electromagnetism, 43–46, 46n, 125, 163–64, 169, 177–78, 309–10, 312–14
and Helmholtz's influence on his research on mechanics, 70–73, 299–300, 315–18
and Hertz-Heaviside form of Maxwell's equations, 53–54, 142–43, 152–53
honors bestowed on, 55–56, 315, 398–99
illness of, 73–76, 78–84
importance of, in history of physics, xiii-xiv, 78–84, 147–48, 257n, 289, 291–92, 293
in Karlsruhe, 24–46, 159–62, 395–98
in Kiel, 19–23, 26, 106–8, 125, 390–95
and Kirchhoff, 91–92, 96, 330n, 347–48
and Lenard, 57–62, 299
letters of, 23n
and Lodge, 56, 80, 165, 268n, 331n
and Mach, 69n, 321n
marriage of, 28–29, 395
and mathematics, 392

and Maxwell's theory, 21–23, 54, 125, 127–45, 183–84, 191–92
as a mentor, 64–65
and meteorology, 27, 107, 391
and microwaves, 40, 179n
military service of, 10, 73–74
Miscellaneous Papers, 16n, 97–108, 125–26, 147–57, 384–95
in Munich, 10–12, 89–91
notes to papers of, 192n
and philosophy, 15, 323–27, 346n
and the photoelectric effect, 37–39, 82, 166, 223–40, 305, 397
and physics in Germany, 83–84
and M. Planck, 81, 383–403
Principles of Mechanics, xv, 65–70, 82–83, 108, 299–300, 317–18, 319–22, 323–64, 365–77, 402
as a researcher, 5–6, 16, 17, 32–33, 36–37, 64, 79–80, 96, 97n, 99, 102–5, 107–8, 182n, 282, 295, 313, 332n, 390, 395–97, 397–98
reviews of his *Collected Works* by G.F. FitzGerald, 147–57, 289–92, 365–77
and the skin-effect, 40, 400
as a teacher, 21, 27, 52, 63–64, 101–2, 403
and wireless telegraphy, 41–43, 80, 178n
and x-rays, 82
young manhood of, 7–12, 383–84
Hertz, Johanna (daughter), 4n
birth of, 29
Hertz, Mathilde (daughter), 4n, 29
birth of, 51
and her father, 23n, 76
Hertz, Melanie (sister), 4n
Hertz, Otto (brother), 20
death of, 20

Hertz, Rudi (brother), 85
Hertzian oscillators, 81, 291
Hertz's (third) representation of mechanics, 347–63, 371–74
advantages of, 353–55
characteristics of, 355–63
comparison with other representations, 363–64
concept of force in, 351
and the ether, 371
exposition of, 347–53
fundamental concepts of, 347–51
fundamental law of, 350–53
and Helmholtz's cyclical systems, 349
and hidden masses and hidden motions, 348–49
incompleteness of, 364
inconsistencies in, 356n
and Kirchhoff's *Textbook of Mechanics*, 347–48
limitation of, to inanimate matter, 361
mathematical form of, 351–53
practical applications of, 363
purpose of, 364
and systems of points, 351–53
Hidden masses, 68–69, 69n, 317, 349, 358n, 362, 371
Hidden motions, 68–69, 69n, 317, 349, 362, 371
High-frequencies, 194
importance of, in Hertz's experiments, 34–35, 180–81
History
of electromagnetic theories, Helmholtz on, 307–12
of physics, Hertz's importance in, 78–84
Huber, Heinrich, 41–42
Humidity
Hertz's letter to his parents on, 101–2
Hunt, Bruce J., 289n, 291n, 369n
Hygrometer
Hertz's research on, 101–2, 389

Index

Hz (hertz)
   as accepted unit of frequency, 42–43

"Images" in Hertz's *Mechanics*, 323n
Induction
   and electric oscillations, 203–4
   in open rectilinear circuits, 201–7
   physiological effects of, 203
   self and mutual, 219–21
Induction coils, 163
Inductive effects, 134, 136
   in electric circuits, 131–32
   possible observation of, in magnetic circuits, 132
Inertia
   and centrifugal force, 328–29
   of electric currents, 13–14, 92–95, 148–50, 309–10, 385–87
   and Galileo, 316, 327

Jefferson, Thomas, 25
Johanneum Gymnasium
   Hertz as student at, 6
Jolly, Philipp von (1809–1884)
   professor of physics in Munich, 11, 91
Jones, Daniel E.
   obituary of Hertz by, 76n
   research student in Bonn, 63n
   translator of Hertz's *Collected Works*, xv
Joule, James Prescott (1818–1889), 119n, 291, 292
Jungnickel, Christa, and Russell McCormmach
   Intellectual Mastery of Nature, xiv, 20n, 24n, 47n, 66

Karlsruhe
   Hertz in, 25–26, 125
Karlsruhe *Technische Hochschule*, 24–27
   as doctoral institution, 28n
   Hertz as professor at, 24–46
Karsten, Gustav (1820–1900)
   professor at Kiel, 19, 395
Kayser, Heinrich (1853–1940), 17, 25
   and death of Hertz, 77–78
   Erinnerungen of, xiv
Kelvin, Lord. *See* Thomson, William.
Kerr, John (1824–1907), 150n
Kerr cell, 150n
Kiel, 20
   Hertz in, 19–23, 26
Kiel University
   Hertz as professor at, 19–23, 26, 106–8, 125, 390–95
Kinetic energy, 338, 344
   of an electric current, 92–97, 100–1, 148–50, 385–87
Kirchhoff, Gustav R. (1824–1887), 17, 25
   and atoms, 341
   and concept of force, 330n
   death of, 47
   and foundations of mechanics, 347–48
   and Hertz's doctoral examination, 15
   and Hertz's *Mechanics*, 347–48
   and position for Hertz in Kiel, 19
   as professor in Berlin, 13
   as teacher of Hertz in Berlin, 91–92, 96, 384–85
   and theory of electrical oscillations, 216
Klein, Martin, 66, 68n
Knochenhauer spirals, 30,32
Koenigsberger, Leo (1837–1921)
   as author of Helmholtz's biography, xiv, 9n
   as mathematics professor in Dresden, 9
Kohlrausch, Rudolf (1809–1858), 278n
Kohlrausch, R., and W. Weber
   measurement of ratio of electrostatic to electromagnetic unit of charge, 278
Kolacek, Frantisek (1851–1913), 264n
König, Walter (1859–1936), 266n

Kummer, E. (1810–1893), 15
Kundt, August (1839–1894), 295n
　director of the Berlin Physics
　Institute, 47

Lagrange, J.L., 327, 329–30
Lange, Wichard, 5, 94
Larmor, Joseph (1857–1942)
　and inertia of electric currents,
　366
Laue, Max von (1879–1960), 79n
Least Action. *See* Principle of Least
　Action
Least Constraint. *See* Principle of
　Least Constraint.
Lecher, Ernst (1856–1926)
　and measurement of velocity of
　electromagnetic waves on
　wires, 175
Lenard, Philipp (1862–1947), 400n
　anti-Semitism of, 61
　as assistant to Hertz, 57, 156
　and cathode-rays, 57–59, 61–62,
　82, 299
　and discovery of x-rays, 61n,
　156n
　as editor of Hertz's
　Miscellaneous Papers, 60,
　87, 147–48
　as editor of Hertz's *Principles of
　Mechanics*, 60, 71, 307n,
　365
　Erinnerungen of, xiv
　FitzGerald on experiments of,
　62n
　and Helmholtz, 71
　and Hertz, 57–62, 299
　on Hertz and the Berlin prize-
　problems, 307n
　Introduction to Hertz's
　Miscellaneous Papers, 87,
　89–108, 147–48
　and H. Kayser, 6n
　and Nobel Prizes, 60–61, 82
　and photoelectric effect, 82
　and the Third Reich, 61n
Lenard window, 59, 299

Lenz, Heinrich (1804–1865), 308
LeSage, George Louis (1724–1803)
　and ether theory of gravitation,
　367n
Leyden jars, 149, 163–64, 204
　damping ratio of discharge
　from, 219, 280–81
　oscillatory nature of discharge,
　193, 280–81
　use in velocity measurements of
　electromagnetic waves,
　280–81
Lichtenberg, Georg (1742–1799),
　165n
Lichtenberg figures, 165
Light
　finite velocity of propagation of,
　280–83
　ultraviolet, 235n, 237–38
　uncertainties about nature of,
　274
　velocity of, 241n, 278–79, 280–84
　as a wave phenomenon, 274
　See also Light and Electricity;
　Photoelectric effect.
Light and Electricity
　direct relations between, 276–80
　Hertz's contributions to, 280–
　86, 313–14
　Hertz's Heidelberg address on,
　273–87
　indirect relations between, 273,
　279n
　intimate connection between,
　285–86
　proof that they have similar
　properties, 181–82, 261–68,
　284–86
Lipschitz, Rudolf (1832–1903), 47,
　321
Lloyd's mirror, 253
Lodge, Oliver J. (1851–1940), 80, 268n
　and electromagnetic waves on
　wires, 165
　and foundation of dynamics,
　331n

Index

and Hertz's 1890 visit to
London, 56
and wireless telegraphy, 41n
London
Hertz's visit to, 56
scientific reputation of, 113
Lorberg, Hermann (1831–1906), 401n
Lorentz-FitzGerald contraction. *See* FitzGerald, George F.
Lorenz, Ludwig (1829–1891)
and electromagnetic theory, 141, 278
Love, Augustus Edward Hough (1863–1940)
review of Hertz's *Mechanics* by, 365n
Ludwig, Hubert
eulogy by, at Hertz's funeral, 76

McCormmach, Russell, 66
Dictionary of Scientific Biography article on Hertz, xiv
See also Jungnickel, Christa.
Mach, Ernst (1838–1916)
and his *Science of Mechanics*, 321
and Hertz's *Mechanics*, 69, 321n
and the principles of mechanics, 331n
Magnetic currents, 134–35
Magnetic field
effect on cathode rays, 57, 304
in electromagnetic waves, 263–64
Magnetic force, 137–39
meaning of, in Hertz's writings, 191n
Magnetic permeability, 144n
Marconi, Guglielmo (1874–1937), 27, 41, 80
and development of wireless communications, 41, 161
Mass
Hertz's definitions of, 347n
W. Thomson's definition of, 329
Masses, hidden. *See* Hidden masses.

Maxwell, James Clerk (1831–1879), 14n, 79, 169–71, 173–74, 177–78, 217–18
and action-at-a-distance theories, 289, 310
disciples of, 279. *See also* Maxwellians.
and displacement current, 152–53
1865 paper on electromagnetic theory, 278–79
and the ether, 278
and models of electromagnetic processes, 318
prediction of transverse electromagnetic waves, 278
relations between light and electricity, 278–79
theory of, confirmed by Hertz's experiments, 283–84
Maxwellians
and Hertz's 1890 visit to London, 56
relationship to Hertz, 126
Maxwell's electromagnetic theory
apparent disagreement of, with experiment, 254n
confirmation of, by Hertz's experiments, 182, 254, 283–84
difficulties in understanding, 183–84, 284
four different views of, 184–91
and Helmholtz's theory, 183
Hertz's views on, 184, 190–92, 277
and Maxwell's equations, 184, 290n
need for experimental proof of, 254
and opposing theories, 144–45, 183
precise meaning of, 183–92
Maxwell's equations, 142–43, 277
acceptance of, in 1879, 14n, 161
advantages of, 144
derived by Helmholtz, 169

derived by Hertz, 137–45
FitzGerald's comments on, 152–53
in the Hertz-Heaviside form, 53–54, 142–43, 152–53
Hertz on relationship of Maxwell's theory to, 184, 290
validity of Hertz's derivation of, 144
Mayer, Julius Robert (1814–1878), 119n
Mechanical models, 125, 318
Mechanics
difficulty in expounding, 329
disagreements about, 329–30, 331–32
Hertz's *Principles of Mechanics*, 65–70, 315–18, 319–22, 323–64, 365–77
primacy of, in physics, 65–66, 319
principles of, 321n, 326–27
representations of, 326–27, 331–32
simple laws of, 319
and structure of physics, 319
three pictures of, 363–64
triumphs of, 331
See also First representation of mechanics; Second representation of mechanics; Hertz's representation of mechanics; Hertz, Heinrich
Metallic layers
passage of cathode-rays through, 301–4
Metaphysical principles
Hertz and, 346–47
Meteorology
Bjerknes as father of modern, 62
Hertz and, 27, 107, 391
Microwaves
Hertz's discovery of, 40, 179n
Millikan, Robert (1868–1953)
and photoelectric effect, 82

Mirror
concave parabolic, 259–60
for producing electromagnetic beams, 259–60
Miscellaneous Papers (H. Hertz), 60, 97–106, 125–26
Lenard's Introduction to, 87, 89–108, 147–48
reviewed by G.F. FitzGerald, 147–57
value of, 147
See also Helmholtz; Hertz; Lenard; Planck.
Models
conformity with things, 324
and *Bilder*, 324n
Mossotti, O.F. (1791–1863), 289
Motions, hidden. *See* Hidden motions.
Multiple resonances, 179–80
explained by Hertz, 179
explained by Poincaré, 179
explained by Sarasin and de la Rive, 179–80
research of V. Bjerknes on, 179–80
Münchener Allgemeine Zeitung, 112
Munich
Hertz in, 10–12
Polytechnic, 10–12
University, 11

Natural motion, 358n
Naturphilosophie, 113n
Neumann, Carl (1832–1925), 308
Neumann, Franz (1798–1895), 143n, 161
and action-at-a-distance theories, 125
and formula for inductance, 217–18
and potential theory of electromagnetism, 21–22n, 130n, 393n
neutrino, 119n
Newton, Isaac (1642–1727)
and action-at-a-distance, 316

Index

FitzGerald on third law of
motion of, 369–70, 376
law of universal gravitation, 275
laws of motion, 68, 319, 320,
321, 327–30
Newton's rings, 254
Nicol prism, 263n
Nodes
and multimode oscillations,
212–13, 214–15
of oscillation, 198, 212–16,
in standing waves, 245–48
Normal, 243
Null-points. *See* Nodes.

Oersted, Hans Christian (1775–1851),
127
O'Hara, James G. and W. Pricha, 14n,
97n
Ophthalmoscope, 115–16
Order-of-magnitude calculations
by Hertz, 216–25
and Fermi, 220n
Oscillations, electric
calculation of frequencies of,
216–19
comparison of experiments with
theory for, 219–21
damping of, 215
effect of capacitance and
inductance on frequency
of, 200n, 216–18
effect of magnetic materials on
frequency of, 201, 211
effect of resistance on frequency
of, 201, 210
Helmholtz on, 193n, 216, 218
Hertz's experiments with, 193–
221
high-frequency, 193–221, 281
induced in open circuits, 206
of Leyden jars, 193
of open induction coils, 193
resonance effects with, 207–11
Oscillator, spark
damping of, 291

Hertz's experiments with, 193–
221, 241–55, 257–71

Parabolic reflector, 35–36
Paris
scientific reputation of, 292
Pasteur, Louis (1822–1895), 292
Period of oscillation
dependence of, on capacitance
and inductance, 200n
lack of direct way to measure,
220–21
Philosophers
reception of Hertz's *Mechanics*
by, 70
Philosophy
Helmholtz's views of, 117–18
Hertz's interest in, 15, 323n,
346n
Hertz's philosophy of science,
323–27
Naturphilosophie, 113n
philosophical content of Hertz's
Mechanics, 323n, 346n, 377
Phosphorescence, 301–4
Photoelectric effect, 162, 216, 279n
discovery of, 166, 305n
explanation of, 38, 240n
Hertz's experiments on, 37–39,
82, 223–40
initial observation of, 223–26
larger magnitude of, at negative
pole, 223–24
light sources used in research
on, 226–29, 238–40
metals used in research on, 227
rectilinear propagation of cause
of, 228–29
reflection and refraction of
cause of, 234–37
results of experiments on, 239–
40
significance of, 166n, 397n
substances tested for
transmission of, 229–33
thoroughness of Hertz's
research, 233n

ultraviolet light as cause of, 235n, 237–38
Physical Review
and Hertz's contributions to physics, 84
Physics
in German universities, 25
Hertz's contributions to, 97–108, 312–14, 386–402
Hertz's importance in history of, 78–84
Institutes, 13, 25–26, 51–52
role of mechanics in, 319
theory and experiment in, 182n
unifying principle of, 70
Physikalisch-Technische Reichsanstalt (PTR), 43–44, 47
Pictures
in Hertz's *Mechanics*. See *Bilder*.
Planck, Max (1858–1947), 147n, 360n
admiration of, for Hertz, 76, 381, 403
and Helmholtz, 381
on relationship of Hertz to Helmholtz, 384–85, 399–400
and Hertzian oscillators, 81
on Hertz's character, 383, 399–400
on Hertz's death, 383, 402–3
on Hertz's research ability, 390, 395–98
on Hertz's research in Berlin, 384–90
on Hertz's research in Bonn, 400–3
on Hertz's research in Karlsruhe, 395–98
on Hertz's research in Kiel, 390–95
on Hertz's studies, 383–86
on Hertz's theoretical ability, 392–94, 400–1
on importance of Hertz's research, 79–80, 398–99

Memorial Address for Hertz by, 383–403
and modern physics, 83
on the role of intellect and chance in Hertz's research, 395–96, 398
Plane of oscillation, 243
Planté accumulators, 240n
Plücker, Julius (1801–1868), 57n
Poggendorff, Johann C. (1796–1877), 14n
Poincaré, Henri (1854–1912), 34n, 191n, 290
and diffraction effects in Hertz's experiments, 34–35, 35n
and error in Hertz's calculations, 171–72, 217n
and explanation of "multiple resonance," 179
Poisson, Simeon Denis (1781–1840), 290
Polarization
of dielectrics, 188–89, 191–92
of electromagnetic waves, 262–64, 265–66, 285
magnetic, 130–31
Polarizing array, 36
Polytechnics, 8n
Dresden Polytechnic, 8–10
Munich Polytechnic, 10–12
and *Technische Hochschulen*, 8n
Popov, A.S. (1859–1906), 80
and development of wireless communications, 40–41
Potential theory
of Franz Neumann, 21–22n
Potentials, 129–31
in Maxwell's equations, 54, 141–43
See also Vector potentials.
Precht, Julius (1871–1942), 63n
Pricha, W. *See* O'Hara, James G.
Principle of Least Action, 70, 72, 119–20, 339–40
Principle of Least Constraint, 68, 350–51, 354–55
Principles of Mechanics (H. Hertz),

and Boltzmann, 69
contents of, 67–68
departure of, from Helmholtz's ideas, 71–72
dependence of, on Helmholtz's work, 70–73, 320–21
dependence on previous research, 320–22
estimate of importance of, 69–70, 322
and the ether, 65–67, 69, 319, 343, 349n, 361n, 364
fundamental law in, 67–68
Helmholtz's Preface to, 299–300, 305–18
Hertz's Introduction to, 67, 299–300, 323–64
Hertz's Preface to, 300, 319–22
Hertz's reasons for writing, 65–67, 319–20, 332, 356
and "hidden masses," 68–69, 348–49, 362, 371
and "hidden motions," 68–69, 348–49, 360, 362, 371
and Mach, 69, 321
philosophical content of, 321n, 323n, 345–47, 377
reception by philosophers of, 70
reception by physicists of, 69–70, 73n
review by G.F. FitzGerald of, 300, 365–77
as a textbook, 320
and Wittgenstein, 70
See also FitzGerald; Helmholtz; Hertz's representation of mechanics; First representation of mechanics; Second representation of mechanics.
Prisms
of asphalt and hard pitch, 159, 266–68, 285
Nicol, 363n
Privatdozent, 390n
Hertz as, 106–8, 125

Prize-problems
1878 prize of Berlin Philosophical Faculty, 13–14, 309–10, 365–66, 385–86
1879 prize of Berlin Academy of Sciences, 14–15, 366–67
work by Hertz on, 14n, 44–45, 163–64, 169–70, 307n, 312, 365–67

Quincke, Georg H. (1834–1924), 61n

Radiation, electromagnetic
Hertz's 1888 article on, 257–71
See also Electromagnetic waves; Hertz, Heinrich.
Radio. *See* Wireless telegraphy.
Radio astronomy, 260n
Refractive index
of hard pitch, 268
Representations (*Darstellungen*)
and *Bilder*, 327n
in electromagnetism, 183–94, 326n
of mechanics, 326, 363–64
Research success
dependence of, on knowledge of equipment, 282
Residual charge, 98, 295, 389
Resistance
effect on frequency of, 200
effect on resonance of, 210
Resonance
between open electric circuits, 207–11
Bjerknes' research on, 40, 62, 210n
experiments to demonstrate, 208–11
graphs of electrical, 211
in Hertz's experiments, 33–34, 39–40, 207–11, 220–21, 396
lesser importance of, at high frequencies, 259
mechanical and electrical, 281n
and observed spark lengths, 220
tuning circuit to, 208–11

See also Syntonic effects.
Resonance curves
  obtained by Hertz and Bjerknes, 211
Reynolds, Osborne (1842–1912)
  and mechanical model for the ether, 367n, 369
Riemann, Bernhard (1826–1866),
  and electromagnetic theory, 141, 278n, 308
Riess, Peter (1804–1883), 194n
  spark-micrometer introduced by, 194, 225
  spirals designed by, 30, 32–33, 164
Righi, Augusto (1850–1920)
  and photoelectric effect, 166
Ring-magnet, 129
Roentgen, Wilhelm C. (1845–1923), 233n
  and discovery of x-rays, 60–61
  importance of research of Hertz and Lenard to, 82
Rolling sphere, 342–43
Ruhmkorff coil, 196, 204–7, 225, 258
Rumford, Count (Benjamin Thompson; 1753–1814), 55
Rumford Medal
  awarded to Hertz, 55–56

Sarasin, Éduard (1843–1917), 34n, 74, 176n
Sarasin and de la Rive,
  measurements in the Rhône waterworks, 176n
  and "multiple resonance," 179–80
  velocity measurements of electromagnetic waves, 175–76
Savart, Nicolas (1790–1853), 253
Schiller, Friedrich, 90
Schleiermacher, August (1857–1953), 26, 28
Schulte, Aloys, 74
Science
  future of, 286–87

methods and values of, xiv
Second (energy) representation of mechanics, 337–47, 370–71
  absence of concept of force, 338–39
  characteristics of, 339–47
  exposition of, 337–39
  Hertz's criticism of, 341–45
  Hertz's conclusions about, 347
  merits of, 341
Seebeck, Thomas (1770–1831), 253n
Side-circuit (*Nebenkreis*), 167, 196
Side-sparks (*Nebenfunken*), 164, 196
Skin-effect
  Hertz and, 40, 400
Sommerfeld, Arnold (1868–1951), 41n
  and Hertz's theoretical work, 178n
  and importance of Hertz-Heaviside form of Maxwell's equations, 54
Spark-detector
  for electromagnetic waves, 258, 282
  secret of its success, 282n
Spark-micrometer, 29n, 33, 194n, 225
  See also Riess, Peter.
Sparks
  production of, 195–96
  size of electric, 258, 282
  visibility of, 198, 282
  See also Oscillations, electric.
Specific inductive capacity, 143–44
Speed of light. See Velocity of light.
Standing-waves
  measurements on, 34, 212–16, 245–50
Stokes, George G. (1819–1903), 61n
"straightest path"
  meaning of, 68n, 350n
Syntonic (resonance) effects, 207n

Tait, Peter G. (1831–1901)
  definition of mass by, 329
  Treatise on Natural Philosophy of, 331n

Index

Tangent galvanometer
  construction of, by Hertz, 12
  use of, by Hertz for research, 94
Technical Institutes, 8n
Technische Hochschulen
  and Polytechnics, 8n
Technische Universitäten, 8n
Third representation of mechanics.
  See Hertz's representation
  of mechanics.
Thompson, Benjamin. *See* Rumford,
  Count.
Thomson, J.J. (1856–1940)
  and cathode rays, 62, 82, 155n
  and dynamical principles, 321
  and 1906 Nobel Prize in physics,
    82
Thomson, William (Lord Kelvin;
    1824–1907), 144n, 152, 318
  definition of mass by, 329
  and oscillatory discharge of
    Leyden jars, 193n, 216,
    281n
  Treatise on Natural Philosophy
    of, 331n
  and vortex-atoms, 318, 349,
    360–61
Three pictures of mechanics (*Drei
    Bilder der Mechanik*)
  comparison of, 363–64
  importance of ether in deciding
    among, 364
Tourmaline, 263, 266
Translations of Hertz's papers, xvii
Trouton, Frederick Thomas (1863–
    1922), 175n

Ultraviolet light
  and photoelectric effect, 235,
    237–38, 253n
Unifying principle of physics, 70
Unity of electric force, 21–22, 393–94
  and action-at-a-distance
    theories, 394
  and Hertz's derivation of
    Maxwell's equations, 128–
    29, 145n

Unity of magnetic force, 127–28, 145,
  393
Universities
  academic year in German, 16
  Berlin, 12–13
  Bonn, 47–48
  doctoral examination in
    German, 15n
  Kiel, 19–20, 125
  most important duty of, 156–57
Uranium glass, 301

Vapor pressure of mercury
  FitzGerald's comments on, 151–
    52
  Hertz's research on, 151–52
Vector potentials, 135
  convergence of series for, 140
  differential equation for, 140–41
  elimination from final
    equations, 141–42
  and the Hertz-Heaviside form of
    Maxwell's equations, 53–54
  infinite series for, 139
  propagation at velocity of light,
    141
Velocity of electromagnetic waves
  and conflict between theory and
    experiment, 169–70, 254
  on wires and in air, 169–70, 260–
    61
Velocity of light, 241n, 278–79
  reciprocal of, 130n, 278–79
Vibrations
  first overtone, 212–14
  fundamental, 212–14
  See also Oscillations, electric.

"Waterdale" on structure of the
  ether, 369
Wave-particle duality, 325n
Wave-plane, 243
Waves, electromagnetic. *See*
  Electromagnetic waves.
Weber, Wilhelm (1804–1891), 143n,
    161, 218, 278n, 290–91
  and action-at-a-distance, 128

and inertia of electric currents, 309, 365
and law for electromagnetic interactions, 275
on nature of electricity, 385, 393n
and theory of electromagnetism, 21–22n, 307–8, 365n
Weierstrass, Karl T. (1815–1897), 19
Wheatstone bridge, 385
Wiedemann, Gustav H. (1826–1899), 14n
Wireless telegraphy, 40–43
   and Hertz, 41–43
   and Hertz-Huber correspondence, 41–42
   use of Hz (hertz) as frequency unit in, 42–43
Wittgenstein, Ludwig
   and Hertz's *Mechanics*, 70

X-rays, 156
   Roentgen's discovery of, 60–61
   Quincke and Stokes on, 61n

Zeller, Eduard (1814–1908), 15n

For Product Safety Concerns and Information please contact our EU representative GPSR@taylorandfrancis.com
Taylor & Francis Verlag GmbH, Kaufingerstraße 24, 80331 München, Germany

www.ingramcontent.com/pod-product-compliance
Lightning Source LLC
Chambersburg PA
CBHW071233300426
44116CB00008B/1021